Psychopharmacology Series 6

Benzodiazepine Receptor Ligands, Memory and Information Processing

Psychometric, Psychopharmacological and Clinical Issues

Editors

I. Hindmarch H. Ott

With 76 Figures

Springer-Verlag Berlin Heidelberg New York
London Paris Tokyo

DR. IAN HINDMARCH
Head, Human Psychopharmacology Research Unit
University of Leeds
Leeds LS2 9JT, Great Britain

DR. HELMUT OTT
Research Laboratories
Department Pharmacopsychology
Schering AG, P.O. Box 65 03 11
1000 Berlin 65, Federal Republic of Germany

Vols. 1 and 2 of this series appeared under the title "Psychopharmacology Supplementum"

Figure on the front cover: "Basic Chemical Structure of the β-Carbolines"

ISBN-13: 978-3-642-73290-4 e-ISBN-13: 978-3-642-73288-1
DOI: 10.1007/978-3-642-73288-1

Library of Congress Cataloging-in-Publication Data.
Benzodiazepine receptor ligands, memory, and information processing. (Psychopharmacology series; 6) "Papers were presented at an international workshop on benzodiazepine receptor ligands, memory, and information processing, held during the 15th CINP meeting in Puerto Rico in December 1986" – Pref. Includes bibliographies and indexes. 1. Benzodiazepines – Side effects – Congresses. 2. Memory – Effect of drugs on – Congresses. 3. Human information processing – Effect of drugs on – Congresses. 4. Benzodiazepines – Receptors – Congresses. I. Hindmarch, I. (Ian), 1944– . II. Ott, H. (Helmut). III. Collegium Internationale Neuro-psychopharmacologicum. Congress (15th: 1986: Puerto Rico). IV. Series. [DNLM: 1. Benzodiazepines – pharmacology – congresses. 2. Cognition – drug effects – congresses. 3. Ligands – congresses. 4. Memory – drug effects – congresses. 5. Receptors, GABA-Benzodiazepine – metabolism – congresses. Q1 PS773J v. 6/QV 77.9 B479016 1986] RM666.B42B42 1988 615'.788 88-6548 ISBN-13: 978-3-642-73290-4 (U.S.)

Softcover reprint of the hardcover 1st edition 1988

2125/3145-543210 – Printed on acid-free paper

Preface

The following papers were presented at an international workshop on benzodiazepine receptor ligands, memory and information processing, held during the 15th CINP meeting in Puerto Rico in December 1986 and organised by the editors and T. Roth. This workshop was aimed at reviewing and reflecting on past experience with benzodiazepines, evaluating the current state of knowledge of the actions of psychotropic drugs (particularly benzodiazepines, β-carbolines and benzodiazepine antagonists), and laying a basis of interest and speculation for future research into the potential use of these drugs in disorders of memory and information processing.

There is much published material regarding the theoretical underpinnings of what psychologists call "memory", and doubtless there will be several more libraries filled with theses on the topic before a consensus is reached as regards basic definitions or even meaningful distinctions between for example, "short"- and "long"-term memory – everyone has his own particular theory. The qualitative and quantitative diversity of the different approaches might seem to present an insurmountable problem for the psychologist seeking a unified conceptual framework. In practice the various theories produce a plethora of pragmatic and experimental techniques for psychopharmacologists and clinicians to use when investigating drugs with putative amnestic or promnestic properties.

Amnesia as a direct consequence of drug administration has not received the clinical attention it deserves, especially as regards elderly patients, since both normal ageing and senile dementia are associated with a broad array of changes in cognitive function. The advent of effective computer-supported test systems for use in older patients will, hopefully, raise the quantity and quality of psychogeriatric investigations.

Research on benzodiazepines has provided a copious source of pharmacological agents which have had a profound impact on the clinical management of anxiety, sleep disorders, skeletomuscular disorders, preoperative stress, epilepsy, panic and depression over the last 28 years. Benzodiazepines have their strengths and weaknesses and, with few exceptions, are more problematic in clinical use than at first thought in the 1960s. Nevertheless, they remain the most widely prescribed class of psychotropic drugs, not only because of their efficacy but also because they are considered to be relatively safe.

Evidence from a large number of studies indicates that benzodiazepines might impair some aspects of cognitive function but not all of them. Investigations of the characteristics of the benzodiazepines predictive of amnesic potency assist in determining the mechanisms underlying benzodiazepine-induced amnesia and are of benefit in the search for drugs such as the β-carbolines that might improve learning. In addition, the pharmacological potential offered by the β-carbolines themselves is matched by the possibility of using them in research-orientated studies to help understand the basic processes underlying anxiety, memory and information processing.

I. Hindmarch
H. Ott

Table of Contents

Benzodiazepines, Memory and Information Processing

β-Carbolines and Benzodiazepine Antagonist Actions on Memory and Vigilance

Résumé

List of Contributors

You will find the addresses at the beginning of the respective contribution

Theoretical Aspects of Memory and Information Processing

Models of Memory: Information Processing

M. W. EYSENCK[1]

Abstract

A complete understanding of human memory will necessarily involve consideration of the active processes involved at the time of learning and of the organization and nature of representation of information in long-term memory. In addition to process and structure, it is important for theory to indicate the ways in which stimulus-driven and conceptually driven processes interact with each other in the learning situation. Not surprisingly, no existent theory provides a detailed specification of all of these factors. However, there are a number of more specific theories which are successful in illuminating some of the component structures and processes.

The working memory model proposed by BADDELEY and HITCH (1974) and modified subsequently has shown how the earlier theoretical construct of the short-term store should be replaced with the notion of working memory. In essence, working memory is a system which is used both to process information and to permit the transient storage of information. It comprises a number of conceptually distinct, but functionally interdependent components.

So far as long-term memory is concerned, there is evidence of a number of different kinds of representation. Of particular importance is the distinction between declarative knowledge and procedural knowledge, a distinction which has received support from the study of amnesic patients. Kosslyn has argued for a distinction between literal representation and propositional representation, whereas Tulving has distinguished between episodic and semantic memories. While Tulving's distinction is perhaps the best known, there is increasing evidence that episodic and semantic memory differ primarily in content rather than in process, and so the distinction may be of less theoretical value than was originally believed.

In sum, recent models of memory have made steady progress in clarifying the major characteristics of the human memory system.

1 Historical Introduction

During the late 1960s and early 1970s there were several attempts made to produce multi-store models to describe the structures and processes of human memory in information-processing terms. Most of these attempts were based to a greater or lesser extent on the pioneering theorizing of BROADBENT (1958), and as a consequence they resembled each other sufficiently for it to be possible to construct a "modal" multi-store model.

What are the advantages of this "modal" model? First, it was assumed that there are three different kinds of memory store: modality-specific stores which hold incoming information in a relatively uninterpreted form for no longer than

[1] Psychology Department, Royal Holloway and Bedford New College, University of London, Egham Hill, Egham, Surrey TW20 OEX, UK.

1–2 s; a short-term store of very limited capacity which stores information for several seconds; and a long-term store of essentially unlimited capacity. Secondly, information is transferred from the modality-specific stores to the short-term store by means of the process of attention, whereas rehearsal plays a major role in the transfer of information from the short-term store to the long-term store. Thirdly, it was argued that the nature of the forgetting process depends upon which memory store is involved. Forgetting from the modality-specific stores occurs primarily via spontaneous physiological decay, forgetting from the short-term store is largely a consequence of displacement or of an attentional shift, and forgetting from the long-term store is often cue-dependent, in the sense that the relevant memory trace is still in the memory system, but is inaccessible to most retrieval cues.

In sum, the major assumption of the "modal" multi-store theory (e.g., Atkinson and Shiffrin 1968) that it is valid to distinguish three different kinds of memory store is broadly supported by the evidence. More specifically, the various memory stores appear to differ in their capacity, in terms of the temporal duration of information within them, and in terms of the forgetting mechanisms involved.

Despite the successes of the multi-store approach, it can no longer be regarded as an adequate theoretical conceptualization of the architecture of the memory system. Why is that so? One major weakness of multi-store theory is that the processes involved in human memory are substantially underspecified. According to some versions of multi-store theory, information in the modality-specific stores is in a relatively uninterpreted form, whereas in the short-term store it is in a phonological form because of verbal rehearsal, and then in the long-term store it is mainly stored in semantic form. These dramatic changes in the nature of the information as it moves from one memory to another remain almost entirely unaccounted for by multi-store theorists.

Another major weakness of the multi-store approach is its emphasis on rehearsal as the key process by which information is transferred from the short-term store to the long-term store. This does not seem plausible at all. We normally store much information in long-term memory every day, and yet this is typically done with little recourse to verbal rehearsal. In other words, the role of rehearsal in producing long-term memories was grossly exaggerated by multi-store theorists.

A final major weakness of the multi-store approach has received considerable attention from theorists attempting to modify the original assumptions of the multi-store theorists so as to create more realistic theoretical models. According to multi-store theorists, there is a single, unitary short-term store, and also a single, unitary long-term store. It has increasingly been argued that this theoretical position is untenable, and that much more complex conceptualizations of both the short-term and the long-term store are required. Recent theoretical developments relating to these two memory stores are considered in the two sections which follow.

2 Working Memory

Baddeley and Hitch (1974) argued that the short-term store is a theoretical contruct of strictly limited usefulness. They put forward two main reasons to support their argument. Firstly, the variety of coding is much greater than was allowed for by multi-store theories with their emphasis on verbal rehearsal. Secondly, if, as was often claimed, the short-term store represents "the contents of consciousness", it would seem very clear that the short-term store should be involved in a great variety of cognitive tasks, not just those explicitly concerned with memory.

In essence, Baddeley and Hitch (1974) proposed replacing the concept of a short-term store with that of a working memory. In the original version the working memory system consisted of three components: a central executive, an articulatory loop, and a visuo-spatial scratch pad. The most important component was the central executive. This was an attention-like system which was modality-free and of limited capacity. The articulatory loop and the visuo-spatial scratch pad were slave systems that could be used by the central executive for specific purposes. The articulatory loop was a verbal rehearsal system, organizing phonemic information in a temporal and sequential fashion. The characteristics of the visuo-spatial scratch pad were less clear, but there was some evidence that it relied mainly on spatial rather than on visual coding.

Baddeley (1986) has recently modified the working memory model in a number of ways. For example, the nature of the central executive is more clearly specified. According to Baddeley (1986), the central executive is called upon under a range of circumstances including "(a) tasks that involve planning or decision making; (b) situations in which the automatic processes appear to be running into difficulties and some form of trouble-shooting is necessary; (c) where novel or poorly learned sequences of acts are involved; (d) where the situation is judged to be dangerous or technically difficult; and (e) where some strong habitual response or temptation is involved". The articulatory loop is a limited-capacity system for retaining speech-based material. Originally, it appeared that the loop was based on articulation, and that it was a time-based loop resembling a closed loop on a tape recorder. However, the increasing complexity of experimental findings has led to a revised conceptualization in which the articulatory loop system consists of a phonological or speech-based store and an articulatory control process. Finally, there is the visuo-spatial scratchpad or sketchpad. The characteristics of this component of working memory remain rather unclear, but it appears to consist of a relatively short-lasting visuo-spatial store which can "hold" and/or manipulate images, and which can be disrupted by concurrent spatial processing.

One of the crucial insights underlying the working memory model is the notion that numerous everyday activities (e.g. comprehension, solving a problem) require a combination of active processing and transformation as well as the transient holding of relevant information. For example, mental solution of a complex, multi-stage problem requires that the outcomes of early stages of processing are readily accessible during the subsequent stages of processing. It is appropriate, therefore, to postulate an integrated system designed to permit both active processing and transient storage, and these are precisely the functions provided by the working memory system. This conceptualization differs profoundly from the previous notion of a short-term store, in that it is explicitly of relevance to numerous situations other than specifically memory tasks.

The role of verbal rehearsal within the working memory model is clearly more acceptable than its role within the multi-store model. In general terms, verbal rehearsal is accorded a much less central role in the working memory model. It is regarded as an optional process that occurs primarily in only one out of the three components of the working memory system rather than as the major process within the short-term store.

What are the problems with the working memory model? Firstly, there is a clear imbalance in terms of our knowledge concerning the three components of the working memory system. The detailed functioning of the articulatory loop is reasonably well understood and documented, whereas that of the central executive is much less so. It is especially unfortunate that rather little is known of the functioning of the central executive, because that is by far the most important component of working memory. Secondly, the distinctions among the components of working memory are sharper at the conceptual level than at the experimental level. For example, if one wanted to decide whether performance on a particular task usually involves the use of the visuo-spatial scratchpad or sketchpad, then the optimal strategy would compare performance of that task on its own with a second task that only made use of the visuo-spatial scratchpad or sketchpad. The presence or absence of a disruption or interference effect under dual-task conditions would provide strong evidence. However, this strategy depends upon the use of a task employing only one of the components of working memory (i.e. the visuo-spatial scratchpad or sketchpad in the above example), and such a "pure" task probably does not exist. The problem would be at least as great if one wanted to use a secondary task which relied exclusively on the central executive.

3 Long-Term Memory

In recent years various theorists have tried to identify functionally separate long-term memory systems. These theorists all agree that the conceptualization of the long-term store as a unitary and homogeneous structure is demonstrably inadequate. The most enduring of the proposed divisions of the long-term store are into episodic and semantic memory (TULVING 1972, 1983) and into procedural and declarative knowledge (e.g. COHEN 1984), and these will be considered in turn.

According to TULVING (1972, 1983), episodic memory is basically autobiographical memory, and episodic memories generally contain spatial and temporal information. On the other hand, semantic memory is a "mental thesaurus", which contains our knowledge of the world, the meanings of words, and so on. TULVING argued that episodic and semantic memory form separate systems, although he did admit that the two systems are usually highly interdependent in their functioning.

At a superficial level, research on amnesic patients (mostly suffering from Korsakoff's syndrome) appears to support this proposed theoretical distinction, since they often demonstrate intact semantic memory but grossly impaired episodic memory. In fact, most of the research involves a confounding of variables, since the usual comparison is between amnesics' good pre-morbidly acquired semantic memory and their poor post-morbidly acquired episodic memory. When the factors of type of memory and time of acquisition are not confounded, then the evidence suggests that amnesia reduces the ability to acquire new information, and this is equally true of semantic and episodic memory. In contrast, most of the information learned pre-morbidly is spared, and again this is equally true for both types of memory. Thus, amnesia research fails to confirm the notion that there are distinguishable episodic and semantic memory systems.

Amnesia research, however, has provided rather strong evidential support for a distinction between declarative knowledge and procedural knowledge. According to COHEN (1984, p. 96), declarative knowledge is represented "in a system quite compatible with the traditional memory metaphor of experimental psychology, in which information is said to be first processed or encoded, then stored in some explicitly accessible form for later use, and then ultimately retrieved upon demand." In contrast, procedural knowledge accrues because "experience serves to influence the organization of processes that *guide* performance without access to the knowledge that *underlies* the performance" (COHEN 1984, p. 96).

The distinction between procedural and declarative knowledge is not very precise. As the above quotations indicate, the distinction overlaps considerably with that between knowledge to which we can and cannot gain conscious access. Despite this definitional problem, the results from most amnesia research are consistent with the hypothesis that amnesics have an impaired declarative system and so experience great difficulties in acquiring new declarative knowledge, but that they have an essentially intact procedural system. Episodic and semantic memory both involve declarative knowledge, and we have seen that amnesics cannot readily acquire and retrieve episodic and semantic memories in the post-morbid phase. So far as procedural learning is concerned, amnesics have shown normal (or nearly normal) rates of acquisition on various tasks such as the pursuit rotor, mirror reading, and the Tower of Hanoi. These tasks seem to rely largely on procedural knowledge, although it may well be that declarative knowledge is involved in the initial stages of skill acquisition, especially on a complex task such as the Tower of Hanoi.

It is undeniable that neuropsychological studies of amnesia have revealed phenomena which necessitate modification to information-processing models of normal memory functioning. However, it has still not been established whether the critical distinction is between declarative and procedural knowledge or between stored information which can and cannot be accessed consciously. It is possible (if unlikely) that the two distinctions are essentially the same for most practical purposes (cf. COHEN 1984). If they are not, then it becomes important to unconfound the two factors and investigate their separate contributions to amnesic deficits in long-term memory.

The distinction between procedural and declarative knowledge may be of fundamental importance, but it should certainly not be assumed that each knowledge system is entirely homogeneous. Within the declarative knowledge system, for example, it seems important to distinguish among different-sized units of information. At one extreme, there are individual concepts or isolated facts, whereas at the other extreme there are relatively large, well-integrated sets of knowledge which have been variously described as schemata, scripts or frames (for review, see Eysenck 1984). One of the uses of schemata is to facilitate comprehension. A complex situation can often be interpreted by accessing a single schema rather than a substantial number of very specific pieces of information.

Schema theories (e.g. Rumelhart 1980) assume that schemata influence what is learned and what is subsequently retrieved, and in so doing they introduce systematic distortions into long-term retention. The fact that such distortions are frequently found provides some support for schema theories. However, most schema theories suffer from the disadvantage that they are rather imprecise in their specification of what constitutes a schema. In addition, most studies of schemata have failed to provide any independent measures of the schemata theoretically involved in acquisition and retrieval.

In sum, the notion of a unitary long-term store must be replaced with more complex conceptualizations. As part of such a reconceptualization, separate declarative and procedural knowledge systems should be identified. Within the declarative (and probably also the procedural) knowledge systems, some information is in the form of integrated chunks whereas other information is in much smaller units.

4 Retrieval From Long-Term Memory

Most theory and research concerned with retrieval from long-term memory has dealt with declarative knowledge. Far and away the most used retention tests to assess retrieval of declarative knowledge have involved either recognition or recall (cued, free, or serial). Since the results from recognition and recall tests typically differ quantitatively (i.e. in terms of the level of performance) and often differ qualitatively (i.e. in terms of the pattern of performance across conditions), it is obviously a matter of some theoretical and practical consequence to consider the processes involved in these two kinds of retention tests.

According to the two-stage or two-process theory (discussed at length by Watkins and Gardiner 1979), different processes are involved in recall and recognition memory tests. Recall requires an initial directed search or retrieval process, followed by a decision or recognition process which operates on the information which has been retrieved. In contrast, recognition memory is rather simpler and more straightforward, since it involves only the second of the two processes involved in recall (i.e. the decision process).

This theoretical approach has been severely criticized in recent years because of the greatly over-simplified view of retrieval processes which it offers. This is especially obvious in the case of recognition memory. Two-stage or two-process theorists claimed that there is no retrieval problem in recognition memory, a claim which is definitely erroneous. The fact that recognition memory is affected by contextual information presented on the recognition test indicates strongly that there can be a retrieval problem in recognition memory. If recall and recognition can both involve a retrieval process and a decision process, then the theoretical distinction between the processing involved in recall and recognition has become so blurred as to be practically meaningless.

A very different approach has been advocated by TULVING (1979, 1983). He has consistently argued that there are rather fundamental similarities between recall and recognition. For example, TULVING (1979) argued that the encoding specificity principle applied equally to recall and to recognition. He expressed this principle in the following way (TULVING 1979, p. 408): "The probability of successful retrieval of the target item is a monotonically increasing function of informational overlap between the information present at retrieval and the information stored in memory".

Experimental tests of the encoding specificity principle require that the degree of "informational overlap" between the memory trace on the one hand and the retrieval environment on the other hand be manipulated in a systematic fashion. This is most readily achieved by varying the contextual information at the time of learning and at the time of test. In general terms, the levels of both recall and recognition memory are higher when the context at study and at test is the same rather than being different. More strikingly, it has been possible on occasion (e.g. TULVING and THOMSON 1973) to make recall memory superior to recognition memory by ensuring that the study and test contexts are the same for recall, while making the study and test contexts as dissimilar as possible for recognition.

TULVING (1983) extended some of these theoretical notions in his General Abstract Processing System. He still argued in favour of the encoding specificity principle, but claimed that the amount of informational overlap was not the sole determinant of memory performance. More informational overlap (or "ecphoric information" as he labelled it) is required for recall than for recognition, because the naming of a previous event which is needed in recall demands more ecphoric information than does the judgement of familiarity involved in recognition.

The notion that the success or failure of retrieval depends upon both the information in the memory trace and the information in the retrieval environment is theoretically fruitful, but it is rather limited in some ways. One limitation is TULVING's (1979, 1983) use of the theoretical construct of "context". He failed to distinguish between interactive context (i.e. context which changes what is stored) and independent context (i.e. context which does not change what is stored; BADDELEY 1982). The distinction between these two kinds of context is important, because it points to a major difference between recall and recognition. Recall appears to be affected by manipulations of either interactive or independent context, whereas recognition memory is affected by interactive context but not by independent context. One of the implications of these findings is that the processes involved in recall and recognition are more different than TULVING's (1983) theoretical model allows for; more specifically, independent context can facilitate (or interfere with) the processes involved in recall but not those involved in recognition.

The most important limitation of TULVING's (1983) theory is that it assumes that the information contained in the retrieval cue is used in a rather automatic and passive fashion. This may, indeed, be the case with the main paradigm used by TULVING, in which pairs of words are presented, with one word in each pair then being used as a retrieval cue for the other word in the pair. However, it seems totally implausible that a retrieval cue such as "Where were you on Thursday evening?" is processed in this fashion. Such a retrieval cue often initiates a series of problem-solving processes which eventually produce the required memory trace.

If recall can actually occur in a number of different ways, then the various strategies which may be involved should be identified at a theoretical level. A useful starting point was provided by JONES (1982). He distinguished between a direct recall route and an indirect recall route. In the direct recall route, the retrieval cue produces immediate access to the target information. In the indirect recall route, in contrast, the retrieval cue produces recall following processing activities such as the making of inferences and the generation of possible responses. In essence, TULVING (1983) emphasized the direct recall route at the expense of the indirect recall route.

There is increasing evidence that recognition memory can also occur in at least two different ways (MANDLER 1980). Recognition memory often occurs on the basis of stimulus familiarity: high familiarity produces rapid recognition, whereas low familiarity produces a rapid decision that the stimulus has not been seen before. When the level of stimulus familiarity is intermediate, then familiarity alone provides insufficient evidence to make a definite recognition decision. Under these circumstances, recognition decisions are based on a retrieval process which recovers relevant contextual information about the stimulus, such as the context or contexts in which the target item has previously been encountered.

In sum, it has been established that neither recall nor recognition depends on a single invariant process. Rather, the processes which intervene between the presentation of a retrieval cue and subsequent recall or recognition are influenced by various factors such as the nature of the retrieval cue and the relevant knowledge possessed by the subject.

5 Relevance to Psychopharmacology

Advances in memory theory and research have led to the introduction of several new theoretical distinctions. Thus, the previous unitary short-term store construct has been replaced by the working memory system with its three components, and the unitary long-term store construct has been extended in various ways to account for important differences among long-term memories. Furthermore, the notion (TULVING 1979) that the retrieval system operates in the same fashion for both recall and recognition has been replaced with the notion that retrieval can involve several different strategies.

The relevance of these theoretical advances to psychopharmacology is that they enable the effects of drugs on the memory system to be assessed more precisely than was previously possible. Instead of merely deciding whether a particular drug is enhancing or reducing short-term or long-term memory, it is now possible to decide which component or aspect of the short-term or long-term memory system is most affected by the drug.

References

Atkinson RC, Shiffrin RM (1968) Human memory: a proposed system and its control processes. In: Spence KW, Spence JT (eds) The psychology of learning and motivation, vol 2. Academic, London, pp 89–195

Baddeley AD (1982) Domains of recollection. Psychol Rev 89:708–729

Baddeley AD (1986) Working memory. Clarendon, Oxford

Baddeley AD, Hitch G (1974) Working memory. In: Bower GH (ed) The psychology of learning and motivation, vol 8. Academic, London, pp 123–175

Broadbent DE (1958) Perception and communication. Pergamon, Oxford

Cohen NJ (1984) Preserved learning capacity in amnesia: evidence for multiple memory systems. In: Squire LR, Butters N (eds) Neuropsychology of memory. Guilford, New York, pp 425–459

Eysenck MW (1984) A Handbook of Cognitive Psychology. Erlbaum, London

Jones GV (1982) Tests of the dual-mechanism theory of recall. Acta Psychol 50:61–72

Mandler G (1980) Recognizing: the judgement of previous occurrence. Psychol Rev 87:252–271

Rumelhart DE (1980) Schemata: the building blocks of cognition. In: Spiro R, Bruce B, Brewer W (eds) Theoretical Issues in Reading Comprehension. Erlbaum, Hillsdale, NJ, pp 221–267

Tulving E (1972) Episodic and semantic memory. In: Tulving E, Donaldson W (eds) Organization of Memory. Academic, New York, pp 333–371

Tulving E (1979) Relation between encoding specificity and levels of processing. In: Cermak LS, Craik FIM (eds) Levels of Processing in Human Memory. Erlbaum, Hillsdale NJ, pp 405–428

Tulving E (1983) Elements of Episodic Memory. Oxford University Press, Oxford

Tulving E, Thomson DM (1973) Encoding specificity and retrieval processes in episodic memory. Psychol Rev 80:353–373

Watkins MJ, Gardiner JM (1979) An appreciation of generate-recognize theory of recall. J Verbal Learning Verbal Behav 18:687–704

Measuring Memory

A. BADDELEY[1]

Abstract

Three broad approaches to the measurement of memory functioning will be described. The first of these involves using memory as a general indicator of any dysfunction in the central nervous system. This approach will be illustrated using STERNBERG's short-term memory scanning paradigm. Its strengths are that such tests are often very sensitive, but they are often very difficult to interpret both theoretically and in practical terms.

A second approach is to use a range of tasks selected so as to tap different aspects of human memory. Such an approach is of considerably more theoretical interest, and is discussed in more detail by EYSENCK (this volume). Its weaknesses are that theories of memory are still changing relatively quickly, and that mapping such results onto memory outside the laboratory is often complex.

A third approach is to attempt a more direct measure of everyday memory. The use of questionnaires for this purpose will be critically discussed, and a new test of everyday memory will be described. This test, the Rivermead Behavioural Memory Test, correlates well with observations of memory lapses in patients, and appears to offer a promising new line of development.

1 Introduction

The capacity to remember is one of the most important and impressive characteristics of the human brain, which provides a memory storage and retrieval system that far exceeds that of the best computer in terms of useful capacity, flexibility and robustness. Memory is however also a very fallible system, and the capacity to store and retrieve information is one of the most sensitive indices of central nervous system (CNS) impairment, whether temporary as produced by drugs, or permanent, following brain damage. Measurement of memory performance is therefore an important component of any form of intellectual assessment.

Attempts to measure performance typically approach memory from one of three perspectives. First of all, memory may simply be regarded as an indicator of the general performance of the CNS, with a drop in performance simply signalling that all is not well. The problem with such an approach is that it treats memory as a single function, which as EYSENCK (this volume) points out, it certainly is not. Consequently, one might come to quite erroneous conclusions if only a single measure is used. The second approach therefore is to attempt to test each of the various types of memory, producing a pattern of results that is more

[1] MRC Applied Psychology Unit, Chaucer Road, Cambridge CB2 2EF, UK.

likely to be informative in terms of the overall functioning of the CNS, and in addition may well be theoretically much more revealing.

Such an approach is scientifically much preferable, but it in turn has limitations if one wishes to extrapolate from the experiment to performance outside the laboratory. As I shall discuss later, it is by no means always the case that a memory test that is sensitive to CNS damage is necessarily a good predictor of memory problems in everyday life. I shall discuss these three approaches to memory testing in turn.

2 Memory as an Index of CNS Functioning

The capacity for new learning and speed of retrieval both tend to be extremely sensitive to many factors that impair the overall functioning of the CNS. For example, the first indicator of the onset of Alzheimer's disease tends to be a decrement in memory performance (MILLER 1977). Similarly, one of the most sensitive indicators of closed-head injury is again memory performance (BADDELEY et al., in press). As many chapters in the present volume will testify, memory is also typically very susceptible to the effects of drugs, and as I, and I suspect one or two others, can testify, difficulty in retrieving known material, such as people's names, from memory is perhaps the first intellectual sign of ageing.

For this reason, it makes sense to include at least one memory test in any general battery assessing CNS function. The criteria for selecting such a test obviously include test sensitivity. This in turn depends on having a test that is reliable, giving the same score when the same subject is tested under equivalent conditions on more than one occasion. Over the years, laboratory studies of memory have produced many paradigms that broadly meet this specification, ranging from sequential recall of nonsense syllables used by EBBINGHAUS (1885) in the first experimental study of human memory, to memory for complex prose passages, analysed using whatever is the currently fashionable story grammar. For the purpose of illustrating my point however, I shall stick to one paradigm that appears to be currently used quite widely in studying the effects of drugs on performance, namely that associated with Saul STERNBERG's serial exhaustive memory scanning model (STERNBERG 1966).

This task involves presenting the subject with a sequence of from one to six numbers, followed by a single probe number. The subject's task is to decide whether the probe had occurred in the prior sequence, pressing a "yes" key as rapidly as possible if it had, and a "no" key if it had not. Typically subjects make very few errors, and their performance is measured in terms of reaction time. This normally increases linearly with the number of its items presented, with "yes" and "no" responses yielding parallel linear functions.

As a measure of performance, the STERNBERG paradigm has a number of virtues. First of all, if given with sufficient care and practice, it can yield reliable results, based on a large number of responses that can be collected reasonably quickly. Secondly, it appears to be relatively sensitive to a number of drugs. Thirdly, it yields two potentially independent measures in the slope of the line relating reaction time to number of items in the stimulus set, and its intercept. Finally, these two measures have been given a theoretical interpretation by STERNBERG, who is a very able and highly respected cognitive psychologist. There is therefore a great deal to be said for this measure as an indicator of CNS functioning.

Unfortunately however, the paradigm is open to two major objections. The first is that STERNBERG's theoretical interpretation is by no means universally accepted. The second concerns the question of what a decrement in performance on this task might mean for performance outside the laboratory.

STERNBERG's own theoretical interpretation concentrates on explaining why the slopes tend to be both linear and parallel. He explains this by assuming that the subject performs a very rapid internal scan of the items that have been presented, using a process that waits until all the items have been processed before responding. The reason for assuming this exhaustive scanning mechanism is that if the subject were to respond as soon as he detected a match, then some of the responses would be very rapid (e.g. when a match is detected on the first item scanned), some rather slow (e.g. when the target is the last item checked), but on average, only about half the items should need to be checked before detecting a match on positive trials. However, negative trials would always require that every location be checked before responding. On average therefore, the slope for positive responses should only be about half as steep as that for negative responses.

There are two major problems with this theoretical interpretation. First of all, there are a number of more detailed aspects of the data that do not fit into the model. For example, if the last item presented is probed, it tends to evoke a particularly rapid response. Furthermore, given a sequence like *5 7 1 7 6*, if the repeated digit, *7* is probed, it again tends to evoke a fast response (BADDELEY and ECOB 1973). STERNBERG himself explains these results by assuming that they reflect components of the task other than memory scanning, but does not explain them in detail (STERNBERG 1975). This raises the question of whether such additional processes could not also explain the central findings of linear and parallel slopes. Consequently, a number of other competing models have been proposed, some assuming parallel testing of trace strength (e.g. BADDELEY and ECOB 1973; CORBALLIS et al. 1972), others assuming that memory operates like a push-down memory stack in a computer (THEIOS 1973), while yet others attempt to interpret the phenomena within much more ambitious parallel computer simulation models of memory (e.g. ANDERSON 1973). I know of no powerful evidence for choosing one of these models rather than the others, leaving the theoretical interpretation of the STERNBERG paradigm extremely uncertain.

A second problem is this. If the task does indeed measure a retrieval process of the type described by STERNBERG, it is difficult to imagine it being used at all generally within recognition memory. For example, when I recognize the face of my wife, do I systematically "scan" a representation of every face I have ever encountered before coming to the conclusion that I know her? Such a view is surely absurd, and has not of course been proposed by STERNBERG, or so far as I know by anyone else. It is therefore not at all easy to see what general memory function might be served by such a scanning procedure.

A final problem concerns the practical interpretation of a decrement in performance on the STERNBERG task. Does it matter that Mr. X now has a slope that is 10 ms per item steeper? Perhaps it is unfair to ask such questions of such techniques, but if it is, then one should avoid generalizing from STERNBERG task performance to performance outside the laboratory.

In conclusion then, the STERNBERG task is an intriguing one that may offer a very sensitive measure of performance, but at present, its theoretical and practical significance are both open to serious question. If we are to continue to use it, we should be much more concerned about its theoretical and ecological significance than is usually the case.

3 Components of Human Memory

Human memory is not a single monolithic system. It represents the operation of at least three separate subsystems, which themselves can be further fractionated. Neuropsychological evidence indicates that any of these three may be impaired independently of the other two (e.g. BADDELEY 1982), suggesting that any adequate assessment of the influence of a given drug on memory is likely to need several tests. Furthermore, as we increase our knowledge of human memory, we tend to fragment it further, producing yet more complexity. I will not go into details here of the subcomponents that appear to be emerging, since this is discussed by EYSENCK (this volume), but will merely add the, I hope reassuring, comment that for practical purposes, some components are both more susceptible than others, and probably of more significance in everyday life. In particular, the tasks involved in most standard laboratory learning and recall tests probably reflect a very important component of human memory, namely the capacity to learn and remember new material. However, this and other components are discussed by EYSENCK and I will therefore devote the rest of this paper to the important but difficult area of assessing memory performance outside the laboratory.

4 Measuring Everyday Memory

When patients complain that they have memory difficulties they do not typically mean that they find it difficult to learn to associate unrelated words, or to remember meaningless designs, and they might justifiably question the use of such tasks to assess their problem. Such scepticism is not of course limited to patients. A couple of years ago I was approached by a drug company which had been investigating a drug that appeared to have promise of alleviating some of the memory problems of early dementia. They had carried out trials using standard psychometric memory tests, together with physicians' ratings of the mental state of their patients. They had heard that we were using questionnaire methods to assess the memory performance of patients, and were interested in the possibility of using our techniques.

Their results already indicated a highly significant though not enormous effect of the drug on psychometric performance, and I therefore asked them why they needed to use questionnaires. It appears that their physicians' ratings showed nothing, other than a consistent tendency for the doctor to rate the patient a little better each time he saw him. As they pointed out, the people they were hoping would prescribe their drugs were doctors, and doctors tend not to be very impressed by small but significant improvements in performance on highly artificial laboratory tasks. Perhaps the doctors have a point.

We ourselves, as experimental psychologists, had become interested in the question of the relationship between our laboratory-based measures and everyday memory, and had started to tackle this in a study concerned with the laboratory and real-world memory problems of patients who had suffered a head injury (SUNDERLAND et al. 1984).

Why should we doubt that there is a relationship between laboratory memory tests and everyday memory? A few years earlier, Arnold WILKINS and I had become interested in the question of absentmindedness, and in particular that of remembering to do things at a specified time. We decided to try to set up an analogue to the task of taking pills four times a day, and did so as follows. The subjects were given a small box and instructed that at four specified times on each day for a period of a week, they should press a button and turn a knob. The box contained a cheap digital watch and a film. When the button was pressed, the LED on the watch was illuminated, registering the time on the film, which was then moved on by the turn of the knob.

We selected from the Applied Psychology Unit subject panel two groups, one that we knew to be particularly good at remembering lists of words, and one that we knew to be particularly poor of the task. If remembering to do things is simply one facet of having a generally good memory, then those who were good at learning word lists should also be good at remembering to "take their pills". Our results showed that there was indeed a difference between the two groups, but not in the expected direction; those who were good at remembering words were *less* good at performing the simulated pill-taking task, a result that, being wise after the event, we labelled the "absentminded professor" effect (WILKINS and BADDELEY 1978).

We decided to explore further the relevance of laboratory measures to everyday memory performance, and chose to tackle the problem using patients who had suffered a mild-to-moderate head injury, typically associated with a traffic accident. Head injury is known to cause substantial impairment in performance on psychometric memory tests, and to be associated with frequent complaints of everyday memory problems by patients (BROOKS 1972). The degree of head injury and hence of memory decrement was likely to vary widely, allowing us to use correlational techniques to assess the relationship between laboratory tests and everyday memory performance.

We had no difficulty in finding a range of well-established tasks that would be expected to reflect impaired memory performance, on the basis of earlier literature. A much more difficult problem however, was that of assessing everyday memory. Ideally, we would have liked to have had objective observational data, but how could one obtain this? It was clearly not feasible to follow our patients around in their everyday lives, and even if it were, memory lapses are by no means always obvious to the observer. We therefore decided to use a combination of structured interviews and diaries to be completed by both the patient and a close relative, usually his wife.

We tested three groups of male patients, one group that had had their head injury within the last few months, but had returned home from hospital. A second head-injury group comprised people who had had their accident from two to eight years before. Finally, as a control group, we used orthopaedic patients who had also been in traffic accidents, but had suffered limb fractures but no head injury. All three groups completed a battery of conventional memory tests; then they and their relatives were interviewed regarding their everyday memory problems.. Finally, both patients and relatives were asked to fill in a checklist concerned with lapses of everyday memory, completing the list every evening for a period of a week.

As expected, both groups of patients showed impaired performance on a range of memory tasks, the degree of impairment being approximately the same for the recently injured subjects and for those who had their head injury several years before, suggesting that the memory decrement is a persisting one. Furthermore, subjective complaints of memory problems were significantly greater for the patients than for the orthopaedic controls, whether based on the responses of the patient or the relative, and whether measured by interview or diary. There was again no evidence that memory lapses became less frequent over time, since the chronic patients reported just as many problems as those who had had their head injury quite recently.

The crucial analysis of course concerns the relationship between the various measures. Some indication of the results we obtained are shown in Fig. 1. This shows two things, the magnitude of the correlation between the various subjective measures of performance and performance on each of the separate memory tasks, and on the other hand the susceptibility of the objective memory tests to head injury, as measured by the statistical reliability of the difference between patients and controls on the measure in question. The data shown are taken from the results of patients who had received their head injury several years before.

The results shown in Fig. 1, together with data from other groups, suggest the following conclusions:

1. The extent to which laboratory-based measures of performance correlate with memory complaints varies greatly.
2. The extent to which a test predicts everyday memory problems is unrelated to its overall sensitivity to head injury. For example, Kimura's repeated figures test, in which the subject identifies the repetition of a meaningless design, is highly sensitive to the effects of head injury, but totally uncorrelated with memory complaints. On the other hand, the capacity to remember a short paragraph of prose is less sensitive to head injury, but offers a much better prediction of everyday memory performance.
3. The various subjective measures vary in their degree of correlation with objective performance. In general, the relatives' interview offers the best correlation, and the patients' interview the worst.
4. Data from subjects who had their head injury quite recently showed the same overall pattern, but were very much more noisy, with the relation between subjective and objective measures being uniformly lower than in the chronic patients.

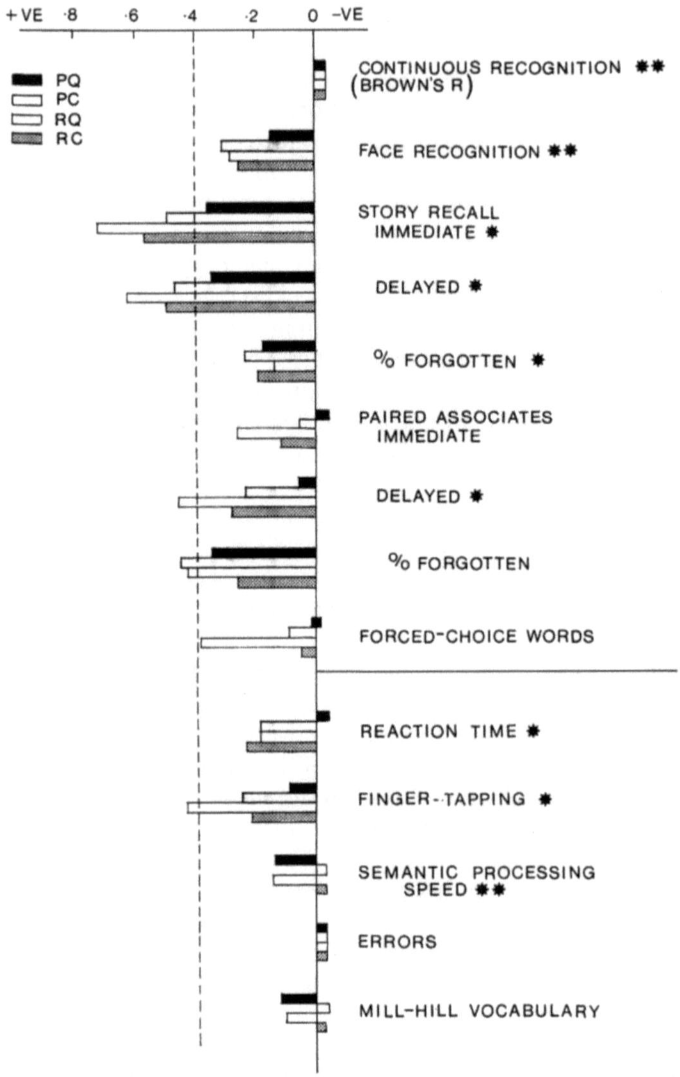

Fig. 1. Relationship between reports of everyday memory problems and performance on a range of laboratory tests. The histograms on the *left* indicate the degree of correlation between performance on the test in question and four estimates of everyday memory lapses: *PQ*, patient's interview; *PC*, patient's checklist; *RQ*, relative's interview; *RC*, relative's checklist. The laboratory tests are listed on the *right*. Those showing significantly poorer performance in patients who suffered a head injury 2–8 years before are marked *($p<0.05$) or **($p<0.01$). Data from SUNDERLAND et al. (1984)

We suspected that this latter point occurred because the recently head-injured patients were often still settling into a routine. Those who had milder deficits would be most likely to be back at work, in a situation where demands on memory and possibility of memory lapses would be greatest. The more serious cases would probably still be at home, with relatively few demands being placed on them, and consequently fewer opportunities for the inadequacy of their memory to show up.

In general then, our results suggest that one should extrapolate from laboratory-based memory tasks to memory performance in everyday life with considerable caution. Unfortunately, our questionnaire measures of memory themselves seem to be relatively unreliable, particularly in the case of patients who are still coming to terms with their memory problems. On the whole however, the relative's estimate of the problems was more satisfactory than the patient's, or indeed than the patient's or the relative's diary.

A subsequent study was carried out using normal elderly subjects instead of head-injured patients (SUNDERLAND et al. 1986). Here, we found even the spouse's interview to be unreliable. Why should that be? The most obvious reason is that such questionnaires and diaries themselves depend on being able to remember one's memory lapses. Consequently the worse a person's memory, the more lapses he or she is likely to make, but also the more likely he or she is to forget them. The poor correlations obtained between objective performance and subjective assessment found with the spouses of the normal elderly presumably reflects the fact that the spouses themselves would be elderly, and hence liable to forget. Overall therefore, our studies demonstrated that there was a problem in generalising from laboratory measures, but offered no very satisfactory alternative solution.

However, while we were concluding our studies in Cambridge, a clinical psychologist at Rivermead Rehabilitation Centre in Oxford, Barbara WILSON, was exploring another approach to the same problem. As a clinician concerned with helping brain-damaged patients cope with their memory problems, she felt the need to find alternatives to the existing laboratory-based memory tests. As she points out (WILSON 1986), such tests are very helpful in assessing whether the patient is or is not performing within the normal range on a given task, but yield little information on the practical problems the patient is likely to encounter, and give clues as to what aspects of memory to concentrate on in treatment. Bearing in mind the limitations of questionnaire and interview measures revealed by our own work, she decided to attempt to develop a type of test that was intermediate between the laboratory-based psychometric tests, and observation. She did this by first of all identifying a range of everyday memory problems on the basis of our work and her own, and then creating a series of objectively scorable tasks that would capture the essence of these problems. The outcome was the Rivermead Behavioural Memory Test.

5 The Rivermead Behavioural Memory Test

The test comprises the following subcomponents, each aiming to tap an aspect of everyday memory that tends to be impaired in patients with memory problems.

1. *Remembering a name*. The subjects are shown a photograph of someone, told their name and asked to remember it, with recall tested later in the session.

2. *Remembering a hidden belonging.* A possession belonging to the subject is borrowed and secreted in a drawer or cupboard. The subject's task is to remember to ask for it at the end of the session.
3. *Remembering an appointment.* A timer is set to ring after 20 min. The subject is instructed that when this occurs, he or she should ask "When is my next appointment?".
4. *Picture recognition.* The subject is shown line drawings of 10 common subjects and required to name them. After a delay, the subject is required to recognize the 10 targets from a set of 20.
5. *Prose recall.* A short passage of prose is read to the subject and recall is required both immediately and after a filled delay.
6. *Face recognition.* The subject is shown five faces which must be categorized according to age and sex. After a delay, the five faces have to be selected from a set of 10.
7. *Remembering a short routine.* The experimenter follows a simple route within the room (e.g. door to window to chair to table and back to door). The subject must repeat this both immediately and after a delay.
8. *Remembering to deliver a message.* While walking the route, the experimenter picks up an envelope and leaves it in a specific place. The subject's capacity to remember to do this both immediately and after a delay is noted.
9. *Orientation.* This involves questions about the current day, week and year, the present location, the age of the subject and date of birth, together with questions about the name of the current British Prime Minister and US President. The date is also requested and scored separately, since this does not necessarily correlate particularly highly with the other measures of orientation.

Performance on the test can be scored in two ways, either by deciding whether the subject passes or fails each subtest, giving a screening score, or if a more fine-grained measure is used, by scoring the number of subcomponents of each test that are successfully completed.

In a preliminary validation study, WILSON et al. (1984) asked the occupational therapists treating a range of brain-damaged patients to categorise them on the basis of whether or not they had memory problems that were severe enough to interfere with therapy. Patients categorised as having difficulties failed a mean of 10.0 out of 12 items, while those who did not had a mean score of 3.76 failed items. In general, most patients proved to have some memory problems, regardless of whether their brain damage resulted from head injury, or from a left- or right-sided stroke, as Table 1 suggests.

We are at present engaged in a more extensive validation and collection of normative data. We have used as our primary validation measure observations over an average of 55 h per patient made by the various therapists treating patients at Rivermead. At the end of each therapy session over a 2-week period, the therapist would note on a checklist the occurrence of any memory lapses. We have also collected ratings of memory problems by the patients and their relatives, but we believe that the therapists' observations have a number of advantages over these. First of all, a therapist has the advantage of seeing many different patients rather than just one, and hence is likely to give a more balanced and reliable assessment.

Table 1. Mean percentage of Behavioural Memory Test items passed by control subjects, and head-injured and stroke (CVA) patients

Subject group	*n*	Mean items passed (%)
Controls		
<50 years	20	100
50–69 years	12	86
70+ years	12	71
Head-injured	28	47
Right CVA	24	60
Left CVA	21	59

The relationship to the various patients is likely to be more objective than that of the relative, and furthermore the therapist is likely to see all patients in broadly similar surroundings. The therapy session is one in which the patient is likely to be stretched intellectually, and hence one in which memory errors are likely to occur.

We obtained ratings for a total of 80 patients, and despite the fact that mean number of lapses reported per session was relatively low, we still obtained an encouragingly high correlation between number of lapses reported and performance on the Behavioural Memory Test ($r = 0.75$). The test has proved easy and reasonably quick to administer, and is well received by patients and therapists, who both seem to like its obvious face validity.

We are still in the process of detailed analysis of the data, in particular looking at the pattern of particular profiles to see if the test is successful in predicting particular types of everyday memory problem. So far we have relied mainly on a simple screening score based on the total number of items successfully completed by the patient. However, a more detailed profile score is also available, based on the number of components of each subtest successfully completed. Since the different subtests have different numbers of components, some form of weighting will be necessary to produce a scaled score. We are currently in the process of finalising our collection of normative data, and will use this in order to determine the weightings necessary for the profile score.

Although the final validation is not quite complete, the degree of interest shown in the test was so great that we decided to publish on the basis of the preliminary evidence of validity (WILSON et al. 1985). It appears to be becoming a popular clinical test, but is also being used in studies examining performance in the normal elderly (COCKBURN and SMITH 1986) and in studies of senile dementia, where preliminary results suggest that it is sensitive to the effects of dementia, but relatively resistant to the effects of depression in the non-demented elderly. Preliminary results suggest that the tests that measure prospective memory, the capacity to remember to do things, may be particularly vulnerable to the effects of ageing (COCKBURN and SMITH 1986). Other current applications include studies of the capacity to learn new technological skills as a function of memory performance, and use of the Behavioural Memory Test to monitor possible side effects of surgery on the patient's memory.

In conclusion, although the Rivermead Behavioural Memory Test is still at an early stage of development, it has established its value clinically, and seems extremely promising as a potentially valid and sensitive indicator of the effects of drugs or stresses on everyday memory performance. It is particularly useful for studying groups such as the elderly, who often find standard psychometric testing rather stressful.

References

Anderson JA (1973) A theory for the recognition of items from short memorized lists. Psychol Rev 80:417–438

Baddeley AD (1982) Implications of neuropsychological evidence for theories of normal memory. Proc R Soc Lond B 298:59–72

Baddeley AD, Ecob JR (1973) Reaction time and short-term memory: implications of repetition for the high-speed exhaustive scan hypothesis. Q J Exp Psychol 25:229–240

Baddeley AD, Harris J, Sunderland A, Watts K, Wilson B (1987) Closed head injury and memory. In: Levin H (ed) Neurobehavioral recovery from head injury. Oxford University Press, pp 295–317

Brooks DN (1972) Memory and head injury. J Nerv Mental Dis 155:350–355

Cockburn J, Smith P (1986) A dissociation between age and intelligence on two memory tasks. Paper presented at the Annual Conference of the B. P. S. Neuropsychology Interest Group, Nottingham, 8 November 1986

Corballis MC, Kirby J, Miller A (1972) Access to elements of a memorized list. J Exp Psychol 94:185–190

Ebbinghaus H (1885) Über das Gedächtnis. Dunker, Leipzig (Translation: Ruyer H, Bussenius CE (1913) Memory. Teachers College, Columbia University, New York)

Miller E (1977) Abnormal ageing. Wiley, New York

Sternberg S (1966) High-speed scanning in human memory. Science 153:652–654

Sternberg S (1975) Memory scanning: new findings and current controversies. Q J Exp Psychol 27:1–32

Sunderland A, Harris JE, Baddeley AD (1984) Assessing everyday memory after head injury. In: Harris JE, Morris PE (eds) Everyday memory, actions and absentmindedness. Academic, London, pp 191–206

Sunderland A, Watts K, Baddeley AD, Harris JE (1986) Subjective memory assessment and test performance in the elderly. J Gerontol 41:376–385

Theios J (1973) Reaction time measurements in the study of memory processes: theory and data. In: Bower GH (ed) The psychology of learning and motivation, vol 7. Academic, New York, pp 43–85

Wilkins AJ, Baddeley AD (1978) Remembering to recall in everyday life: an approach to absentmindedness. In: Gruneberg MM, Morris PE, Sykes RN (eds) Practical aspects of memory. Academic, London, pp 27–34

Wilson BA (1986) The Rehabilitation of Memory. Guilford, New York

Wilson BA, Baddeley AD, Hutchins H (1984) The Rivermead Behavioural Memory Test: a preliminary report. Rivermead Rehabilitation Centre, Oxford (Technical Report 84/1)

Wilson BA, Cockburn J, Baddeley AD (1985) The Rivermead Behavioural Memory Test. Thames Valley Test Co., 34 The Square, Titchfield, Hants, England

Drugs and Information Processing in Skilled Performance

A. F. SANDERS[1] and C. H. WAUSCHKUHN

Abstract

A concise review is presented of recent research on task aspects with respect to determining the effects of drugs on human information processing. It is concluded that progress in this area is hampered by lack of a theoretical basis to most behavioral tasks, preventing firm conclusions about the effects of drugs on either well-defined mental functions or on real-life performances. It is argued that the effects of drugs should only be tested in behaviorally well-researched tasks. Some proposals are discussed with an emphasis on perceptual-motor skills.

1 Introduction

Notions about the nature of the effects of organismic and environmental stressors on human performance are gradually changing from the hypothesis that stressors have rather general and nonspecific effects on arousal and mood to the hypothesis that they have quite specific points of application on aspects of human information processing. This shift in emphasis is a natural consequence of the change in conceptualization about the relation between energetic and information processing aspects of human performance. As long as the hypothesis prevailed that energetics and information processing were relatively independent, it could only follow that stressors affected a generalized drive state or arousal mechanism (e.g. HEBB 1955). Even if this simple notion is abandoned and replaced by multiple energetic systems (PRIBRAM and MCGUINNESS 1975), the effects on the level of performance may still be nonspecific; there may be no specific ties between energetic and processing mechanisms.

The next step, however, is the idea that performance is based upon the operations of highly specific cognitive-energetic structures (HOCKEY 1979; RABBITT 1979; SANDERS 1983), which implies that stressors have quite specific effects depending on the type of stressor as well as on the type of task. This development has also been seen in the concept of stress, which changed from a situation-independent physiological response of the body (SELYE 1956) to quite situation-specific stress patterns (e.g. MASON 1975; URSIN et al. 1978).

Psychotropic drugs have traditionally been subsumed under the category "organismic stressors" and the most simple notion has been that their operation is largely limited to either stimulation or sedation. Although their operation is still usually tied to the energetics of behavior, the current emphasis on cognitive-en-

[1] Institute of Psychology RWTH Aachen, Jägerstrasse 17/19, 5100 Aachen, FRG.

ergetic structures has blurred the sharp distinction between energetics and cognition. In addition, the notion is gaining ground that psychotropic drugs may also have direct effects on cognitive processing mechanisms involved in perceptual, memoryal, reasoning, and motor functions. The search for a "learning pill," or at least for a drug affecting memory functions (e.g. GAILLARD 1980), is a case in point.

At the same time this development means that there is a considerable interest in developing standard performance test batteries, the tests of which reflect certain mental functions and can be assumed to "measure" these functions or at least the quality of their functioning. The recent growth in interest in composing such batteries (e.g. FLEISHMAN and QUAINTANCE 1984; SANDERS et al. 1986) is at least partly based on dissatisfaction with assessing the effects of psychotropic drugs on arbitrary tasks and generalizing from there to presumed underlying mechanisms or real-life tasks.

2 Task Versus Process Orientation

Following SCHMIDT (1982), the aims of behavioral research are either task or process oriented. I may add that I have never seen a positive result from attempts to combine both orientations in one study. Usually neither aim is served.

In a task orientation one chooses a task or a set of tasks as representative of a skill or a relevant aspect of a skill. Thus, in the area of motor behavior, tests of pursuit and compensatory tracking, stability, complex coordination (i.e., matching light sequences through hand and foot controls) and, more recently, the reference tests specific for Fleishman's motor abilities, have been quite popular (FLEISHMAN and QUAINTANCE 1984). It is assumed, however, that aspects of, say, flying, car driving, manual and supervisory control, and sport skills, can be properly simulated so that a set of relatively simple tests could serve for selecting personnel as well as for determining the human factors involved in complex real-life tasks. As such, the task-oriented approach is fairly atheoretical but has the advantage of preventing ivory-tower attitudes. It ought to be admitted that the need for more adequate models is gradually being realized (e.g., SANDERS 1986) since the correlations between the test tasks and the criterion measures of the actual skills (e.g., flying) are not impressive. It is noteworthy that the most encouraging results still always stem from the performance tests such as those designed by MELTON (1947) for pilot selection (0.3–0.4). Various reasons can be mentioned for this disappointing state of affairs, including differences in practice between the selection test and the real-life skill, the possibility that as practice proceeds, originally separate skills are transformed into a new integrated skill, which is not reflected by the simple tests in the laboratory, differences in perceptual-motor load between the laboratory tests and the real-life skill, and finally, differences in motivational state.

It is, of course, not out of the question that more satisfactory measures can be developed by using tasks which have been thoroughly analyzed with regard to the mechanisms and processes that play a role in their performance. In fact, I have recently argued in favor of developing a set of laboratory tests for driving (SANDERS 1986). Yet the problems that are encountered should not be underestimated and they should warn any researcher to refrain from using laboratory tasks as so-called simulations of, say, driving skills. In the same way one should be extremely careful about generalizing from performance in simple laboratory tests to that in complex real-life skills. In particular, this error is frequently made in research on effects of psychotropic drugs (e.g., WILLUMEIT et al. 1984); the dangers will be discussed in more detail in the next section.

With a process orientation the same types of task may be used as with a task orientation, but with very different rationales. Real-life simulation is not the primary aim but rather the determination of the internal processes underlying performance. The experiments are designed to test predictions based on different theoretical views. Applied to psychotropic drugs one would not generalize the results to the real world but rather to some internal mechanism or process. Again, this should not be taken too simply, in that a task should not be thought to reflect a single function such as "memory," "speed," or "perceptual encoding." It is clear that, even in the most simple laboratory settings, performance is based upon a composite of internal processes. Laboratory research has been carried out in the hope that attempts towards uncovering these composite processes through variations of interesting independent variables would lead to sets of theoretical principles, the value of which go beyond the specific settings and paradigms that had been used originally. Whether this hope is justified has been seriously questioned (e.g., ALLPORT 1980; TURVEY et al. 1978) and the issue will undoubtedly continue to be one of the major methodological questions of the years to come. GOPHER and SANDERS (1984) and SANDERS (1984) have proposed so-called "back-to-back" experimentation. In this approach twin experiments are carried out which use similar paradigms but of a different level of complexity. The idea is to investigate whether theoretical principles underlying the more simple paradigm are still relevant in explaining performance at a more complex level. It should be noted, though, that analysis of real-life skills in terms of process models is feasible but becomes increasingly difficult as the task becomes more complex.

In the field of psychotropic drugs and performance the matter of the meaning of a task is often underestimated or hardly considered at all (e.g., HINDMARCH 1980). Yet process-oriented research, incomplete as its models may be, could provide relevant tools for locating effects of psychotropic drugs and thus serve as a bridge between "brain and behavior." It is quite relevant, therefore, to consider well-researched paradigms with respect to their usefulness as standard tasks for evaluation effects of psychotropic drugs. In the next sections some recent research on drugs (especially the benzodiazepines) and performance will be reviewed; it will be concluded that the tasks that are used in most research are extremely arbitrary, and that the types of problems outlined so far, have hardly been touched upon. This section is followed by some proposals and preliminary considerations with regard to potentially useful paradigms and standard reference tests.

3 Benzodiazepines and Performance Measurement

In 1980, HINDMARCH presented a broad synopsis of the experimental tasks commonly used in drug research. Most of the studies he mentioned seem to be neither task nor process oriented, but rather claim to tap some vaguely defined psychomotor function.

To check whether this is still true for recent studies, a sample of the literature (1981–1986) on the effects of benzodiazepines on performance and memory in normal subjects has been reviewed. This review is neither exhaustive nor representative. The literature sample was taken more or less from the *Index Medicus* with the help of a literature search computer program. Our main focus of interest has been on the tasks used and their theoretical basis.

In general, the research strategy is much the same as before 1980. Most studies have to be regarded as superficial with regard to critical evaluation of the tasks. They neither aim at real-life skills nor intend to investigate basic psychological mechanisms underlying drug effects. They merely show that benzodiazepines impair, say, "memory," but they are not interested in analyzing which aspects or processing stages are involved. Explicit task or process orientation is found only in a small number of studies.

Apart from convention, in these "surface-oriented" studies the main criterion for selecting a specific task seems to be the task's "sensitivity to drug effects." Further theoretical considerations are seldom reported. Results are either interpreted on a quasi-operational level – which as such is no problem; it only prevents integration of results into a broader theoretical context – or they are supposed to measure specific psychomotor functions. This latter approach may well be promoted by articles like HINDMARCH's (1980), which provides a rather tempting scheme in which a set of drug-sensitive tasks is related to a set of psychomotor functions (Fig. 1). In the course of his paper HINDMARCH reports evidence that most of these tasks are indeed also especially affected by benzodiazepines (see Table 1 for a concise summary). Claiming a fictitious one-to-one relation between specific tasks and processes means a rather dangerous simplification: any task affects a number of mental functions (SANDERS 1986), since at least sensory input, central processing, and motor output functions are always involved.

HINDMARCH explicitly refers to his scheme as a "scheme for conducting an investigation on the effects of a psychoactive drug on human psychomotor functions ..." and "Any study which utilises measures from each of the above divisions of human psychomotor performance will produce relevant, valid and reliable results only if the experimental conditions, methodology and selection of subjects are carefully controlled" (HINDMARCH 1980, p. 202).

This relevant and certainly necessary (e.g., see MEWALDT et al. 1983) warning has probably often been misinterpreted as an assurance that the use of measures from the scheme combined with control of experimental errors is sufficient for relevant, reliable, and valid results. Sensitive standard measures and experimental error control are certainly necessary conditions, but whether they are also sufficient depends on theoretical considerations.

Table 2 clearly shows that most of the popular tasks used since 1981 are included in HINDMARCH's task set. Table 3 summarizes the main empirical results.

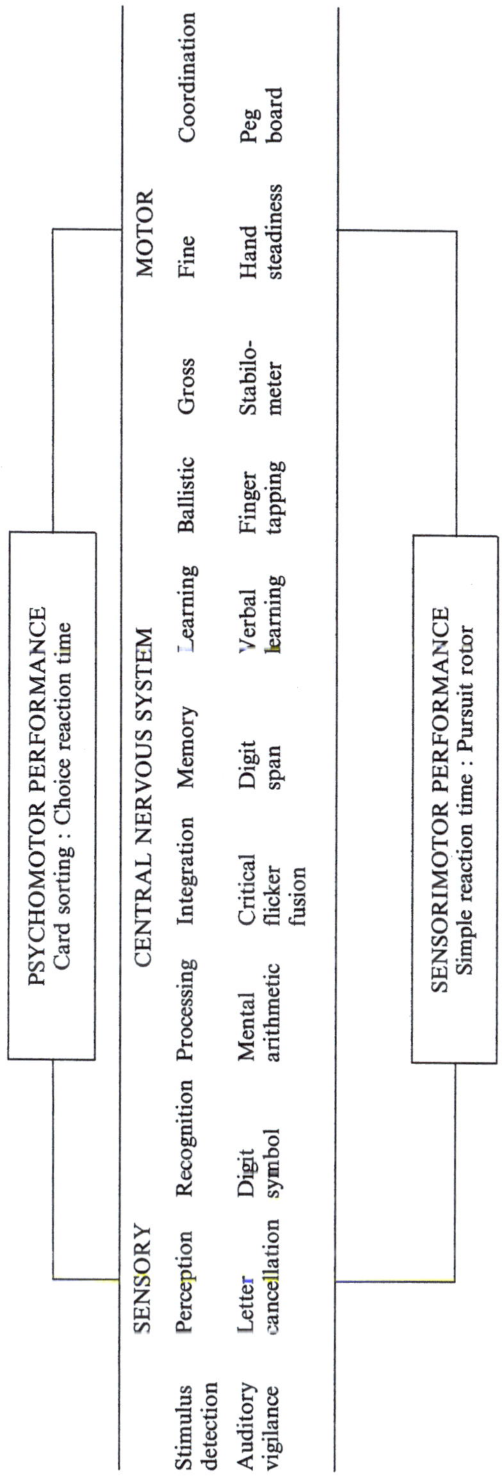

Fig. 1. A summary of tests of psychomotor function which have been shown to be sensitive to a wide range of psychoactive drugs (from HINDMARCH 1980)

Table 1. Summary of the effects of benzodiazepine reported by Hindmarch (1980)

Card sorting	
# ZIMMERMANN-TANSELLA et al. (1976)	Chlordesmethyldiazepam 10, 20
# TANSELLA et al. (1974)	*N*-desmethyldiazepam 20
# MALPAS and JOYCE (1969)	Nitrazepam 10
# MALPAS (1972)	Nitrazepam 10
# VELDKAMP et al. (1974)	Triazolam 0.5, 1
Choice reaction time	
See HINDMARCH (1980, Table 4)	
Vigilance	
# HART et al. (1976)	Diazepam 2.5
# PECK et al. (1977)	Nitrazepam 10
/ PECK et al. (1977)	Nitrazepam 5
Symbol cancellation	
# ZIMMERMANN-TANSELLA et al. (1976)	Chlordesmethyldiazepam 1, 2
# FILE and BOND (1979)	Lorazepam 1, 2.5
# LAWTON and CAHN (1963)	Diazepam 5
# BOND and LATER (1972)	Nitrazepam 5, 10
# STITT et al. (1977)	Diazepam 5
# JONES et al. (1978)	Diazepam 5
/ BOND and LADER (1975)	Flunitrazepam
/ HINDMARCH and CLYDE (1980a)	HR 158
DSST	
See HINDMARCH (1980, Table 1)	
Mental arithmetic	
# MASUDA and BAKKER (1966)	Diazepam 10, 20
# FROSTAD et al. (1966)	Diazepam 10–20
# HINDMARCH (1977)	Flunitrazepam 1
# HINDMARCH (1977)	Flurazpam 15
# HINDMARCH and PARROTT (1979)	Clorazepate 15
# HINDMARCH and GUDGEON (1980)	Lorazepam 1
# HINDMARCH and CLYDE (1980b)	Triazolam 0.5
# HINDMARCH and CLYDE (1980b)	Nitrazepam 10
# HINDMARCH and CLYDE (1980a)	HR 158
Critical flicker fusion frequency	
See HINDMARCH (1980, Table 2)	
Digit span	
# JONES et al. (1979)	Nitrazepam 5
# HAFFNER et al. (1973)	Diazepam 10, 20
/ OGLE and DITMAN (1966)	Chlordiazepoxide 10, 25
Tapping	
# PECK et al. (1977)	Nitrazepam 10
# CERNY et al. (1973)	Diazepam 10
# GHONHEIM et al. (1975)	Diazepam 10
# SALKIND and SILVERSTONE (1975)	Flurazepam 30
# DIMASCIO and BARRETT (1965)	Oxazepam 10
# BOND and LADER (1972)	Nitrazepam 5, 10
* ZIMMERMANN-TANSELLA et al. 1976	Chlordesmethyldiazepam 1
Stabilometer	
* HINDMARCH (1979)	Nitrazepam 5
* HINDMARCH (1979)	Clobazam 20

Table 1 (continued)

* WITTENBORN et al. (1979)	Clobazam 10
/ WITTENBORN et al. 1979	Diazepam 5
# ORR et al. 1976	Diazepam 10, 20
Handsteadiness	
* DOONGAJI (1979)	Clobazam 10
Pegboard	
# ROTH et al. (1977)	Flurazepam
# ROTH et al. (1977)	Triazolam
/ HINDMARCH and GUDGEON 1980	Lorazepam 1
/ HINDMARCH and GUDGEON (1980)	Clobazam 10
Simple reaction time	
See HINDMARCH (1980, Table 3)	
Pursuit Rotor	
# SALKIND and SILVERSTONE (1975)	Flurazepam 30
# OGLE et al. (1976)	Diazepam 5
# OGLE et al. (1976)	Lorazepam 2

impairment of performance; * improvement; / no influence on performance.
The numbers following the drug names give doses in mg for references, see HINDMARCH (1980).

In recent studies the tasks have usually been interpreted according to HINDMARCH's scheme. There are only a few attempts which aim at the investigation of underlying psychological mechanisms through systematic variation of task parameters. One example is FILE and LISTER's (1982) work about variation of rehearsal in a verbal learning task; another example is a study by MORGAN (1984), who varied target size in a tapping task, and found that benzodiazepines affect accuracy rather than speed.

Memory is an essential area in which research activities are not restricted to a single standard task. Although the individual models of memory vary, most studies claim to investigate effects of benzodiazepines on memory functions quantitatively as well as qualitatively. Table 4 summarizes the tasks used. The majority of studies have used immediate or delayed free recall in order to assess the effects on short- and long-term retention. Some studies not only used the proportion of correct recall as a dependent variable, but additionally analyzed the serial position curve of free recall in terms of asymptote and recency (e.g., MEWALDT et al. 1983).

Experiments on single-list memorization preceding or following drug administration showed that neither retention nor retrieval was directly impaired but that the acquisition of new information was directly impaired by benzodiazepines. Indirectly, the negative effect on acquisition even had the effect of facilitating retrieval of material learned before drug administration. Less acquisition of new material presumably causes less interference with material learned earlier (e.g., HINRICHS et al. 1984; GHONHEIM et al. 1984a). This effect is quite opposite to the phenomenon of state-dependent memory.

Table 2. Tasks used in the reviewed studies (except long-term memory tasks); numbers refer to literature references

Hindmarch's drug-sensitive tasks						
Card sorting	8					
Choice reaction time	2					
Auditory vigilance	22					
Symbol cancellation	5	8	9	13		
DSST	1	6	22			
Mental arithmetic	2	7	8	9	13	21
Critical flicker fusion frequency	3	6	19			
Digit span	3	10	17	19	22	
Tapping	1	8	9	13	20	
Stabilometer	13					
Hand steadiness						
Pegboard	2	10	21			
Simple reaction time	2	3	19	22		
Pursuit rotor	20	21				
Other tasks						
Symbol copying	1	22				
Tracking	2	21				
Divided attention	22					
Sustained attention	6					
Spatial rotation	13					
Sequence completion	13					

1 File and Lister (1983).
2 McManus et al. (1983).
3 Pomara et al. (1984b).
4 Scharf et al. (1983).
5 Kleindienst-Vanderbeke (1984).
6 Wesnes and Warburton (1984).
7 File and Lister (1982).
8 Mewald et al. (1986).
9 Ghonheim (1984a, b).
10 Brown et al. (1983).
11 Hinrichs et al. (1984).
12 Ghonheim (1984a).
13 Ghonheim et al. (1984b).
14 Subhan and Hindmarch (1984).
15 Scharf et al. (1984).
16 Desai et al. (1983).
17 Lister and File (1984).
18 Mewaldt et al. (1983).
19 Pomara et al. (1984a).
20 Morgan (1984).
21 Ott (1984).
22 Roehrs et al. (1984).
23 Subhan (1984).

The majority of memory studies implicitly proceed from Atkinson and Shiffrin's memory model. In contrast, Kleindienst-Vanderbeke's (1984) study on the influences of benzodiazepines on storage strategies is based on Craik and Lockhart's "levels of processing" approach. The reproduction deficit following administration of benzodiazepines is ascribed to decreased congruence between strategies of storage and retrieval.

Apart from tests of span and free recall in assessing memory effects of benzodiazepines, recent studies have used memory search (Sternberg 1969), which allows differentiation between effects on different processing stages, such as stimulus encoding, serial comparison, and response organization. Thus, a study by Subhan and Hindmarch (1984) showed differential effects on these processing

Table 3. Summary of the benzodiazepine effects reported in recent literature (1981–1986)

Card sorting	
# MEWALDT et al. (1986)	Diazepam 0.3/kg
/ MEWALDT et al. (1986)	Oxazepam 1.2/kg
Choice reaction time	
/ MCMANUS et al. (1983)	Loprazolam 1
Vigilance	
# ROEHRS et al. (1984)	Flurazepam 30
/ ROEHRS et al. (1984)	Temazepam 30
/ ROEHRS et al. (1984)	Lormetazepam 1.5
Symbol cancellation	
/ MEWALDT et al. (1986)	Diazepam 0.3/kg
/ MEWALDT et al. (1986)	Oxazepam 1.2/kg
# GHONHEIM et al. (1984a)	Diazepam 0.2/kg
# GHONHEIM et al. (1984b)	Diazepam 0.1–0.3/kg
# KLEINDIENST-VANDERBEKE (1984)	Clobazam 30
# KLEINDIENST-VANDERBEKE (1984)	Lorazepam 3
DSST	
# FILE and LISTER (1983)	Lorazepam 2.5
/ WESNES and WARBURTON (1984)	Temazepam 40
/ WESNES and WARBURTON (1984)	Flurazepam 30
# ROEHRS et al. (1984)	Flurazepam 30
# ROEHRS et al. (1984)	Temazepam 30
/ ROEHRS et al. (1984)	Lormetazepam 1.5
Mental arithmetic	
# MCMANUS et al. (1983)	Loprazolam 1
# FILE and LISTER (1982)	Lorazepam 1, 2.5
# MEWALDT et al. (1986)	Diazepam 0.3/kg
# MEWALDT et al. (1986)	Oxazepam 1.2/kg
# GHONHEIM et al. (1984a)	Diazepam 0.2/kg
# GHONHEIM et al. (1984b)	Diazepam 0.1–0.3/kg
# OTT (1984)	Lormetazepam 1
Critical flicker fusion frequency	
/ POMARA et al. (1984b)	Diazepam 2.5, 5, 10
/ WESNES and WARBURTON (1984)	Temazepam 40
/ WESNES and WARBURTON (1984)	Flurazepam 30
/ POMARA et al. (1984a)	Diazepam 2.5, 5, 10
Digit span	
/ POMARA et al. (1984b)	Diazepam 2.5, 5, 10
/ BROWN et al. (1983)	Lorazepam 2.5
/ LISTER and FILE (1984)	Lorazepam 2.5
/ POMARA et al. (1984a)	Diazepam 2.5, 5, 10
/ ROEHRS et al. (1984)	Flurazepam 30
/ ROEHRS et al. (1984)	Temazepam 30
/ ROEHRS et al. (1984)	Lormetazepam 1.5
Tapping	
# FILE and LISTER (1983)	Lorazepam 2.5
/ MEWALDT et al. (1986)	Diazepam 0.3/kg
/ MEWALDT et al. (1986)	Oxazepam 1.2/kg
# GHONHEIM et al. (1984a)	Diazepam 0.2/kg
# GHONHEIM et al. (1984b)	Diazepam 0.2/kg, 0.3/kg
/ GHONHEIM et al. (1984b)	Diazepam 0.1/kg

Table 3 (continued)

/ MORGAN et al. (1984)	Lormetazam 1
# MORGAN et al. (1984)	Nitrazepam 5
Stabilometer	
# GHONHEIM et al. (1984b)	Diazepam 0.1–0.3/kg
Pegboard	
# MCMANUS et al. (1983)	Loprazolam 1
# BROWN et al. (1983)	Lorazepam 2.5
/ OTT (1984)	Lormetazepam 2
Simple reaction time	
/ MCMANUS et al. (1983)	Loprazolam 1
# POMARA et al. (1984)	Diazepam 2.5, 5, 10
# POMARA et al. (1984a)	Diazepam 2.5, 5, 10
# ROEHRS et al. (1984)	Flurazepam 30
/ ROEHRS et al. (1984)	Temazepam 30
/ ROEHRS et al. (1984)	Lormetazepam 1.5
Pursuit rotor	
/ MORGAN (1984)	Lormetazepam 1
/ MORGAN (1984)	Nitrazepam 5
# OTT (1984)	Lormetazepam 2
Symbol copying	
/ FILE and LISTER (1983)	Lorazepam 2.5
/ ROEHRS et al. (1984)	Flurazepam 30
/ ROEHRS et al. (1984)	Temazepam 30
/ ROEHRS et al. (1984)	Lormetazepam 1.5
Tracking	
# MCMANUS et al. (1983)	Loprazolam 1
# OTT (1984)	Lormetazepam 2
Sustained attention	
/ WESNES and WARBURTON (1984)	Temazepam 40
# WESNES and WARBURTON (1984)	Flurazepam 30
Divided attention	
# ROEHRS et al. (1984)	Flurazepam 30
/ ROEHRS et al. (1984)	Temazepam 30
/ ROEHRS et al. (1984)	Lormetazepam 1.5
Spatial rotation	
# GHONHEIM et al. (1984b)	Diazepam 0.1–0.3/kg
Sequence completition	
/ GHONHEIM et al. (1984b)	Diazepam 0.1–0.3/kg

# indicates impairment of performance; * improvement; / no influence on performance. The numbers following the drug names give doses in mg.

stages for benzodiazepines and a pharmacologically and clinically similar non-benzodiazepine.

All the above-mentioned tasks refer to episodic memory. Studies on semantic categorization show that benzodiazepines do not affect semantic memory. Impaired performance in such tasks, as measured by longer reaction times, can be

Table 4. Memory tasks used in the reviewed literature (numbers refer to literature references, see Table 2)

	Verbal						Visual	
Immediate free recall	4	5	7	8	9	10	3	19
	11	12	15	17	18			
Delayed free recall	4	8	9	10	11	12	3	19
	15	17	18					
Recognition	8	9	10	11	17	19	5	17
Running memory	16							
Paired associates	1							
List-learning	2	3	8	9	12	19		
Semantic category task	3	9	10	12				
Sternberg task	14	23						
Backward reading	17							

ascribed to speed rather than to memory factors (GHONHEIM et al. 1984a; BROWN et al. 1983).

Finally, it should be mentioned that multitrial tests appear to be more sensitive to effects of benzodiazepines than single-trial procedures, because benzodiazepines seem primarily to block acquisition of new information or skills (GHONHEIM et al. 1984a), or to delay improvement of performance. The size of the benzodiazepine effect is considered to be a direct function of the task's learning component (e.g., GHONHEIM et al. 1984a). This is in contrast to results from experiments aiming to study procedural knowledge. For example, FILE and LISTER (1983) found that benzodiazepines do not impair the acquisition of backwards-reading skills.

Considering the results on memory, learning, and psychomotor performance together, it should be noted that the empirical data do not necessarily imply an effect of benzodiazepines on learning as such. Less rapid acquisition or improvement of performance could also be due to changes in motivation (SANDERS and HOOGENBOOM 1970).

Apart from the above-mentioned tasks there are also simple "task-oriented" memory tasks. ROTH and coworkers "tried to develop a set of tasks with direct clinical relevance. The tasks were chosen to mimic real life situations which might be encountered by patients during a night-time awakening after having used a benzodiazepine at bedtime" (ROTH et al. 1984), like pill-taking, etc. Using such a task battery, ROTH and his coworkers found differential amnesic effects for various benzodiazepines (e.g., ROTH et al. 1984; ROEHRS et al. 1984).

Results from this wide range of studies seem to be consistent, but due to the diversity and the eclectic character of the tasks, as well as to the lack of theoretical considerations, it is hard to reach a more psychologically relevant conclusion than that impairment of performance and memory are functions of drug, dose, time, application form, and population. The problem is that the experiments aim to find effects on tasks rather than on the mental functions underlying the tasks.

4 Some Possible Standard Paradigms

4.1 The Additive Factor Approach

One way of going beyond task effects is to be concerned with interactions rather than with main effects. Thus if, a drug constitutes one independent variable (drug vs. placebo) and a task factor another (e.g., stimulus quality), an interaction between the effects of these variables suggests that the drug affects perceptual encoding. For example, the effect of the drug might be stronger as signals are more degraded. For a long time the position has been defended that such an approach should be strongly preferred to a task battery, the structural properties of which are virtually unknown (SANDERS 1977).

The relevance of this notion has been clearly demonstrated by the work of FROWEIN (1981 a, b), who observed highly selective effects of pentobarbital und phentermine HCl on choice reaction time variables. The barbiturate had a stronger effect on the response to degraded than to intact signals, while its effects were not selectively related to any other variations, such as signal intensity, stimulus-response compatibility, foreperiod duration, or complexity of the response. This result was essentially replicated by LOGSDON et al. (1984). SUBHAN (1984) also found interactive effects of some benzodiazepines and the variables stimulus quality and memory set size in a STERNBERG memory-scanning task. Using a memory-scanning task, SUBHAN and HINDMARCH (1984) also investigated the effects of zopiclone, flunitrazepam, and triazolam, which were found to be located in the memory search as well as in the response selection stages of the reaction process.

In tests with phentermine HCl, the effects of the drug were largely located in the output stages of the reaction process. Interactions were found between the effects of phentermine and the effects of foreperiod duration and of movement variables. The relations were particularly pronounced after 24 h sleep loss (FROWEIN et al. 1981 b).

Establishing patterns of interactions and additive effects of drugs with variables affecting choice reaction time is within the framework of the additive factor logic (STERNBERG 1969; SANDERS 1980). This method merely claims to describe clusters of interacting variables, thus defining stages of processing. Hence, the additive factor logic is not a process model of choice reaction time; rather, it describes substructures of choice reaction processes which in turn require individual process models. Yet classification of drug effects in terms of which stages are and are not affected carries the interpretation one step beyond simple task effects.

4.2 The Factor-Analytic Approach

Application of the additive factor method is limited to the analysis of choice reaction times and hence is concerned with a limited range of behavioral phenomena. In this respect the factor-analytic approach which – similarly to the additive factor method – merely claims to describe clusters of performance relations, may cover a much wider area. This approach forms the basis of the research of FLEISHMAN and associates (e. g., FLEISHMAN and QUAINTANCE 1984) who have defined some 50 mental abilities and accompanying reference tests. Various abilities are related to sensory and motor functions (e.g., arm-hand steadiness, manual dexterity, wrist-finger speed, near vision, night vision) while some 15 factors concern central functions including comprehension, reasoning, category flexibility, and memorization. As remarked, the abilities merely reflect the outcomes of factor-analytic studies and do not constitute process models of information processing. Yet the abilities are based upon more than simple factor-analytic studies. In the projects of FLEISHMAN and associates, the quality and boundary conditions of factors were subsequently checked, by way of further factor-analytic studies in which variations of the tests were used to determine in which cases a factor was also involved in such variations. For example, correlations between various types of tracking task led to the definition of an ability called "rate control." Later studies showed that, when subjects had to time movements in relation to different stimulus rates but did not have to follow a target or compensate for its movement, the ability "rate control" was still involved. Actual movements appeared to play a role, however, since the correlation disappeared in tests that asked merely for judgements about the future location of stimuli. While most of the work in the perceptual-motor area stems from FLEISHMAN's own research, the primary sources for the cognitive and perceptual domains come from GUILFORD (1967) and from FRENCH and associates (FRENCH 1951; FRENCH et al. 1963).

It has been clearly realized that the subset of abilities which plays a role in a given task is not static but depends on variables such as practice. In a now classic paper, FLEISHMAN and HEMPEL (1955) used a multichoice reaction task in which subjects responded to light patterns. The correct response depended on the combination of lights presented. Early in practice the task loaded considerably on abilities concerned with spatial relations and verbal skills. Later in practice the contribution of motor factors became larger while the verbal component almost disappeared (see also HEUER 1984). This is consistent with FITTS' (1964) hypothesis that skill acquisition starts with a verbal-analytic phase and results in almost complete automatic perceptual-motor control.

It is interesting to note that the effects of drugs and of environmental stressors such as noise (THEOLOGUS et al. 1974) on subsets of reference tasks (i.e., tasks loading specifically on one ability and not on others) have shown selective rather than general effects. This is consistent with the general trend in FROWEIN's work. Another relevant observation is that most of the processing stages, as uncovered by additive factor logic in standard choice reaction tests, can be readily equated to FLEISHMAN's abilities (see Table 5), although validation in a wider factor-analytic study including FLEISHMAN's reference tests has not yet been attempted.

In conclusion, it is undoubtedly true that FLEISHMAN's reference tests deserve closer scrutiny despite a number of shortcomings and traditional criticisms of the factor-analytic approach. These criticisms include (a) the extreme task orientation, (b) the reduction of behavioral theory to a number of independently operating abilities or traits, and (c) the common doubts about the factor-analytic technique with respect to rotation, unaccounted and specific variance, and exhaustiveness. It has been seriously questioned whether performance in more complex

Table 5. Relations between additive factor variables and factor-analytic abilities

Additive factor variable	Factor analytic ability
Signal intensity	Visual color discrimination
Signal quality	Flexibility of closure
Signal discrimination	Visualization/perceptual speed
Memory set size	Category flexibility
Stimulus – response compatibility	Response orientation
Motor programming	Multilimb coordination
Motor adjustment	Reaction time

tasks can be described in terms of a multiple regression analysis with different weights on a constant set of abilities. There is obviously a total neglect of strategic aspects of performance and of hierarchical principles in mental functioning. A further criticism is that the abilities are largely limited to perceptual-motor and physical skills. Yet the criticisms do not remove the fact that FLEISHMAN and associates have certainly contributed to constructing a more general task taxonomy and its accompanying reference tests. It is even more relevant that, despite their limitations, his procedures suggest a way forward.

4.3 Process-Oriented Tasks

Process-oriented experimental paradigms do not provide simulations or even attempt to approach reality. Rather, they suggest tests which distinguish between types of processes that are thought to be affected by a drug or by an experimental stressor.

4.3.1 Reactive Inhibition

One example concerns the so-called "reactive inhibition" paradigm in sensorimotor skills like tracking and discrete choice reactions. A common finding is that when such a skill is practised uninterruptedly over a prolonged period of time, hardly any improvement is observed in comparison with the condition where brief periods of practice and rest periods alternate (massed vs. spaced practice). It has been well established (e.g., ADAMS and REYNOLDS 1954; SANDERS and HOOGENBOOM 1970) that massed and spaced practice schedules do not basically differ with regard to the actual degree of skill acquired. If a group of subjects is assigned to massed practice but shifts to spaced practice after a number of sessions, the same performance level is observed in the spaced practice trials as for the group that had spaced practice throughout the experiment. Again, the level of performance hardly differs between subjects who receive either massed or spaced practice on a common after-test (Fig. 2). The usual interpretation is that subjects learn equally well with massed and spaced practice but suffer from brief periods

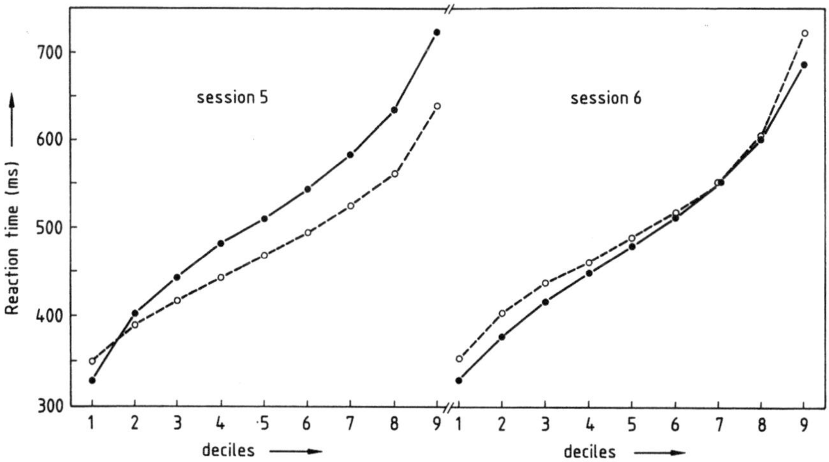

Fig. 2. Mean reaction times as a function of deciles. In sessions 1–5 subjects received either massed (*continuous line*) or spaced practice (*dotted line*). In session 6, all subjects had massed practice. The massed vs. spaced indication refers to the condition in sessions 1–5 (from SANDERS and HOOGENBOOM 1970)

of inefficiency in the massed condition. These lead to a performance loss which counteracts the learning effect. At a higher level of skill, there is less learning but there are also fewer moments of inefficiency due to increasing automatization of performance and probably also due to experience in working uninterruptedly over longer periods of time. Consistent with this interpretation, SANDERS and HOOGENBOOM (1970) found that with massed practice, the highest deciles of the distribution of choice reaction times increased as a function of time on task, while the lowest deciles showed practice effects of a size comparable to those observed with spaced practice.

This paradigm was employed in experiments that aimed at distinguishing effects of the peptide adrenocorticotrophic hormone (ACTH) 4–10 on either learning – skill acquisition – or motivation (GAILLARD and SANDERS 1975). It was found that when using a massed practice schedule in which self-paced discrete choice reactions were carried out uninterruptedly for 30 min, followed by a break and then by a 2-min after-test, that the ACTH and placebo groups did not differ significantly in the after-test, although the ACTH group showed superior performance during the 30-min work spell. More detailed analyses confirmed that this advantage of ACTH was due to less frequent periods of inefficiency during the 30-min work period. This result, which was subsequently confirmed by GAILLARD and VAREY (1979), suggests that learning was not affected by ACTH 4–9, but that the drug had a stimulating effect on maintaining motivation in a self-paced serial reaction task. It is perhaps interesting to note that tests of the effect of ACTH 4–9 on learning and retention of paired associates did not show any effect of the drug. In verbal learning there is usually very little evidence for effects of high motivation, perhaps because the test sessions are relatively short and contain frequent feedback at the test trials (SCHMIDT 1982). It is interesting that effects of sedatives on tests of memory span (TALLAND and QUARTON 1965) have been found to depend on presentation speed. This also suggests that there is no direct effect of the drug on learning (acquisition or storage) but rather on rehearsal activity. Sedative drugs had a relatively stronger negative effect on short-term retention under a condition with a longer presentation time and in particular when the material was presented visually.

4.3.2 Strategy Changes

RABBITT and HOCKEY (1979) have strongly criticized the earlier discussed abilities approaches. They maintain that effects of stressors are not reflected by way of suboptimal processing in a subset of abilities or processing stages but rather by changes of strategy in allocating processing resources to certain aspects of the tasks. Processing under stress changes qualitatively and not quantitatively. Research should employ tasks sufficiently complex to allow strategy changes to become apparent.

Although HOCKEY's research (HOCKEY 1979) has usually been concerned with stressors like noise and sleep loss, his main argument applies equally to drug research. Two of his experimental tasks will be briefly described. The first is a traditional running span task in which subjects receive a long list of consonants which ends at some unexpected point in time. The instruction is to recall as many items as possible in the correct order of presentation from the end of the list. The main finding was that, under intensive noise, recall of the final items improved while recall of the less recent items declined. The reverse was found after sleep loss. In that case the less recent items were recalled better while the final items were recalled less well. The conclusion is that noise promotes a strategy of "immediate throughput" of information, while after sleep loss subjects are more inclined to consider earlier-stored items at the expense of recall of the most recent ones. Another strategy shift concerns phenomena of "levelling vs. sharpening" of attention in a divided-attention test. HOCKEY's task consisted of centrally located tracking together with monitoring of peripheral signal lights. Under noise, tracking improved at the expense of more frequent failures in peripheral monitoring. By contrast, sleep loss had its main effect on tracking (HOCKEY 1970, 1973). The strategic emphasis of attending to the center or the periphery was affected but no general aspecific type of decrement was found. It should be noted that, despite the general value of these studies with regard to emphasizing the strategic nature of performance, the results described above do not appear easily replicable. Hence, they need further scrutiny before the tasks can be seriously considered as candidates for a standard battery. If any, the running span task seems to be the most well-researched and backed by satisfactory process models.

4.3.3 Measuring Multiple Task Aspects

One of the issues raised by the strategy proponents is that studies of stressor effects should measure a variety of different variables in a particular task in order to describe the nature of the strategy changes. Rather than the simple measures of speed and accuracy on which the additive factor and factor-analytic approaches are usually based, there is a search for more composite measures of performance. For instance, SMITH (1985) used a serial choice reaction task in which signals differed in relative frequency. Under noise the reaction time to the most probable signal decreased while the speed of reaction increased for the improbable signals. One average performance measure would not have shown this effect.

BROADBENT (1985) has recently proposed an attention task in which a variety of measures can be obtained and which has the further advantages of consuming relatively little time and possessing a reasonable theoretical basis. Subjects view a screen on which three fixation points appear which are either a little or more widely separated (0.5° or 2°). The first task is to react to a subsequently appearing letter *A* with a left-hand response and to a letter *B* with a right-hand response, but only if the letter appears at the central fixation position. The irrelevant outer fixation points may contain either a pair of *A*s, a pair of *B*s, a pair of asterisks, or nothing. This task has been well researched (e.g., ERIKSEN and ERIKSEN 1974) and is a standard paradigm for research on focused selective attention. In a second part of the test, the central fixation light is removed and the task is now to react to the *A*s and *B*s irrespective of their position on the screen. This is a standard paradigm for studying divided attention: the location is irrelevant and it is merely the category (*A*, *B*) which counts. Distractors can be asterisks, digits, etc., but obviously no *A*s or *B*s. As BROADBENT suggests, the difference between the two parts of this task has clear links to theoretical analyses of attention (BROADBENT 1982) and it allows a wealth of different and subtle performance measures. As such it certainly qualifies for inclusion in a standard battery.

A similar plea for more fine-grained measures may apply to most commonly used performance tests for psychotropic drugs. It has been argued that the types of tasks described in HINDMARCH's (1980) timely review of performance research usually lack a good theoretical basis as well as subtlety of measurement. The previous section of this paper has shown that since 1980 not much has changed. For instance, the Digit-Symbol Substitution Test (DSST) is a code transcription task appearing as a subtest in the Wechsler-Bellevue Intelligence Scale. This is presumably the major reason for its popularity as a test for assessing drug effects on "cognitive functioning." In addition, the usual measures are limited to the number of correct and erroneous transcriptions during a certain period of time. This can be considerably improved, for example by separately assessing eye movements, fixation times of the code key for the transcription, and fixation times at the actual transcriptions. In the early phase of learning, subjects usually consult the code key prior to entering a unit, but the acquisition of digit-code combinations should gradually eliminate the need for consulting the key. Experiments on this issue by DEBUS and SCHROIFF (1984) suggest that this is indeed the case, but also that some subjects keep consulting the key for a much longer period of time than do other subjects. Besides registering different rates of mastering the associations, this finding could reflect individual differences in risk taking. Further research in this direction might provide worthwhile additional measures of the DSST, in addition to a better insight into what actually happens when performing this task.

4.4 Motor Skills

A variety of motor tasks may be considered for inclusion in a standard set, among which *tracking* is certainly a most prominent candidate. Tracking has been a traditional research area from the point of view of task orientation (e.g., piloting) and consequently it has also aroused considerable process model activity to the extent that satisfactory computer simulations of manual control are available, which account for the major variables playing a role in tracking. Tracking activity is a prime example of closed-loop control to which principles of control theory of feedback, error correction and transfer functions can readily be applied (e.g., Poulton 1974). In addition, tracking has been also found to be quite sensitive to the effects of psychotropic drugs (e.g., Payne and Hauty 1954).

Acquisition of tracking skill takes place according to two major sets of variables. One set concerns acquisition of the action-reaction rules of the controlling elements such as gain and control rate, while the second set is concerned with building an internal model about the system's future behavior, allowing perceptual and cognitive anticipation and the development of feedforward actions relating to anticipated changes of the signal. There is a basic difference between actions which are the result of error feedback and actions which aim at adapting to an anticipated future development. The latter type of activity is dominant in more complex motor skills and is the result of a well-developed reference trace of the system. Feedforward activities seldom occur in standard tracking tests, which usually employ a fairly irregular signal and therefore are largely based on error feedback. Yet from an interest in drug effects on memory skills, a feedforward system is the more relevant albeit more time-consuming. Besides skills relying strongly on feedback control, many motor skills have quite pronounced elements of open-loop control, presumably governed by schemata consisting of motor programs (Schmidt 1982). While a motor program is being carried out, errors arising from external circumstances are usually not considered, but "internal feedback" (i.e., lower-level control of the actual proceedings of the program) could well occur (e.g., Sternberg et al. 1978). Open-loop control is predominant in all types of rapid-flow movement such as piano playing, typing and handwriting. In particular, the latter skill has recently been the subject of considerable research activity (Thomassen et al. 1984).

The distinction between closed- and open-loop motor performance is a major dimension in theorizing on motor functions (e.g., Adams 1971; Schmidt 1975).

It is evident that the development of motor programs implies a great deal of skill acquisition in order to go from the production of single controlled units (say in typing) to integrated command structures allowing a rapid serial flow of movements. It should be noted that a motor program of some complexity is usually viewed as a hierarchically ordered system at a fairly abstract level, which controls a subsequent program loading in which the movement parameters are further specified. Finally, there is activation of the muscles which are involved in executing the action. The relevance to drug research is that each of the above-mentioned stages has its own parameters that might be selectively affected by psychotropic drugs.

A second major principle which is rapidly gaining currency is the notion of hierarchical control in motor performance (e.g., Pew 1984). Well practised actions may be merely controlled at a lower level, which has its own open- und closed-loop mechanisms, but which is hardly cognitive. Higher cognitive levels only get involved in the case of problems that cannot be solved at the lower level. It is even fair to say that higher-level control negatively affects motor performance, since it is not capable of taking over the exact functions of the lower level. Instead, it can only interfere with gross and slow actions at the cost of considerable capacity investment.

Although the principle of hierarchical control is generally accepted (e.g., BROADBENT 1977), it is not yet possible to describe the levels in greater detail – neither their relative roles in existing motor tasks, nor their modes of interaction. This is unfortunate since the action of psychotropic drugs might be related to the level of control.

A significant example comes from the work of RIEMERSMA et al. (1977) on the effects of long-term night driving on performance. It was found that keeping a straight course on a motorway deteriorated as a function of time until some critical variance was reached. This variance, then, remained about constant during the rest of the 10-h spell. It corresponded to a considerable probability of occasionally moving from one traffic lane into the adjacent one. The effect of accumulating sleep loss on course keeping may not be surprising; the question of why the variance did not further deteriorate beyond this critical level is of great interest. RIEMERSMA et al. (1977) suggested that it marks a transition from a lower-level fine-grained control of the straight course to a higher-level control. Presumably the higher level detects and counteracts only larger deviations, but it is still capable of preventing a further increase in variance. The higher level is presumably characterised by cognitive, capacity-demanding, and controlled processing – in fact, subjects might consciously note deviations from the straight course at this level – while the lower level is based upon largely automatic control that demands little capacity. In turn, this would mean that the higher level is less sensitive to sleep loss than the lower level. It is interesting that O'HANLON et al. (1982) observed approximately the same critical variance in course keeping after intake of diazepam, which again suggests that the low level is more sensitive to "sedative" types of stressors. This is obviously no support for models that proceed from the assumption that such stressors primarily affect processing capacity (e.g., KAHNEMAN 1973; SANDERS 1983). The prime effect of stressors on motor skills might rather be on well-developed and acquired subtle control.

One consequence of this analysis is that, when considering a set of reference tests that are characteristic for human performance, one should be careful to choose tests that reflect lower-level as well as higher-level control. For instance, a conventional pursuit rotor task would probably fail to tap the lower-order control level since even small deviations are easily detected. As I have argued elsewhere (SANDERS 1986), choosing a tracking task with vague safety boundaries, such as simulating course keeping on a straight road, would be more desirable.

4.4.1 *Knowledge of Results*

Although motor learning is described by various rival theories, there is one common set of parameters concerned with the performer's use of feedback to acquire the appropriate motor behavior. Feedback about the outcome of the action is usually called knowledge of results (KR), while augmented feedback about the movement itself is called knowledge of performance (KP). Finally, it is possible to provide knowledge about what would have been the correct action in a certain situation, rather than about the outcome (e.g., SCHUFFEL 1986). It is generally acknowledged that KR, as well as KP, may have motivational effects as well as effects on skill acquisition; these factors are easily confounded in actual performance tests (e.g., SCHMIDT 1982; SALMONI et al. 1984). The best technique for separating effects of learning and motivation is by having both an acquisition and a retention phase in the experimental design. For instance, one may use an experimental and a control group which receive KR or no KR respectively (a between-subjects design is always necessary in learning studies) for, say, 50 trials on a complex movement pattern. This is followed by additional trials in which neither group receives KR, in order to evaluate the effect of KR on acquisition, presumably uncontaminated by motivation. Subjects are usually blindfolded in motor learning studies to avoid effects of nonmotor factors. Prior to the experimental session the desired movement pattern is demonstrated visually or through passively moving the hand.

One common notion is that, when carrying out a movement pattern, there is response-produced feedback which is coded, stored, and subsequently compared with KR about the movement outcome or with the internal reference acquired by the preexperimental demonstration. The divergence between the response-produced feedback and KR leads to modifications in the plan of action for the next trial. The end result of learning is a well-organized motor scheme as well as a reference trace with which the movement can be compared. This finally eliminates the need for KR (see SCHMIDT 1982).

Although there are various issues of debate in the area of motor acquisition (e.g., the role of response-produced feedback vs. initially existing action plans; MARTERNIUK 1986), this general paradigm of motor learning has sufficient basis to qualify for the study of the effects of drugs on motor acquisition and retention. The same can be said of paradigms concerned with tests of short-term motor memory (STMM), the most common of which consists of reproduction of a movement pattern that is demonstrated in advance. The time elapsing between this criterion movement and the recall trial is one of the major variables in this paradigm.

5 Conclusions

One straightforward and perhaps trivial conclusion is that tests of human memory and performance should be either task or process oriented in order to qualify as tools for assessing effects of psychotropic drugs. In the case of task orientation, there should be evidence that the test is related to some real-life performance. Usually the correlations between laboratory tests and real-life tasks are small or moderate. It is obvious, therefore, that tests with mere surface validity of measuring, say, "driving-related skills" are fully unwarranted. In the case of process orientation, there should be a well-researched process model providing satisfactory rationales for assigning effects of parameters to certain mental processes. The use of procedures which are merely postulated as "proper measures" of some postulated cognitive or sensorimotor skill is clearly unwarranted. The same is true for tasks such as the DSST, "choice reaction time," and "digit span," which are derived from existing intelligence tests or are otherwise popular as means of clinical assessment, but which as such are usually void of theoretical content. The finding that a test is "sensitive to drug effects" is in itself no sufficient rationale for its usefulness as a tool for assessing valid effects.

One of the major arguments of this paper is that a "task" as such is always performed by way of a whole score of processing structures – sensory, central, and motor alike. Hence, it is through proper variation of task variables or through consistently high loading on one and only one ability or factor that individual processing stages can be tapped. Typical tests for certain processing stages (e.g., HINDMARCH 1980) may exist, but should be the result of research and not of intuition.

Therefore, the results derived from application of additive factor logic to choice reaction times (FROWEIN 1981) and from FLEISHMAN's factor-analytic research have been emphasized. In addition there are a number of emerging experimental paradigms with fairly elaborate process models. These include massed vs. spaced practice schedules for separating acquisition from motivation; tracking with either vague or well-defined boundaries as the most promising task-oriented tool; running memory for studying the extent of storing bias as affected by stressors; the ERICKSON paradigm and its divided attention extension as a summary measure of visual attentional functioning; and finally, motor acquisition procedures. It is important to note that, in order to allow interpretation, parameter variation is almost always necessary.

The above list has obviously no claims to being exhaustive. First, it is largely limited to motor functions, although it extends to some general aspects of human information processing. Ultimately a task battery should be much larger and include tests of perception, memory, and other higher cognitive functions. Even within the area of motor behavior the proposed paradigms should be considered merely as examples of how to proceed. Even if they were suitable for inclusion in a future "testicon" (i.e., a "laundry list" of tests suitable for assessment of stressors) a lot of work remains to be done before their selection can be justified. As yet, none of the tests has been properly standardized to allow comparison between results of different laboratories. There are usually no norms, and there is lack of effort towards establishing validation.

Why is progress so slow in this field? I think that this is due to the usually short-term goals of applied research, and a lack of interest in standardization in basic research. A major problem in drug research is that the pharmaceutical industry is solely interested in testing a drug and usually ignores problems relating to the validity of the performance tasks or, more generally, to the methodology of behavioral research (e.g., GAILLARD 1985). Again, basic research is not sufficiently well coordinated to fill this gap. Moreover, it suffers from present-day dwindling research funds and hence from diminished research effort. Continuing research funding and strong international cooperation are needed to establish the "testicon," even in a preliminary version. If the present paper has contributed to spreading the idea that such a development is greatly needed, it has served its aim.

References

Adams JA (1971) A closed-loop theory of motor learning. J Mot Behav 3:111–149

Adams JA, Reynolds B (1954) Effect of shift in distribution of practice conditions following interpolated rest. J Exp Psychol 47:32–36

Allport DA (1980) The state of cognitive psychology: a critical notice of W. G. Chase (ed) Visual information processing. Q J Exp Psychol 27:141–152

Broadbent DE (1977) Levels, hierarchies and the locus of control. Q J Exp Psychol 29:181–201

Broadbent DE (1981) Task combination and selective intake of information. Acta Psychol 50:253–290

Broadbent DE (1985) Reaction time with distractors: some possibilities for drug assessment. Paper at the symposium Drugs and reaction time, Aachen, FRG, September 1985

Brown J, Brown MW, Bowes JB (1983) Effects of lorazepam on rate of forgetting, on retrieval from semantic memory and on manual dexterity. Neuropsychologia 21:501–512

Debus G, Schroiff H-W (1985) Paradigmen zur differenzierten Leistungsanalyse beim „Zahlen-Verbindungs-Test" (ZVT) und beim „Zahlen-Symbol-Test" (ZST). Institut für Psychologie der RWTH Aachen, FRG, Arbeitsbericht no I/19

Desai N, Taylor-Davies A, Barnett DB (1983) The effects of diazepam and oxprenolol on short term memory in individuals of high and low state anxiety. Br J Clin Pharmacol 15:197–202

Eriksen BA, Eriksen CW (1974) Effects of noise letters upon identification of targets in a non-search task. Percept Psychophysics 16:143–149

File SE, Lister RG (1982) Do lorazepam-induced deficits in learning result from impaired rehearsal, reduced motivation or increased sedation? Br J Clin Pharmacol 14:545–550

File SE, Lister RG (1983) Does tolerance to lorazepam develop with once weekly dosing? Br J Clin Pharmacol 16:645–650

Fitts PM (1964) Perceptual-motor skill learning. In: Melton JW (ed) Categories of human learning. Academic, New York

Fleishman EA, Hempel WE (1955) The relation between abilities and improvement with practice in a visual reaction discrimination task. J Exp Psychol 49:301–312

Fleishman EA, Quaintance MK (1984) Taxonomies of human performance. Academic, New York

French JW (1951) The description of aptitude and achievement tests in terms of rotated factors. Psychometric Monographs, no 5

French JW, Ekstrom RB, Price LA (1963) Kit of reference tests for cognitive factors. Educational Testing Service, Princeton

Frowein HW (1981 a) Selective effects of barbiturate and amphetamine on information processing and response execution. Acta Psychol 47:105–115

Frowein HW (1981 b) Selective drug effects on information processing. Doctoral Thesis, Tilburg, Netherlands

Gaillard AWK, Sanders AF (1975) Some effects of ACTH 4–10 on performance during a serial reaction task. Psychopharmacologia 42:201–208

Gaillard AWK, Varey CA (1979) Some effects of an ACTH 4–9 analog on human performance. Physiol Behav 23:79–84

Gaillard AWK (1980) ACTH analogs and human performance. In: Martinez JL et al. (eds) Endogenous peptides in learning and memory processes. Academic, New York

Gaillard AWK (1985) The evaluation of drug effects in laboratory tasks. Paper at the symposium Drugs and reaction time, Aachen, FRG, September 1985

Ghonheim MM, Hinrichs JV, Mewaldt SP (1984a) Dose-response analysis of the behavioral effects of diazepam: I. Learning and memory. Psychopharmacology 82:291–295
Ghonheim MM, Mewaldt SP, Hinrichs JV (1984b) Behavioral effects of oral versus intravenous administration of diazepam. Pharmacol Biochem Behav 21:231–236
Gopher D, Sanders AF (1984) "S-Oh-R": Oh stages! Oh resources! In: Prinz W, Sanders AF (eds) Cognition and motor processes. Springer, Berlin Heidelberg New York, pp 231–253
Guilford JP (1967) The nature of human intelligence. McGraw Hill, New York
Hebb DO (1955) Drives and the conceptual nervous system. Psychol Rev 62:243–254
Heuer H (1984) Motor learning as a process of structural constriction and displacement. In: Prinz W, Sanders AF (eds) Cognition and motor processes. Springer, Berlin Heidelberg New York
Hindmarch I (1980) Psychomotor function and psychoactive drugs. Br J Clin Pharmacol 10:1898–209
Hinrichs JV, Ghonheim MM, Mewaldt SP (1984) Diazepam and memory: retrograde faciliation produced by interference reduction. Psychopharmacology 84:158–162
Hockey GRJ (1970) Signal probability and spatial location as possible bases for increased selectivity in noise. Q J Exp Psychol 22:37–42
Hockey GRJ (1973) Changes in attention allocation in a multi component task under loss of sleep. Br J Psychol 61:473–480
Hockey GRJ (1979) Stress and the cognitive components of skilled performance. In: Hamilton V, Warburton DM (eds) Human stress and cognition. Wiley, New York, pp 141–177
Kahneman D (1973) Attention of effort. Prentice-Hall, Englewood Cliffs
Kleindienst-Vanderbeke G (1984) Information, processing and benzodiazepines. Neuropsychobiology 12:238–243
Lister RG, File SE (1984) The nature of lorazepam-induced amnesia. Psychopharmacology 83:183–187
Logsdon R, Hochhaus L, Williams H, Rundell DH, Marwell D (1984) Secobarbital and perceptual processing. Acta Psychol 55:179–193
Mason JW (1975) Emotion as reflected in patterns of endocrine integration. In: Lavi L (ed) Emotions, their parameters and measurement. Raven, New York
Marteniuk RG (1986) Information processes in movement learning: capacity and structural interference effects. J Mot Behav 18:55–76
McManus IC, Ankier SI, Norfolk J, Phillips M, Priest RG (1983) Effects on psychological performance of the benzodiazepine, loprazolam, alone and with alcohol. Br J Clin Pharmacol 16:291–300
Melton AW (1947) Apparatus test. Government printing office, Washington, DC (AAF Aviation Psychology Programme. Research Report 4)
Mewaldt SP, Hinrichs JV, Ghonheim MM (1983) Diazepam and memory: support for a duplex model of memory. Memory Cognit 11:57–564
Mewaldt SP, Ghonheim MM, Hinrichs JV (1986) The behavioral actions of diazepam and axazepam are similar. Psychopharmacology 88:165–171
Morgan K (1984) Effects of two benzodiazepines on the speed and accuracy of perceptual-motor performance in the elderly. Psychopharmacology (Suppl) 1:79–83
O'Hanlon JF, Haak T, Blaauw GJ, Riemersma JBJ (1982) Diazepam impairs lateral position control in highway driving. Science 217:79–80
Ott H (1984) Are electroencephalographic and psychomotor measures sensitive in detecting residual sequelae of benzodiazepine hypnotics. Psychopharmacology (Suppl) 1:133–151
Payne RB, Hauty GT (1954) The effects of experimentally induced attitudes upon task proficiency. J Exp Psychol 47:267–273
Pew RW (1984) A distributed processing view of human motor control. In: Prinz W, Sanders AF (eds) Cognition and motor processes. Springer, Berlin Heidelberg New York
Pomara N, Stanley B, Block R, Guido J, Russ D, Bercou R, Stanley M, Greenblatt DJ, Newton RE, Gershon S (1984a) Adverse effects of single therapeutic doses of diazepam on performance in normal geriatric subjects: relationship to plasma concentrations. Psychopharmacology 84:342–346
Pomara N, Stanley B, Block R, Guido J, Stanley M, Greenblatt DJ, Newton RE, Gershon S (1984b) Diazepam impairs performance in normal elderly subjects. Psychopharmacol Bull 20:137–139

Poulton EC (1974) Tracking skill and manual control. Academic, New York
Pribram KH, McGuiness D (1975) Arousal, activation and effort in the control of attention. Psychol Rev 82:116–149
Rabbitt PMA (1979) Current paradigms and models in human information processing. In: Hamilton V, Warburton DM (eds) Human stress and cognition. Wiley, New York
Riemersma JBJ, Sanders AF, Wildervanck C, Gaillard AWK (1977) Performance decrement during prolonged night driving. In: Mackie R (ed) Vigilance. Plenum, New York
Roehrs T, McLenaghan, Koshorek G, Roth T (1984) Amnesic effects of lormetazepam. Psychopharmacology (Suppl) 1:165–172
Roth T, Roehrs T, Wittig R, Zorick F (1984) Benzodiazepines and memory. Br J Clin Pharmacol 18:45S–49S
Salmoni AW, Schmidt RA, Walter CB (1984) Knowledge of results and motor learning: a review and critical reappraisal. Psychol Bull 95:355–386
Sanders AF, Hoogenboom W (1970) On the effects of continuous active work on performance. Acta Psychol 33:414–431
Sanders AF (1977) Structural and functional aspects of the reaction process. In: Dornic S (ed) Attention and performance, vol VI. Erlbaum, Hillsdale, NJ
Sanders AF (1980) Stage analysis of reaction processes. In: Stelmach G, Requin J (eds) Tutorials on motor behavior. North Holland, Amsterdam
Sanders AF (1983) Towards a model of stress and human performance. Acta Psychol 53:61–95
Sanders AF (1984) Ten symposia on attention and performance: some issues and trends. In: Bouma H, Bouwhuis D (eds) Attention and performance, vol X. Erlbaum, Hillsdale NJ
Sanders AF (1986) Drugs, driving and the measurement of performance. In: O'Hanlon JF, de Gier JJ (eds) Drugs and driving. Taylor and Francis, London
Sanders AF, Haygood RC, Schroiff H-W, Wauschkuhn CH (1986) Standardization of performance tests: a proposal for further steps. United States Air Force Scientific Report
Scharf MB, Khosla N, Brocker N, Goff P (1984) Differential amnestic properties of short- and long-acting benzodiazepines. J Clin Psychiatry 45:51–53
Scharf MB, Khosla N, Lysaght R, Brocker N, Moran J (1983) Anterograde amnesia with oral lorazepam. J Clin Psychiatry 44:362–364
Schmidt RA (1975) A schema theory of discrete motor skill learning. Psychol Rev 82:225–261
Schmidt RA (1982) Motor control and learning: A behavioral emphasis. Human Kinetics Publishers, Champaign
Schuffel H (1986) Human control of ships in tracking tasks (report). Institute for Perception TNO, Soesterberg, Netherlands
Selye H (1956) The stress of life. McGraw Hill, New York
Smith AP (1985) Noise, biased probability and serial reaction. Br J Psychol 76:89–96
Sternberg S (1969) The discovery of processing stages: extentions of Donders' method. In: Koster WG (ed) Attention and performance, vol II. North Holland, Amsterdam, pp 27–315 (Acta Psychologica 30)
Sternberg S, Monsell S, Knoll RJ, Wright CE (1978) The latency and duration of rapid movement sequences. Comparisons of speech and typewriting. In: Stelmach GE (ed) Information processing in motor control and learning. Academic, New York
Subhan Z (1984) The effects of benzodiazepines on short-term memory and information processing. Psychopharmacology (Suppl) 1:173–181
Subhan Z, Hindmarch I (1984) Effects of zopiclone and benzodiazepine hypnotics on search in short-term memory. Neuropsychobiology 12:244–248
Talland GA, Quarton GC (1965) The effects of metamphetamine and pentobarbital on the running memory span. Psychopharmacologia 7:379–382
Theologus GC, Wheaton GR, Fleishman EA (1974) Effects of intermittent, moderate intensity noise-stress on human performance. J Appl Psychol 59:539–547
Thomassen AWJM, Keuss PJG, Galen GV (eds) Motor-aspects of handwriting (Acta Psychol 54)
Turvey MT, Shaw RE, Mace M (1978) Issues in the theory of action: Degrees of freedom, coordinative structures and coalitions. In: Requin J (ed) Attention and Performance, vol VII. Erlbaum, Hillsdale NJ, pp 557–595

Ursin H, Baade E, Levine S (1978) Psychobiology of stress: a study of coping man. Academic, New York

Wesnes K, Warburton DM (1984) A comparison of temazepam and flurazepam in terms of sleep quality and residual chages in performance. Neuropsychobiology 11:255–259

Willumeit HP, Ott H, Neubert W (1984) Simulated car driving as a useful technique for the determination of residual effects and alcohol interaction after short- and long-acting benzodiazepines. Psychopharmacology (Suppl) 1:182–192

Measures of Memory and Information Processing in Elderly Volunteers

B. AUFDEMBRINKE, H. OTT and A. ROHLOFF

Abstract

A screening study was carried out in normal elderly volunteers to determine

1. their scores in measures of memory and information processing
2. how the measures correlate with sample characteristics
3. how the measures correlate with each other and with the EEG

We report here on 111 subjects, 56 women and 55 men aged 53–77 years (63.9 ± 6.8) with average intelligence (Coloured Progressive Matrices CPM 26.7 ± 5.1).

In the *C*-normed measures of the word list from the Nuremberg Geriatric Inventory (NGI) our sample scored better than average (immediate recall of words $C = 7.5$, delayed recall of words $C = 8.2$; mean $C = 5$). The sample even attained ceiling scores in the test of recognition. The sample likewise had a better than average speed in a color-word test (CWT) of focused attention similar to the Stroop test (e.g., reading $C = 7.5$, color-naming $C = 6.7$, interference task $C = 6.3$) and in the ZVT trail-making test ($C = 7.0$). In contrast, speed-free focused attention was only average ($C = 5.2$ and $C = 5.5$).

The following sample characteristics correlated with test measures of memory and information processing ($p < 0.01$): age negatively with immediate recall, trail making, incidental episodic memory, focused attention, speed in the CWT; intelligence correlated positively with 10-min adding which relies on working memory (Pauli test) and immediate recall; education correlated positively with delayed recall, visual memory performance in a figures test, immediate recall, and reading speed in the CWT; occupation correlated positively with figures test performance; exercise of profession correlated positively with incidental episodic memory and immediate recall.

Age and occupation correlated more with motor measures than with cognitive measures (pegboard, 1 min tapping at maximal speed). Sex had no bearing on memory characteristics, but did on tracking (men better) and color-naming performance in the CWT (women better). Of the semiluxuries, only regular alcohol consumption showed a relationship with the test measures (negatively with color-naming speed, immediate recall, and pegboard performance).

Analysis of memory and information processing measures revealed no correlations ($r \geqq 0.40$) between memory measures from the word list, Pauli test, figures test, and learning. Measures of the CWT correlated with Pauli test and pegboard. ZVT score was the variable with the most correlations (CWT, Pauli test, pegboard). A factor analysis with a reduced set of variables should further clarify the inner structure of our subject's performance. All EEG measures (frequency of stage shifts, epoch number of the first stage shift, two multiple sleep latency criteria, number of epochs of wakefulness) were highly correlated with each other, but not with the psychometric measures.

On balance, the conclusion is that our samples were a positive selection compared with the reference samples of the NGI. In our sample, which was also positively selected from a medical point of view, cognitive measures correlated with sociodemographic characteristics, but not with affective and somatic complaints. There was no clear relationship between measures of the various memory tests, nor was there a relationship with EEG measures.

[1] Research Laboratories, Psychopharmacology Section, Schering AG, 1000 Berlin 65, FRG.

1 Introduction

Pharmacological research in healthy volunteers is not normally concerned with the representativeness of its samples. As our adopted strategy is not to restrict phase I studies to young volunteers but to design special studies even in elderly volunteers, we considered it useful to examine the structure of such an elderly sample.

We report here on a study in which we screened elderly volunteers for participation in an ongoing pharmacological trial. This screening study was not designed to investigate memory and information processing prospectively, but rather to record the sociodemographic, medical, psychometric, and electroencephalographic (EEG) variables of our sample population. Further, the presented data are purely descriptive. We shall relate our data to findings from the literature, being fully aware of the flaws inherent in such an approach.

We shall report on:

1. the sample values in measures of memory and information processing
2. their relationships to the sample characteristics
3. their relationships to each other as well as to EEG measures.

2 Sample

Our sample comprised 111 volunteers, 56 women and 55 men, aged 53–77 years (63.9 ± 6.8 years), nearly all of whom were right-handed Caucasians (Table 1). Just over half were married. One-third were living alone. In a nonverbal test of logical reasoning (Coloured Progressive Matrices, CPM: Raven 1976), our subjects attained scores normal for their age group (Rudinger 1976). Sixty-two (= 56%) subjects had an elementary school education, corresponding to 8 years' schooling. Most of the remainder had had an intermediate or trade school education. Eight subjects were university graduates. Sixty-three (= 57%) subjects were skilled workers, 15% were unskilled workers, and 14% were engaged in lower middle class employment. Less than half of the subjects (46 = 41%) were still practising their profession.

The subjects in our sample were relatively healthy: their physical condition was rated by a physician as good in 75%, and fair in 25% of cases (Table 2). All subjects could still look after themselves. No subject was suffering from a serious disease, as this was a criterion for exclusion. The most common complaints were vegetative symptoms, followed by skeletomuscular disorders, disturbances of affect, and sleep disturbances. The disturbances of vision, the hearing defects (which were twice as common), and the few cases of motor disturbances did not interfere with the subjects' performance in the tests. All the subjects were regular drinkers of coffee or tea. Almost one-third were smokers, and two-thirds drank alcohol, but only 25 (= 23%) did so regularly.

Table 1. Sample characteristics

	n
Sample size	111
Sex	
female	56
male	55
Race	
Caucasian	110
Asian	1
Age (years)	$\bar{x}=63.9$ SD=6.8 median=63.2 range=53–77
Handedness	
right	109
left	1
ambidextrous	1
Marital status:	
single	9
married	64
seperated/divorced	21
widowed	17
Living situation[a]	
single	38
with spouse	70
with family	2
Coloured Progressive Matrices score	$\bar{x}=26.7$ SD=5.1 median=27.0 range=11–36
Education	
elementary school	62
intermediate school	21
high school	7
trade school	13
university	8
Highest occupational level attained	
houswife	4
unskilled worker	17
skilled worker	63
lower middle class occupation	15
upper middle class occupation	9
academic	3
Present employment	
none	65
partial or full	46

[a] Missing.

3 Methods

The psychometric part of our test battery consisted of tests from the Nuremberg Geriatric Inventory (NGI), the Pauli test, and psychomotor tests. The Pauli and the psychomotor tests are our own developments. The NGI (Oswald and Fleischmann 1986) is the first German battery of performance tests and rating scales for assessing changes with age of intellectual functioning, general well-being, and individual instrumental activities of daily living. The various tests were

Table 2. Health indicators

	n	$\bar{x}$	SD	Median	Range
Blood pressure (mmHg): systolic		138.4	15.2	140.0	110–180
diastolic		86.9	8.9	85.0	70–110
Heart rate (bpm)		71.8	6.0	72.0	48– 88
Nuremberg Geriatric Questionnaire		5.7	4.4	5.0	9– 26
Disturbances of affect	45				
Sleep disturbances	41				
Defective vision	14				
Defective hearing[a]	33				
Cardiovascular disorders	27				
Respiratory disorders	19				
Gatrointestinal disorders	12				
Urogenital disorders	12				
Metabolic disorders	14				
Skeletomuscular disorders	71				
Vegetative disorders	76				
Chronic pain[a]	17				
Global state of physical health[a]					
good	83				
fair	27				
Motor disorders[a]	7				
Sensory disorders[a]	1				
Impaired coordination[a]	5				

[a] Missing.
No psychiatric history, serious neurologic or physical disorders, cardiac infarction, cardiac arrhythmia, angina pectoris, hypertension, chronic consumptive processes, malignant growth, brain stem symptoms, cerebral attacks, cerebrovascular disorders, diabetes mellitus, nursing needs.

designed to provide a differential picture for subjects with more or less intact cognitive functioning and those with advanced pathological aging. Its main advantage as far as this study was concerned is that we could compare our sample's scores with values from reference samples of $n \leqq 1240$.

The terms memory and information processing are used as labels. They are not meant to be mutually exclusive, as memory is understood to also encompass learning.

3.1 Measures of Memory

We applied the eight-item word list from the NGI, which provides eight measures: recall, intrusions (false recalls), recognition, and false positives determined (a) immediately after stimulus presentation and (b) after a delay of ca. 10 min, in which time the CPM were presented. A further measure of memory was the number of figures recognized in the NGI figures test. The tenth and eleventh measures of memory were the number of incidentally learned names of the five tests of the session and the number of tests that subjects recalled being administered. Incidental or automatic learning proceeds without instructions and is presumably determined by psychobiological mechanisms other than effortful learning such as of word lists (NEWMAN et al. 1984).

3.2 Measures of Information Processing

The first measure of information processing was the time taken to complete the ZVT-G trail-making test from the NGI, which assesses the speed of information processing. In our modification of the Pauli test (PAULI and ARNOLD 1951), the subject presses a button and a number 0–4 appears on a display. This number has to be added to a previous number that is no longer displayed. The subject has then to enter the sum (maximum 8). The test gives two measures: the number of attempted and the number of correctly solved tasks. The test calls for processing of material in working memory (BADDELEY and HITCH 1974).

The NGI color-word test (CWT), which is similar to the Stroop test (STROOP 1935a, b), deals with focused attention. It provides three measures of speed: (a) for word reading, (b) for color naming, and (c) for color naming when printed names do not correspond to colors (color-word interference). Combining these three measures of overall performance speed results in five derived measures: the color difficulty score (quotient of word reading and color naming), the speed score (sum of word reading and color naming), the interference score (difference between color-word interference and color naming), and two scores which are logs of expectancy values computed from the reference sample: the nomination score (color naming adjusted to exclude reading speed), and the selectivity score (attention adjusted to exclude color naming). Color-word interference, interference score, and selectivity score are regarded as measures of focused attention, and the last two scores are relatively speed-free.

3.3 Psychomotor Measures

Subjects were requested to tap at maximum speed for 1 min and at their own speed for 4 min. The number of taps and time between taps was then recorded for both test runs. These four measures were complemented by three measures from the pegboard test: total actions (i.e., sum of all movements), number of pins set (i.e., removed from the first row but not shifted to the second row), and number of pins shifted. The eighth motor measure was a measure of inaccuracy in a visual tracking task.

3.4 EEG Measures

From a 20-min (=60 epochs) recording of EEG when subjects were resting with closed eyes, the following seven measures were determined visually by an EEG expert: frequency of stage shifts, epoch number of first stage shift, epoch number of first four uninterrupted epochs of sleep stage 1, epoch number of first scored sleep stage 2, frequency of epochs of wakefulness, frequency of epochs of sleep stage 1, frequency of epochs of sleep stage 2 (RECHTSCHAFFEN and KALES 1968; RICHARDSON et al. 1978).

4 Results

4.1 Sample Values

The tables show the sample's results in the memory (Table 3), information processing (Table 4), psychomotor (Table 5), and EEG (Table 6) measures.

The measures from the word list showed high mean values relative to the test ceiling and at the same time a low number of intrusions or false positives (Table 3). Apart from the number of incidentally learned test names, the test ceiling was attained in all measures of memory, particularly in the recognized words.

Color naming in the CWT took more time than word reading but less time than the color-word interference task (Table 4).

Compared with the *C* values of the NGI reference sample, our sample attained better scores in nearly all power- and speed-oriented measures of memory and information processing (Fig. 1). Exceptions were focused attention measures, i.e., the interference score and the selectivity score in the color-word test, where the sample scored only average *C* values. Scores were also above average in the measures of self-rated ageing, Nuremberg Geriatric Questionnaire, and Nuremberg Geriatric Self-Rating Scale.

The EEG measures showed the subjects to have their first shift from wakefulness to a state of subvigilance mostly after less than half of the recording time (me-

Table 3. Memory measures

	n	$\bar{x} \pm SD$	Median	Range	
				actual	possible
NGI word list scores					
immediate recall (no.)	111	5.6±1.1	6.0	3– 8	0– 8
intrusions (no.)	111	0.1±0.3	0.0	0– 1	
recognition (no.)	111	7.3±0.9	8.0	5– 8	0– 8
false positives (no.)	111	0.2±0.7	0.0	0– 4	0– 8
delayed recall (no.)	110	4.6±1.8	5.0	0– 8	0– 8
intrusions (no.)	110	0.4±0.7	0.0	0– 3	
recognition (no.)	110	7.2±1.0	7.5	4– 8	0– 8
false positives (no.)	110	0.1±0.5	0.5	0– 4	0– 8
NGI figures test					
recognitions (no.)	111	8.2±1.7	8.0	4–12	0–12
Incidental learning					
test names (no.)	107	0.7±0.8	1.0	0– 3	0– 5
tests given (no.)	107	4.0±0.8	4.0	2– 5	0– 5

Table 4. Information processing measures

	n	x̄±SD	Median	Range
NGI ZVT-G				
trail making (s)	111	21.1 ± 6.4	19.2	12.5 - 49.3
Pauli test				
10-min adding				
tasks attempted	100	159.4 ±48.8	157.0	89 –288
tasks solved	100	127.1 ±49.9	121.0	33 –257
NGI CWT				
1 word reading (s)	111	14.1 ± 2.9	13.5	10.0 - 31.0
2 color naming (s)	111	22.0 ± 4.6	21.0	11.8 - 11.2
3 color word interference (s)	110	46.7 ±18.3	42.8	25.8 –172.0
color difficulty score (1÷2)	111	0.652± 0.109	0.640	0.412– 0.912
speed score (1+2) (s)	111	36.1 ± 6.8	35.0	22.0 - 65.0
interference score (3–2) (s)	110	24.8 ±16.3	22.4	7.0 –145.0
nomination score	111	8.0 ± 6.9	8.5	−12.3 - 24.4
selectivity score	110	− 9.2 ±12.1	− 8.4	−62.6 - 18.1

Table 5. Psychomotor measures

	n	x̄±SD	Median	Range
Tapping				
maximum speed (no./min)	100	347.8± 38.4	341.5	274– 442
tap-to-tap-interval (ms)	100	174.7± 19.1	175.5	136– 219
personal speed (no./4 min)	101	1004.9±283.0	1040.0	242–1727
tap-to-tap-interval (ms)	101	265.8±111.7	231.0	139– 995
Pegboard				
total actions (no.)	100	197.3± 20.5	200.0	146– 241
pins set (no.)	100	98.6± 10.3	100.0	73– 120
pins shifted (no.)	100	92.4± 9.4	94.0	67– 114
Video tracking test, 5 min (arbitrary distance)	85	818.7±452.8	725.0	242–2014

Table 6. EEG measures

	x̄±SD	Median	Range
1. Stage shifts	10.8± 9.3	9.5	0–31
2. 1st stage shift (epoch no.)	29.5±21.2	23.0	1–60
3. 1st 4 uninterrupted epochs sleep stage 1 (epoch no.)	45.9±21.1	60.0	2–60
4. 1st scored sleep stage 2 (epoch no.)	46.1±17.6	60.0	6–60
5. Epochs awake	44.2±16.3	51.0	3–60
6. Epochs sleep stage 1	11.0±11.1	7.5	0–36
7. Epochs sleep stage 2	4.8± 8.0	0.0	0–37

Measures are derived from a 20-min (=60 epochs) resting recording ($n=106$).

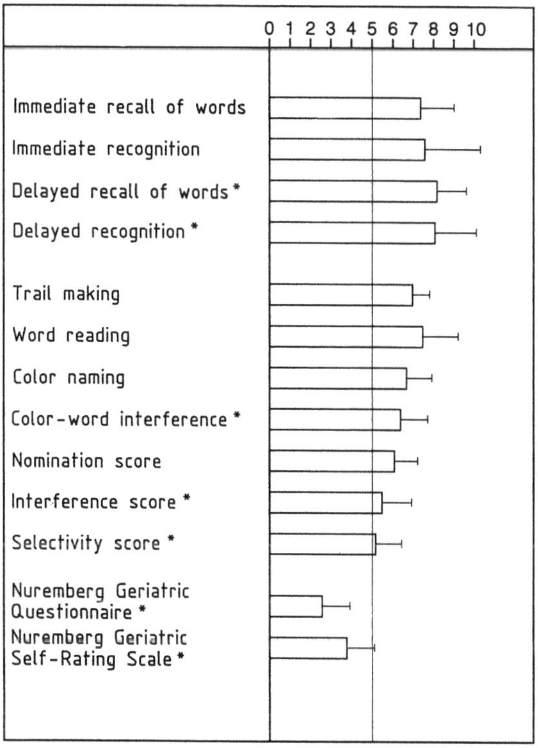

Fig. 1. *C* values in NGI measures ($n=111$; $^*n=110$)

dian epoch no. = 23) but they achieved only a few stages of subvigilance (median no. of epochs of stage 1 = 7.5, stage 2 = 0) and practically did not fulfill the multiple sleep latency criteria (4 uninterrupted epochs stage 1, 1 epoch stage 2; Table 6).

In the following a restricted set of the measures described here will be used. Intrusions and false positives in the word-list, tap-to-tap intervals in the tapping test, and total actions and pins set on the pegboard test are discarded because it is felt that they do not add enough information to justify additional multiple testing.

4.2 Sample Characteristics and Psychometric Measures

Four of a total of seven memory measures were significantly related ($p \leq 0.01$) to the different sample characteristics (Table 7). It should be noted, however, that the low error probabilities were related to the sample size; this and the matter of multiple testing means that data can only be evaluated descriptively. The number of words recalled immediately declined with increasing age. This variable correlated positively with the measure of logical reasoning and increased with higher educational and occupational status. It was greater in subjects still practising their

Table 7. Sample characteristics and psychometric measures ($p \leqq 0.1$)

	Sex	
	female	male
Tracking	969.9±459.4	656.3±389.1
Color naming (s)	20.6±3.4	23.4±5.2
Color difficulty score	0.681±0.098	0.622±0.113

	Age (years)			
	53–57	58–63	64–70	71–78
Pins shifted	97.9± 7.7	93.5± 9.9	88.0± 7.4	87.8± 8.4
Maximum tapping speed (no./min)	371.1±40.0	344.9±38.1	345.5±32.0	323.1±24.3
Immediate recall of words (no.)	6.0± 1.1	5.6± 1.0	5.7± 1.1	4.7± 1.0
Trail making (s)	18.7± 4.9	19.8± 6.3	22.3± 6.7	24.7± 6.4
Color – word interference (s)	41.0± 9.4	41.5±10.0	52.0±19.7	55.4±27.8
Interference score (s)	20.7± 7.8	19.7± 7.7	28.5±16.9	32.8±26.2
CWT speed score (s)	33.4± 4.0	36.5± 8.2	38.3± 7.2	36.7± 6.6

	Education	
	elementary school	intermediate school or higher
Delayed recall of words (no.)	4.0± 1.5	5.3± 1.8
Figure recognition (no.)	7.8± 1.6	8.8± 1.6
Immediate recall of words (no.)	5.3± 1.1	5.9± 1.1
Adding tasks attempted	147.1±48.0	174.0±46.0
Word reading (s)	14.8± 3.2	13.3± 2.3

	Highest occupational level attained		
	≦unskilled	skilled	≧lower middle class
Maximum tapping speed (no./min)	314.3±22.8	347.5±36.6	370.0±35.4
Figure recognition	7.5± 1.6	8.0± 1.7	9.2± 1.4

	Present employment	
	None	Partial/full
Incidental learned test names	3.8±0.8	4.3±0.7
Immediate recall of words	5.3±1.1	5.9±1.1

	Coloured Progressive Matrices
	r
Adding tasks solved	0.40
CWT nomination score	0.33
Immediate recall of words	0.29
Maximum tapping speed (no./min)	0.26
Adding tasks attempted	0.26

profession. The number of figures recognized also rose with increasing educational and occupational status. The number of words recalled after 10 min delay increased with higher education, and similarly the number of test names learned incidentally was greater in subjects still practising their profession.

All the nine investigated measures of information processing correlated with sample characteristics, yet not uniformly. Women responded better to color stimuli (color naming, color difficulty score). Scores on tasks with a speed component, such as trail making, CWT speed score, and focused attention (color-word interference, but also speed-free interference score), became worse with increasing age. Cultural techniques such as calculating (adding tasks attempted) and reading (word reading) improved with higher education. Moreover, scores on the self-paced adding tasks (solved and attempted) correlated with the measure of logical reasoning.

There were considerably fewer relationships between sample characteristics and psychomotor measures. Maximum speed of tapping correlated negatively with age and positively with logical reasoning scores and occupational status. Men performed better in the tracking task. Performance in the pegboard test decreased with increasing age.

Of the semiluxury stimulants, only regular alcohol consumption had an influence (negative) on psychometric measures (color naming, immediately recalled words, CWT speed score, pins shifted; Table 8).

There were hardly any relations between psychomotor measures and indices of physical health. Only maximum speed of tapping correlated negatively with systolic blood pressure, and subjects with sleep disturbances indicated a higher measure of self-rated aging (Table 9).

Table 8. Semiluxuries and psychometric measures ($p \leqq 0.01$)

	Alcohol		
	none	occasional	regular
Color naming (s)	20.9±4.7	21.3±3.7	25.1±4.9
Immediate recall of words (no.)	5.4±1.0	6.0±1.0	5.0±1.2
CWT speed score (s)	34.7±7.4	35.3±5.6	39.9±6.5
Pins shifted	92.6±9.3	94.6±8.7	87.5±9.2

Table 9. Health indicators and psychometric measures ($p \leqq 0.01$)

Maximum tapping	Systolic blood pressure (mmHg) $r = -0.2580$	
	Sleep disturbances	
	no	yes
Nuremberg Geriatric Questionnaire (score)	4.7±3.2	7.5±5.6

Table 10. Intercorrelation of psychometric measures ($p \leqq 0.01$, $r \geqq 0.40$, $n \leqq 111$)

	1	2	3	4	5	6	7	8	9	10	11	12	13	14	15	16	17
1 Immediate recall of words																	
2 Immediate recognition of words																	
3 Delayed recall of words	0.52	0.41															
4 Delayed recognition of words		0.58	0.46														
5 Trail making (s)																	
6 Adding tasks attempted					−0.45												
7 Adding tasks solved					−0.44	0.87											
8 Word reading (s)					0.52												
9 Color naming (s)					0.66	−0.47		0.59									
10 Color – word interference (s)					0.48	−0.44			0.63								
11 Color difficulty score								0.42	−0.46								
12 CWT speed score (s)					0.67	−0.47		0.84	0.94	0.48							
13 Interference score (s)										0.97							
14 Nomination score									−0.75		0.93	−0.48					
15 Selectivity score										−0.78			−0.87				
16 Maximum tapping (no./min)							0.48										
17 Pins shifted (no.)					−0.51				−0.45			−0.47					
18 Age (years)					0.41											−0.40	−0.42

$r < 0.40$ for figure recognition, incidental learning and tracking test.

4.3 Measures of Memory and Information Processing and EEG Measures

For the sake of parsimony, Table 10 contains only correlation coefficients significant at the 1% level that are 0.40 or more. It is at first glance evident that different measures from the word list and the CWT are intercorrelated. Immediate recall and recognition correlated with an r of less than 0.40, and the correlation between delayed recall and recognition was only 0.46. The measures from the figures test and incidental learning do not appear in the intercorrelation matrix. They reflect relatively independent characteristics.

There were also no correlations of 0.40 or greater between measures of memory and measures of information processing or psychomotor performance. Of the basic measures of the CWT two pairs correlated: word reading with color naming and color naming with color-word interference. In addition, there were correlations between the basic measures and the derived measures. Color naming, color-word interference, and speed score were also related to measures from other tests (trail making, adding tasks attempted, pins shifted), which in turn correlated with other measures. The tracking task correlated with no other measure at 0.40 or greater.

The various EEG measures all correlated significantly with one another and, with one exception, with an r of 0.40 or greater (Table 11). In contrast, there was no significant relationship with the measures of memory and information processing. This was also reflected in a factor analysis of 28 psychometric, EEG, and other variables (principal components extraction, varimax rotation). The eight-factor solution, which explained 70% of the variance, showed a pure EEG factor as the first factor (15% of the variance explained; Table 12). The second factor was characterized by word list measures (=10%), while the number of recognized figures and remembered tests given were found on different factors.

The number of figures recognized loaded together with maximum tapping speed, the measure of logical reasoning, and age on the third factor, which could thus be interpreted as a general age-related performance factor (=10%). The fourth factor was composed of the speed variables pins shifted, CWT speed score, trail making, and word reading speed. The fifth factor was made up of the variables adding tasks attempted and solved, interference score, number of incidentally learned test names, and – surprisingly and with a negative sign – daily con-

Table 11. Intercorrelation of EEG measures ($p \leqq 0.01$, $r \geqq 0.40$, $n = 106$)

	1	2	3	4	5	6
1 Stage shifts						
2 1st stage shift (epoch no.)	−0.82					
3 1st 4 uninterrupted epochs sleep stage 1 (epoch no.)	−0.57	0.64				
4 1st scored sleep stage 2 (epoch no.)	−0.70	0.64	0.53			
5 Epochs awake	−0.87	0.78	0.75	0.76		
6 Epochs sleep stage 1	0.79	−0.79	−0.87	−0.55	−0.90	
7 Epochs sleep stage 2	0.66	−0.49		−0.77	−0.79	0.44

Table 12. Factor analysis of 28 variables ($n=73$)

	Factor 1	Factor 2	Factor 3	Factor 4	Factor 5	Factor 6	Factor 7	Factor 8
EEG epochs awake	**0.95**							
EEG stage shifts	**−0.90**							
EEG 1st stage shift (epoch no.)	**0.85**							
EEG 1st 4 epochs sleep stage 1 (epoch no.)	**0.81**							
EEG 1st sleep stage 2 (epoch no.)	**0.79**							
Delayed recognition		**0.74**		−0.25				
Delayed recall of words		**0.73**	0.34					
Immediate recognition		**0.67**						
Immediate recall of words		**0.66**	0.48					
Maximum tapping speed (no./min)			**0.73**					
Coloured Progressive Matrices			**0.62**					
Figure recognition			**0.61**					
Age (years)			**−0.55**	0.31		−0.25	0.41	
Pins shifted				**−0.79**				−0.31
Speed score (s)				**0.75**		0.29		
Trail making (s)			−0.35	**0.74**				
Word reading (s)		−0.28		**0.64**		0.64		
Adding tasks attempted			0.32	−0.26	**0.75**			
Adding tasks solved			0.46		**0.70**			
Cups coffee/tea per day			0.31		**−0.58**		−0.31	
Color difficulty score		−0.25			0.31	**0.72**		
Nuremberg Geriatric Questionnaire score						**0.66**		
Tracking			−0.34				**0.70**	
Personal tapping speed (no./4 min)		0.31					**0.63**	
Systolic blood pressure (mmHg)			−0.27					0.78
Incidental learning, tests given		0.37					−0.31	−0.58
Interference score (s)					**−0.48**		0.36	
Incidental learning, test names		0.39		0.26	**0.39**	−0.29		−0.30
70% of variance explained:	15%	10%	10%	10%	8%	6%	6%	5%

Principal components extraction, varimax rotation, eigenvalues >1.0 Factor loadings >0.25.

Table 13. Factor analysis of 14 variables ($n = 80$)

	Factor 1	Factor 2	Factor 3
Adding tasks solved	**0.79**	−0.27	
Tracking	**−0.68**		
Adding tasks attempted	**0.67**	−0.43	
Maximum tapping speed (no./min)	**0.64**		0.33
Interference score (s)	**−0.54**	0.28	
Figure recognition	**0.51**		0.31
Pins shifted		**−0.82**	
Trail making (s)	−0.38	**0.75**	
Speed score (s)	−0.29	**0.73**	
Immediate recall of words	0.34		**0.72**
Delayed recall of words	0.38		**0.63**
Personal tapping speed (no./4 min)		−0.43	**0.59**
Color difficulty score			**−0.48**
Coloured Progressive Matrices	**0.48**		
54% of variance explained	24%	18%	13%

Principal components extraction, varimax rotation, eigenvalues >1.0 Factor loadings >0.25.

sumption of coffee or tea. It can be described as a combined factor of attention and memory (=8%). The remaining three factors were characterized only by two high loadings and are accordingly difficult to interpret.

Since inclusion of a number of variables in the factor analysis cannot be strictly justified due to their skewed distribution, we carried out a second factor analysis with a reduced set of variables of only 14 measures whose three factors explained 54% of the variance (Table 13). In this analysis, factor 4 ("speed") of the previous analysis became the new factor 2 (=18%). The previous factor 2 ("word list") also took on the variables tapping at personal speed and color difficulty and now formed the new factor 3 (=13%). The old factors 3 and 5 fused to become the new factor 1, a complex performance factor (=24%). Furthermore, in this factor analysis the memory variables were found on different factors. It was also apparent in both analyses that the psychomotor measures did not form a separate factor, but loaded together with some measures of memory and information processing.

5 Discussion

The subjects in our sample were all in fair to good health. This relative homogeneity explains why there were no uniform relationships between characteristics of health and measures of memory and information processing as reported elsewhere in the literature. For instance, several authors have reported that cardiovascular disease has a negative influence on cognitive performance (Hertzog et al. 1978; Spieth 1964; Wilkie and Eisdorfer 1971). Even subjects who are apparently healthy attain poorer scores when they subjectively rate their physical condition as poor (McDONALD and SUCHY 1980). In contrast, our sample rated their condition (Nuremberg Geriatric Questionnaire) as better than average. We assume that elderly subjects who come to an unknown laboratory to participate in a pharmacological investigation, as is the case with our subjects, feel somewhat more robust than those who are visited in their own homes by the investigators, as in the study of McDONALD and SUCHY.

The *C* norms of the NGI show that our sample scored better than the reference sample in most measures of memory and information processing. The poor scores of the reference group could be attributed to the fact that it comprised not only relatively healthy elderly persons living in their own homes, but also residents of old people's homes and even patients with organic brain syndrome. To make the realistic appraisal of cognitive performance of relatively healthy elderly patients possible, the authors of the NGI should furnish reference-sample scores separated according to state of health.

A look at the mean scores for delayed memory performance reveals lower scores in the tests of delayed recall than in tests of delayed recognition. This could be an artifact, since the majority of subjects reached the test ceiling in the delayed recognition task, with the result that there was no recognizable drop in performance when compared with immediate recognition. A further analysis of the data, however, showed that there were 26 subjects with lowered scores in the test of delayed recognition, but also 17 subjects with increased scores. Therefore, it cannot be ruled out that the processes underlying the recognition task are more stable with time than the processes responsible for the recall task.

Of particular interest was the gradient of the various *C* normed indices of the CWT. Our sample scored better than the reference sample in an overlearned task like word reading but not in a task calling for more demanding performance, as illustrated by the interference score and the selectivity score. Here the two samples did not differ. It is tempting to speculate that pathological aging as represented in the reference sample impairs speed performances but leaves more cognitive performances in this test unchanged. The fact that both CWT measures, the interference score and the speed score, deteriorated with increasing age however seems to refute the assumption that performance in tasks of focused attention is more robust with chronological aging than performance in speed tasks.

The association of 13 of the total 16 investigated measures of memory and information processing with sample characteristics underlines the importance of drawing up a comprehensive sample description. This dependence is sufficiently known for the variable "age." Yet before this study we were only partly aware of the relationships in our positively selected sample between memory measures and logical reasoning, education, occupational status, and practising of profession. Education might contribute to explicit memory strategies whereas the practising of ones profession also favors incidental learning that is independent of explicit instructions. OSWALD and FLEISCHMANN (1986) similarly report correlations of incidental learning with activities of daily living. We see here a starting point for constructing memory tests which are valid for everyday life.

Our data on sample characteristics and memory and information processing, although purely descriptive, are supported by other studies that show, for example, that social status correlates with intellectual performance (SCHAIE 1983) and that occupational status correlates with intellectual decline in later life (LEHR 1980).

Given the arbitrary criterion of $r \geqq 0.40$, correlations between measures of memory and information processing were mainly found between different measures from the same test (word list, color-word). In our sample the memory tests (word list, figures test, and incidental learning) correlated neither with one another, nor with other tests at $r \geqq 0.40$. In contrast, there were correlations with $r \geqq 0.40$ between measures of the CWT, trail making, and the pegboard. Trail making showed the most correlations with the other measures and thus assumes a key role in the intercorrelations.

The EEG measures were relatively isolated from the psychometric measures. In a subsample of 33 subjects who showed at least 19 epochs of subvigilance and thus qualified for participation in the above pharmacological study, performance in the psychometric tests did not differ from that of the remaining 78 subjects. The number of epochs of subvigilance found in a clinical EEG under resting conditions was thus of no consequence to test performance in an activated state. This discrepancy is possible due to the fact that the sleep stage measures are relatively coarse: with a finer resolution, as in power spectral analysis, there may indeed be covariation between the EEG and psychometric variables. OTT et al. (1982) found, however, quite low correlations (max. $r = \pm 0.28$) between factor-analytically derived EEG frequency bands and various psychometric measures. It may therefore be difficult to establish relationships between measures determined under resting conditions and those determined in an activated state.

Factor analyses by principal components extraction (as well as by maximum-likelihood extraction, not described here) resulted in a separate EEG factor, while the memory and information processing measures intermingled with each other as well as with psychomotor measures. This suggests a surprising confounding of psychomotor tasks, performance on which is shown to deteriorate with age, and cognitive tasks whose structure in the elderly is to be studied. We plan to analyze the internal structure of some measures of information processing by studying the changes from the beginning to completion of the tasks, thereby possibly diminishing the effects of the motor aspects of the tasks.

In conclusion, our sample was a positive selection compared with the reference sample of the NGI. In our sample, which was also positively selected from a medical point of view, cognitive measures correlated with sociodemographic characteristics, but not with affective or somatic complaints. There was no uniform relationship between measures of the various memory tests. Nor was there a uniform relationship to other psychometric measures. Motor measures were not independent of memory and information processing measures. It is felt that there is a need for an analysis of these measures that is not contaminated by motor aspects. Clinical EEG measures had no bearing on the psychometric measures.

References

Baddeley AD, Hitch GJ (1974) Working memory. In: The psychology of learning and motivation, vol 8. Academic, New York, pp 47–90

Hertzog C, Schaie K, Gribbin K (1978) Cardiovascular disease and changes in intellectual functioning from middle to old age. J Gerontol 33:872–883

Lehr U (1980) Möglichkeiten der Prävention und Intervention im Alter. In: Schütz R-M (ed) Zum Problemkreis „Geriatrisch-rehabilitative Tagesklinik". Klinik für Angiologie und Geriatrie der Medizinischen Hochschule Lübeck, Lübeck

McDonald RJ, Suchy I (1980) Der Einfluß subjektiver Beschwerden auf Leistung und Befindlichkeit im Alter. Z Gerontologie 13:346–358

Newman RP, Weingartner H, Smallberg S, Calne D (1984) Effortful and automatic memory processes: effects of dopamine. Neurology 34:805–807

Oswald WD, Fleischmann UM (1986) NAI-Kurzmanual. Universität Erlangen, Nuremberg

Ott H, McDonald RJ, Fichte K, Herrmann WM (1982) Interpretation of correlations between EEG power spectra and psychological performance variables within the concepts of "vigilance", "attention" and "psychomotoric impulsion". In: Herrmann WM (ed) EEG in drug research. Fischer, Stuttgart

Pauli R, Arnold W (1951) Der Pauli-Test. Seine sachgemäße Durchführung und Auswertung. Munich

Raven J (1976) The coloured progressive matrices. Lewis, London

Rechtschaffen A, Kales A (eds) (1968) A manual of standardized terminology, techniques and scoring system for sleep stages of human subjects. BIS/BRI, University of California, Los Angeles

Richardson GS, Carskadon MA, Flagg W, van den Hoed J, Dement WC, Mitler MM (1978) Excessive daytime sleepiness in man: multiple sleep latency measurement in narcoleptic and control subjects. Electroenceph Clin Neurophysiol 45:621–627

Rudinger G (1976) Correlates of changes in cognitive functioning. In: Thomae H (ed) Patterns of aging, findings from the Bonn longitudinal study of aging, vol 3. Karger, Basel, pp 20–35

Schaie KW (ed) (1983) Longitudinal studies of adult psychological development. Guilford, New York

Spieth W (1964) Cardiovascular health status, age, and psychological performance. J Gerontol 19:277–284

Stroop JR (1935a) The basis of Ligon's theory. Am J Psychol 47:499–504

Stroop JR (1935b) Studies of interference in serial verbal reactions. J Exp Psychol 18:643–662

Wilkie F, Eisdorfer C (1971) Intelligence and blood pressure in the aged. Science 172:959–962

Psychometric Assessment of Drug Effects

Assessment of the Effects of Drugs on Memory

J. R. Wittenborn[1]

Abstract

This survey is based on 70 controlled investigations of the effects of drugs on memory in healthy volunteers. Although detracting effects were predominant, enhancements were reported as well. The effects of 29 well-known drugs on 15 familiar tests are summarized in a way that permits comparisons of the effects of different drugs and of test sensitivities. There is a discussion of factors that could bias or obscure investigations of the effects of drugs on learning and remembering.

The present report provides a review of the effects of psychotropic substances on learning, memory, forgetting, and related aspects of human behavior, and is limited to an examination of the responses of normal subjects. The investigators observed the effects of medication over a relatively short period of time, compared these effects with the effects of placebo, and reported the statistical significance of the comparisons. The review includes only those inquiries that showed at least one significant contrast between drug and placebo effects and is based upon 70 published reports appearing after 1980.

Learning is inferred from behavior change, and a proper study of learning must consider both the conditions under which the behavior change is believed to have occurred and the conditions under which the behavior change is revealed. Conceivably, psychotropic drugs could have their effects as one of the conditions contributing to the behavior change or as one of the conditions under which the change is revealed. The pertinence of any method of describing learning phenomena depends upon the investigators' interests and purposes.

One dimension of general interest is the interval separating the stimulus complex necessary for the learning from the behavioral responses indicative of learning. In some investigations the interval may be *very* short indeed. Some of the tests of very short term memory, such as the digit-symbol substitution test (DSST), symbol copying, and recognizing a series of numbers in a sequence, involve an appreciable perceptual component and may not be considered as involving an important memory component by those who are disinterested in very short-term memory phenomena. Short-term learning is represented by familiar learning tests. Often the tests require the subject to recall or recognize recently presented syllables or numbers from a list. Such material is stored temporarily and may not be retrievable without intervening review. There are other situations in which the learned material is maintained in relatively long-term, if not permanent storage, e.g., names for common objects.

---

[1] Rutgers University, New Brunswick, New Jersey, USA.

Since most investigators choose to examine several different aspects of learning and since there are always limits to the duration of the effect of the drug, to the subject's cooperation, and to the investigator's schedule, time is always at a premium. Popular tests are relatively brief and convenient. Nevertheless, investigators differ greatly in the manner in which they choose to examine the effects of drugs on learning and memory. In addition to the temporal aspects, investigations of learning differ with respect to the amount and complexity of the behavior to be learned, the familiarity or meaningfulness of the material, the sensory modality through which the material is presented, and the behavioral modality by which the learned responses are expressed.

Many investigators have a battery of preferred tests, and the preferences of frequent contributors are a major determiner of the frequency with which tests appear in summaries such as those offered by Tables 1, 2. Only a few of the published reports provide any explicit indication that the tests were selected on the basis of the anticipated nature of the effects of a drug on memory.

Interindividual differences in learning can be reflected in the accuracy with which the learned material is recalled, recognized, or reproduced after either single or multiple presentations. In addition, some investigators assess learning in terms of the number of trials required before a certain criterion of accuracy is attained; other investigators consider the number of trials required before the material is relearned to a given level of accuracy.

The findings of this survey are summarized in Tables 1, 2. All the studies from which data were obtained are listed in the reference list. Since many investigations included the effects of several drugs and may have involved numerous procedures to test the effects, there is no definite correspondence between the number of studies reviewed and the number of times any drug was tested, or between the number of studies and the number of tests applied to a given drug. In Tables 1, 2 the tests are grouped according to the symbols on which they rely, i.e., verbal, numerical, and visual-spatial.

In Table 1, a column is reserved for each drug and a row is reserved for each test. In this way, Table 1 gives a cell to show the effect of each drug on each test. If a given drug impaired the performance on a particular test, a minus sign is entered, if test performance was facilitated by the drug, a plus sign is entered, and if the drug had no significant effect on the response, a zero is entered. The number of instances that a drug was tested is shown at the foot of the column. Some of the tests were used by relatively few investigators, and some of the drugs were also used infrequently. In order to provide some basis for generalization, drugs that were used fewer than twice and tests that were used fewer than three times are not represented in the summary tables.

Most of the drugs were benzodiazepines, and these results are summarized in the left-hand part of Table 1; the effects of various other drugs are summarized in the right-hand part. Although the investigators tested the effects of some drugs much more frequently than others, they rarely indicated the basis of the choice of drug.

In general, the left-hand part of Table 1 shows detracting effects on memory. There were some notable exceptions, however. Tofisopam and clobazam did not have a detracting effect on memory. Flunitrazepam had no detracting effect in

Table 1. Effects of various benzodiazepines and nonbenzodiazepine drugs on performance of tests of memory

	Diazepam	Lorazepam	Triazolam	Oxazepam	Flurazepam	Tofisopam	Nitrazepam	Midazolam	Loprazolam	Flunitrazepam	Meclamezapam	Clobazam	Alcohol	Caffeine	Scopalamine
Verbal recall	– – – – – – – – –	– – – –	–	–0					–				+ – –	0	–
Verbal recall (delayed)	– – – – – – –			– –										0	
Paired associates	– –		–			0	–	–					–		–
Verbal recognition (delayed)	– –0		–	– –										0	
Category recall	–0			0											–
Verbal recognition	–0	–													
Arithmetic	– – –0	–	–	– –					–				–0	0	
Number recall	– – – –			–0									0	0	–
Digit span	0	–0	–		–					0				0	
Numbers in sequence														+	– –
Digits backwards												0	0		
Digit – symbol substitution test	– – – – – – – – –0	– – – –	– – –	–	–0	0				–	–	0	– – –	+	0
Symbol copying	– – –	–0	– –		– –	0				–	–			+	
Visual recall	–0	–	–		–		–	–							
Picture recognition (delayed)		–											+		
Total –	41	14	11	9	5	0	2	2	2	2	2	0	7	0	6
Total 0	7	2	0	3	1	3	0	0	0	1	0	2	3	6	1
Total +	0	0	0	0	0	0	0	0	0	0	0	0	2	3	0
No. of times tested	48	16	11	12	6	3	2	2	2	3	2	2	12	9	7
Percentage 0	15	13	0	25	17	100	0	0	0	33	0	100	25	67	14

–, Drug-related impairment; 0, no significant effect; +, drug-related enhancement.

Table 1 (continued)

	Nicotine	Propranolol	Triprolidine	Barbiturates	Amitriptyline	Terfenidine	Cimetidine	Buspirone	Nomifensine	Ranitidine	Tazifylline	Mianserin	Fluvoxamine	Marijuana
Verbal recall	++	0												
Verbal recall (delayed)	+													
Paired associates														
Verbal recognition (delayed)														
Category recall														−
Verbal recognition				−										
Arithmetic		+												
Number recall														
Digit span														
Numbers in sequence	+00													
Digits backwards					0				0					
Digit – symbol substitution test		0	− − −	−	−0	00	−0	0	0	0	0	−	0	
Symbol copying		−0	− −	−		0	0	−		0	0	−	0	
Visual recall														
Picture recognition (delayed)														0
Total −	0	1	5	3	1	0	1	1	0	0	0	2	0	1
Total 0	2	3	0	0	2	3	2	1	2	2	2	0	2	1
Total +	4	1	0	0	0	0	0	0	0	0	0	0	0	0
No. of times tested	6	5	5	3	3	3	3	2	2	2	2	2	2	2
Percentage 0	33	60	0	0	67	100	67	50	100	100	100	0	100	50

−, Drug-related impairment; 0, no significant effect; +, drug-related enhancement.

Table 2. Summary of effects of benzodiazepines and nonbenzodiazepines on performance of tests of memory, and percentage of trials on which the tests revealed no effects

	Benzodiazepines					Nonbenzodiazepines					Combined				
	–	0	+	Total	% 0	–	0	+	Total	% 0	–	0	+	Total	% 0
Verbal recall	16	1	0	17	6	3	2	3	8	25	19	3	3	25	12
Verbal recall (delayed)	8	0	0	8	0	0	1	1	2	50	8	1	1	10	10
Paired associates	5	1	0	6	17	1	0	1	2	0	6	1	1	8	13
Verbal recognition (delayed)	5	1	0	6	17	0	1	0	1	100	5	2	0	7	29
Category recall	1	2	0	3	67	2	0	0	2	0	3	2	0	5	40
Verbal recognition	2	1	0	3	33	1	0	0	1	0	3	1	0	4	25
Arithmetic	8	1	0	9	11	1	2	1	4	50	9	3	1	13	23
Number recall	5	1	0	6	17	1	2	1	4	50	6	3	0	9	33
Digit span	3	3	0	6	50	0	1	0	1	100	3	4	0	7	57
Numbers in sequence	0	0	0	0	0	2	2	2	6	33	2	2	2	6	33
Digits backwards	0	1	0	1	100	0	3	0	3	100	0	4	0	4	100
Digit – symbol substitution test	20	4	0	24	17	10	11	1	22	50	30	15	1	46	33
Symbol copying	10	2	0	12	17	6	6	1	13	46	16	8	1	25	32
Visual recall	6	1	0	7	14	0	0	0	0	0	6	1	0	7	14
Picture recognition (delayed)	1	0	0	1	0	0	1	1	2	50	1	1	1	3	33

–, Number of trial with drug-related impairment of behavior; 0, number of trials with no significant effect on behavior; +, number of trials with drug-related enhancement of behavior.

one-third of the instances, and oxazepam had no detracting effect in three of the 12 instances in which it was tested.

The right-hand part of Table 1, concerned with nonbenzodiazepine drugs, is organized in the same manner as the left-hand part. In general, the tests appearing most commonly in the left-hand part are those appearing most commonly in the right-hand part. The exceptions are visual recall, delayed verbal recall, and digit span, which were rarely used in the study of nonbenzodiazepine drugs, and recognizing a series of numbers in a sequence, which was not used in the study of benzodiazepines.

In contrast with the benzodiazepines, the nonbenzodiazepines vary greatly in their nature and use. Four of the drugs, scopolamine, triproladine, barbiturates, and alcohol, have a generally detracting effect on memory as tested. The other drugs included have few detracting effects on the memory tests. Nicotine is particularly interesting because, in the present review, there were more instances of it having an enhancing than a detracting effect on memory. Caffeine and alcohol can also have an enhancing effect under appropriate conditions. Table 1 shows that for some drugs verbal recall and recognizing a series of numbers in a sequence can indicate a detracting effect, while for other drugs they can show an enhancement.

A general summary of the use and sensitivity of the test behaviors may be found in Table 2. Verbal recall, delayed verbal recall, and visual recall were among the most discriminating in the present series of studies. The most frequently used tests for very short term memory, i.e., DSST and symbol copying, were quite sensitive to the detracting effect of benzodiazepines, as were two numerical tests, arithmetic and number recall. Approximately one-half of the tests were used too infrequently to support generalizations concerning their sensitivity to psychotropic substances.

The reliability and validity of a test accrue from the context in which it is used. Validity must be stated with respect to some particular use, and if the test does not produce the expected distinction, it may be declared invalid for that use. The manner in which the test is administered, the characteristics of the sample of subjects, and the conditions under which the test peformance is assessed can greatly influence de facto validity. These generalizations are well illustrated in some of the current studies of the effect of psychotropic drugs on various aspects of learning and remembering. For example, DESAI et al. (1983) found that recall of consonants presented in a series was enhanced by diazepam in normal subjects whose anxiety score was above the sample median. The recall of subjects whose anxiety score was low was impaired by diazepam. Diazepam can facilitate learning in anxious subjects and impair it in nonanxious subjects. It is possible that, in samples heterogeneous with respect to anxiety, diazepam might appear to be without significant consequence or the memory test might appear to be invalid.

Another kind of situation where heterogeneity can be important was illustrated by KOHNEN and LIENERT (1980). The conditions under which an object-recall test was administered were manipulated experimentally so that the subjects might have been exposed to one of three levels of situational stress: low, medium, or high. Among subjects assigned cloxazolam, the benefit was most apparent in those subjects experiencing a high degree of situational stress, and learning was poorest among the low-stress subjects. This finding suggests that unrecognized or inadvertent situational stress could lead one to question either the effect of the drug or the sensitivity of the test.

Several investigators have distinguished between the initial input or encoding aspects of learning and the subsequent recovery phase. WARBURTON et al. (1986) have shown that the recovery of learned material can be state dependent. Subjects who smoked prior to being exposed to some Chinese ideograms had a somewhat better recognition score than subjects who did not smoke before the exposure, but who did smoke before they were asked to recognize the ideograms. If, however, the subjects smoked before viewing the stimulus material and subsequently before the recognition phase, recognition was maximal; it was better than that of subjects who smoked initially before exposure but not before recognition, of those who did not smoke at all, and of those who did not smoke before exposure but smoked before the recognition trials. Thus, the apparent effect of smoking is somewhat dependent upon whether the involvement with nicotine is consistent between initial exposure and subsequent recognition.

The positive effects of smoking on the input phase of learning are confirmed by a study by PEEKE and PEEKE (1984), who compared the effect on verbal recall when smoking occurred before the presentation material with the effects when smoking occurred after the presentation of material but before delayed recall. The effect of smoking was dose dependent, but it was found that prepresentation smoking had a more beneficial effect on delayed recall than postpresentation smoking. Much information and many habits of everyday life may have been acquired while the individual was under the influence of such substances as caffeine, nicotine, alcohol, and possible others. Perhaps some of the distress experienced during substance withdrawal may be viewed as a consequence of state-dependent learning.

A study by PARKER et al. (1980) provides further illustration of the importance of a distinction between the acquisition and the recovery phases of learning. They showed that alcohol can enhance subsequent recall if it is taken immediately after the acquisition phase and before the recall phase. Such an enhancing effect would not be expected on the basis of those studies that show that when alcohol is consumed before the acquisition phase, subsequent recall is impaired.

HINRICHS et al. (1984) used a somewhat similar kind of study to illustrate a retrograde facilitating effect for diazepam. They found that if diazepam were given immediately after the acquisition phase and before the presentation of any subsequent material, the recall of the material presented before diazepam was enhanced. This finding, like that of PARKER et al. (1980) for alcohol, is particularly interesting because diazepam, like alcohol, is commonly found to have a detracting effect on memory. The explanation offered by HINRICHS et al. (1984) is that diazepam given immediately after presentation of input material prevents the acquisition of other material which would interfere with the memory of the input material.

Whether learning is found to be enhanced, unaffected, or impaired by a drug may also depend upon dosage level. In a study of clobazam, NICHOLSON and STONE (1982) found that at 20 mg clobazam enhanced DSST performance, while at 40 mg performance was impaired. Presumably at some intermediate dosage

Table 3. A comparison of the impairment of Memory and Psychomotor Behaviors caused by various psychotropic substances

Drug	Memory					Psychomotor					P_{m-}/P_{p-}
	–	0	+	Total	Fraction $-(P_{m-})$	–	0	+	Sum	Fraction $-(P_{p-})$	
Triprolodine	5	0	0	5	1.00	3	1	0	4	0.75	1.33
Scopolamine	6	1	0	7	0.86	4	2	0	6	0.67	1.28
Diazepam	41	7	0	48	0.85	34	12	0	46	0.74	1.15
Flurazepam	5	1	0	6	0.83	4	1	0	5	0.80	1.04
Lorazepam	14	2	0	16	0.88	18	3	0	21	0.86	1.02
Barbiturates	3	0	0	3	1.00	6	0	0	6	1.00	1.00
Alcohol	7	3	0	10	0.70	20	2	0	22	0.91	0.77
Amitriptyline	1	2	0	3	0.33	6	2	0	8	0.75	0.44
Propranolol	1	3	1	5	0.20	5	3	0	8	0.63	0.32
Caffeine	0	6	3	9	0.00	2	8	0	10	0.20	0.00

–, 0, +, see Table 2.

clobazam could have been found to be without effect. Whether a test is a valid indication of an effect of a drug must be stated in highly qualified terms and does not always lend itself to a simple answer.

For many important aspects of human behavior, the interval between the material to be learned and the application of the learning is far too short to compare the effects of drugs on memory as distinguished from learning. Nevertheless, some investigators have used procedures in which the delay between learning and remembering is sufficiently long to enable the two effects to be studied separately. An increasing number of investigators use computers to automate both the presentation of the material to be learned and the recording of responses indicative of memory. In addition to the obvious economies, automation can contribute to refinements in procedures as well as to the reliability of test scores.

Prior to the present survey of the effects of psychotropic drugs on various memory behaviors, a quite similar survey had been made of the effects of psychotropic substances on psychomotor behavior (WITTENBORN 1987). Some drugs that were included in the survey of memory effects were also included in the study of psychomotor effects. Where the behavioral effects of a drug were included in both the memory and psychomotor surveys in a sufficient number of studies, it is possible to compare the proportion of memory losses with the proportion of psychomotor losses (Table 3). Some drugs, such as barbiturates and benzodiazepines, were characterized by a prevalence of detracting effects on both memory and psychomotor behaviors. One drug, caffeine, caused very few detracting effects (these were for tremor, a psychomotor function).

The likelihood of a memory detraction compared with the likelihood of a psychomotor detraction is expressed in Table 3 as a ratio of the proportion of memory detractions (P_{m-}) to the proportion (P) of psychomotor detractions (P_{p-}), i.e., P_{m-}/P_{p-}. From these ratios it can be seen that triprolidine and scipolamine are more likely to have a detracting effect on memory than to have a detracting effect on psychomotor behavior. Other drugs, such as amitriptyline and propranolol, were found to have a detracting effect on psychomotor behavior in a higher proportion of trials than memory.

Therapeutic effects and detracting behavioral effects are recognized as mutually confounding (WITTENBORN 1978). It may be premature to expect therapeutically effective drugs to involve no detracting behavioral effects. The time may be approaching, however, to compare therapeutic effects with various behavioral detractions and thereby provide the clinician with an explicit guide for selecting treatment on the basis of both the behavioral requirements of the patient and the therapeutic requirements for symptom remission. The interaction between the beneficial learning effects observable in patients because of drug-related symptom remission and drug-related behavioral impairment remains an area that invites exploration.

Acknowledgment. The preparation of this manuscript was supported in part by the Cape Branch Foundation, Dayton NJ, USA.

Material Reviewed

Block RI, Wittenborn JR (1985) Marijuana effects on associative processes. Psychopharmacology 85:426–430

Bond A, Lader M (1982) A comparison of the psychotropic profiles of tofisopam and diazepam. Eur J Clin Pharmacol 22:137–142

Bond A, Lader M, Shrotriya R (1983) Comparative effects of a repeated dose regime of diazepam and buspirone on subjective ratings, psychological tests and the EEG. Eur J Clin Pharmacol 24:463–467

Borland RG, Rogers AS, Nicholson AN, Pascoe PA, Spencer MB (1986) Performance overnight in shiftworkers operating a day-night schedule. Aviat Space Environ Med 57:241–249

Breimer DD, Jochemsen R, Kamphuisen HAC, Nicholson AN, Spencer MB, Stone BM (1985) Central effects during the continuous osmotic infusion of a benzodiazepine (triazolam). Br J Clin Pharmacol 19:807–815

Caine ED, Weingartner H, Ludlow CL, Cudahy EA, Wehry S (1981) Qualitative analysis of scopolamine-induced amnesia. Psychopharmacology 74:74–80

Curran HV, Lader M (1986) The psychopharmacological effects of repeated doses of fluvoxamine, Mianserin and placebo in healthy human subjects. Eur J Clin Pharmacol 29:601–607

Darragh A, Lambe R, Kenny M, Brick I, Taaffe W (1982) RO 15–1788 antagonises the central effects of diazepam in man without altering diazepam bioavailability. Br J Clin Pharmacol 14:677–682

Drew PJT, Barnes JN, Evans SJW (1985) The effect of acute B-adrenoceptor blockade on examination performance. Br J Clin Pharmacol 19:783–786

Ellinwood Jr EH, Linnoila M, Easler ME, Molter DW (1981) Onset of peak impairment after diazepam and after alcohol. Clin Pharmacol Ther 30:534–538

Ellinwood Jr EH, Heatherly DG, Nikaido AM, Bjornsson TD, Kilts C (1985) Comparative pharmacokinetics and pharmacodynamics of lorazepam, alprazolam and diazepam. Psychopharmacology 86:392–399

File SE, Lister RG (1982) Do lorazepam-induced deficits in learning result from impaired rehearsal, reduced motivation or increased sedation? Br J Clin Pharmacol 14:545–550

File SE, Lister RG (1983) Does tolerance to lorazepam develop with once weekly dosing? Br J Clin Pharmacol 16:645–650

File SE, Lister RG (1985) A comparison of the effects of lorazepam with those of propranolol on experimentally-induced anxiety and performance. Br J Clin Pharmacol 19:445–451

Ghoneim MM, Mewaldt SP, Hinrichs JV (1984) Dose-response analysis of the behavioral effects of diazepam: II. Psychomotor performance, cognition and mood. Psychopharmacology 82:296–300

Ghoneim MM, Hinrichs JV, Mewaldt SP (1986) Comparison of two benzodiazepines with differing accumulation: Behavioral changes during and after 3 weeks of dosing. Clin Pharmacol Ther 39:491–500

Godtlibsen OB, Jerke D, Gordeladze JO, Bredesen JE, Matheson I (1986) Residual effect of single and repeated doses of midazolam and nitrazepam in relation to their plasma concentrations. Eur J Clin Pharmacol 29:595–600

Golombok S, Lader M (1984) The psychopharmacological effects of premazepam, diazepam and placebo in healthy human subjects. Br J Clin Pharmacol 18:127–133

Gorenstein C, Gentil V (1983) Residual and acute effects of flurazepam and triazolam in normal subjects. Psychopharmacology 80:376–379

Greenberg BD, Moore PA, Letz R, Baker EL (1985) Computerized assessment of human neurotoxicity: Sensitivity to nitrous oxide exposure. Clin Pharmacol Ther 38:656–660

Griffiths RR, McLeod DR, Bigelow GE, Liebson IA, Roache JD (1984) Relative abuse liability of diazepam and oxazepam: behavioral and subjective dose effects. Psychopharmacology 84:147–154

Hamilton MJ, Smith PR, Peck AW (1983) Effects of buproprion, nomifensine and dexamphetamine on performance, subjective feelings, autonomic variables and electroencephalogram in healthy volunteers. Br J Clin Pharmacol 15:367–374

Henauer KA, Hollister LE, Gillespie HK, Moore F (1983) Theophylline antagonizes diazepam-induced psychomotor impairment. Eur J Clin Pharmacol 25:743–747

Hommer DW, Matsuo V, Wolkowitz O, Chrousos G, Greenblatt DJ, Weingartner H, Paul SM (1986) Benzodiazepine sensitivity in normal human subjects. Arch Gen Psychiatry 43:542–551

Idzikowski C, Oswald I (1983) Interference with human memory by an antibiotic. Psychopharmacology 79:108–110

Jubis RMT (1986) Effects of alcohol and nicotine on free recall of relevant cues. Percept Mot Skills 62:363–369

Karacan I, Orr W, Roth T, Kramer M, Thornby J, Bingham S, Kay D (1981 a) Dose-related effects of flurazepam on human sleep-waking patterns. Psychopharmacology 73:332–339

Karacan I, Orr W, Roth R, Kramer M, Thornby J, Bingham S, Kay D (1981 b) Dose-related effects of phenobarbitone on human sleep-waking patterns. Br J Clin Pharmacol 12:303–313

Kupietz SS, Richardson E, Gadow KD, Winsberg BG (1980) Effects of methylphenidate on learning a "beginning reading vocabulary" by normal adults. Psychopharmacology 69:69–72

Lader M, Melhuish A, Harris P (1982) Residual effects of repeated doses of 0.5 and 1 mg flunitrazepam. Eur J Clin Pharmacol 23:135–140

Liljequist R (1981) Codeine-induced memory changes: nature and relationship to opiate system. Eur J Clin Pharmacol 20:99–107

Lister RG, File SE (1984) The nature of lorazepam-induced amnesia. Psychopharmacology 83:182–187

Loke WH, Hinrichs JV, Ghoneim MM (1985) Caffeine and diazepam: Separate and combined effects on mood, memory, and psychomotor performance. Psychopharmacology 87:344–350

McManus IC, Ankier SI, Norfolk J, Phillips M, Priest RG (1983) Effects in psychological performance of the benzodiazepine, loprazolam alone and with alcohol. Br J Clin Pharmacol 16:291–300

McNair DM, Frankenthaler LM, Czerlinsky T, White TW, Sasson S, Fisher S (1982) Simulated public speaking as a model of clinical anxiety. Psychopharmacology 77:7–10

Madden DJ, Blumenthal JA, Ekelund L-G, Krantz DS, Light KC, McKee DC (1986) Memory performance by mild hypertensives following beta-adrenergic blockade. Psychopharmacology 89:20–24

Mewaldt SP, Ghoneim MM, Hinrichs JV (1986) The behavioral actions of diazepam and oxazepam are similar. Psychopharmacology 88:165–171

Myrsten A-L, Rydberg U, Idestrom C-M, Lamble R (1980) Alcohol intoxication and hangover: modification of hangover by chlormethiazole. Psychopharmacology 69:117–125

Naranjo CA, Sellers EM, Kaplan HL, Hamilton C, Khouw V (1984) Acute kinetic and dynamic interactions of zimelidine with ethanol. Clin Pharmacol Ther 36:654–660

Nicholson AN, Stone BM (1983) The Hl-antagonist mequitazine: Studies on performance and visual function. Eur J Clin Pharmacol 25:563–566

Nicholson AN, Stone BM (1984) The H_2-antagonists, cimetidine and ranitidine: Studies on performance. Eur J Clin Pharmacol 26:579–582

Nicholson AN, Stone BM (1986) Antihistamines: impaired performance and the tendency to sleep. Eur J Clin Pharmacol 30:27–32

Nuotto E (1983) Psychomotor, physiological and cognitive effects of scopolamine and ephedrine in healthy man. Eur J Clin Pharmacol 24:603–609

O'Boyle C, Lambe R, Darragh A (1985) Central effects in man of the novel schistosomicidal benzodiazepine meclonazepam. Eur J Clin Pharmacol 29:105–108

Parrott AC, Hindmarch I, Stonier PD (1982) Nomifensine, clobazam and HOE 8476: Effects of aspects of psychomotor performance and cognitive ability. Eur J Clin Pharmacol 23:309–313

Pascoe PA, Stone BM (1984) Ascorbic acid and performance in man. Psychopharmacology 83:376–377

Pomara N, Stanley B, Block R, Guido T, Russ D, Berchou R, Stanley M, Greenblatt DJ, Newton RE, Gershon S (1984) Adverse effects of single therapeutic doses of diazepam on performance in normal geriatric subjects: Relationship to plasma concentrations. Psychopharmacology 84:342–346

Roth T, Hartse KM, Saab PG, Piccione PM, Kramer M (1980) The effects of flurazepam, lorazepam, and triazolam on sleep and memory. Psychopharmacology 70:231–237

Salen SAM, McDevitt DG (1984) Central effects of single oral doses of propranolol in man. Br J Clin Pharmacol 17:31–36
Sellers EM, Naranjo CA, Giles HG, Frecker RC, Beeching M (1980) Intravenous diazepam and oral ethanol interaction. Clin Pharmacol Ther 28:638–645
Seppälä T, Palva E, Mattila MJ, Korttila K, Shrotriya RC (1980) Tofisopam, a novel 3,4,-benzodiazepine: multiple-dose effects on psychomotor skills and memory, comparison with diazepam and interactions with ethanol. Psychopharmacology 69:209–218
Shader RI, Dreyfuss D, Gerrein JR, Harmatz JS, Allison SJ, Greenblatt DJ (1986) Sedative effects and impaired learning and recall after single oral doses of lorazepam. Clin Pharmacol Ther 39:526–529
Spinweber CI, Johnson LC (1982) Effects of triazolam (0.5 mg) on sleep, performance, memory, and arousal threshold. Psychopharmacology 76:5–12
Stromberg C, Mattila MJ (1985) Acute and subacute effects on psychomotor performance of femoxetine alone and with alcohol. Eur J Clin Pharmacol 28:641–647
Subhan Z, Hindmarch I (1983) The effects of lormetazepam of aspects of sleep and early morning performance. Eur J Clin Pharmacol 25:47–51
Subhan Z, Hindmarch I (1985) Psychopharmacological effects of vinpocetine in normal healthy volunteers. Eur J Clin Pharmacol 28:567–571
Walsh JK, Schweitzer PK, Parwatikar S (1983) Effects of lorazepam and its withdrawal on sleep, performance, and subjective state. Clin Pharmacol Ther 34:496–500
Wesnes K, Revell A (1984) The separate and combined effects of scopolamine and nicotine on human information processing. Psychopharmacology 84:5–11
Wesness K, Warburton DM (1984 a) The effects of cigarettes of varying yield on rapid information processing performance. Psychopharmacology 82:338–342
Wesnes K, Warburton DM (1984 b) Effects of scopolamine and nicotine on human rapid information processing performance. Psychopharmacology 82:147–150
Williams HL, Rundell Jr OH, Smith LT (1981) Dose effects on secobarbital in a Sternberg memory Scanning Task. Psychopharmacology 72:161–165
Winokur A, Lindberg ND, Lucki I, Phillips J, Amsterdam JD (1986) Hormonal and behavioral effects associated with intravenous L-tryptophan administration. Psychopharmacology 88:213–219
Wolkowitz OM, Tinklenberg JR (1985) Naloxone's effect on cognitive function in drug-free and diazepam-treated normal humans. Psychopharmacology 85:221–223

References

Desai N, Taylor-Davies A, Barnett DB (1983) The effects of diazepam and oxprenolol on short term memory in individuals of high and low state anxiety. Br J Clin Pharmacol 15:197–202
Hinrichs JV, Ghoneim MM, Mewaldt SP (1984) Diazepam and memory: retrograde facilitation produced by interference reduction. Psychopharmacology 84:158–162
Kohnen R, Lienert GA (1980) Defining tranquilizers operationally by non additive effect in experimental stress situations. Psychopharmacology 68:291–294
Nicholson AN, Stone BM (1982) Hypnotic activity and effects on performance of lormetazepam and camazepam-analogues of temazepam. Br J Clin Pharmacol 13:433–439
Parker ES, Birnbaum IM, Weingartner H, Hartley JT, Stillman RC, Wyatt RJ (1980) Retrograde enhancement of human memory with alcohol. Psychopharmacology 69:219–222
Peeke SC, Peeke HVS (1984) Attention, memory and cigarette smoking. Psychopharmacology 84:205–216
Warburton DM, Wesnes K, Shergold K, James M (1986) Facilitation of learning and state dependency with nicotine. Psychopharmacology 89:55–59
Wittenborn JR (1978) Behavioral toxicity in normal humans as a model for assessing behavioral toxicity in patients. In: Lipton MA, DiMascio A, Killan KF (eds) Psychopharmacology: a generation of progress. Raven, New York
Wittenborn JR (1987) Psychomotor tests in psychopharmacology. In: Hindmarch I, Stonier PD (eds) Human psychopharmacology, vol 1. Wiley, Chichester, pp 69–78

Information Processing, Critical Flicker Fusion Threshold and Benzodiazepines: Results and Speculations

I. HINDMARCH[1]

Abstract

There is evidence to suggest that the rate of information processing, as measured by the critical flicker fusion threshold (CFFT), is slower following some benzodiazepines than others. Changes in CFFT brought about by benzodiazepine administration are usually, but not always, correlated with changes in other measures of cognitive performance and memory. However, the drug-induced changes in information processing and memory cannot be fully explained by simple postulates regarding alterations in the overall level of CNS arousal. Results from a series of studies of the effect of benzodiazepines on measures of CFFT and memory will be reviewed and the utility of CFFT in evaluating the amnesic or mnemonic potential of CNS-active drugs will be assessed.

1 Introduction

The majority of psychological theories of memory, especially those relating to "short-term" or "working" functions, adopt information-processing models (see BADDELEY, this volume; EYSENCK, this volume). Indeed, theories of cognition in general see the prime role of the brain and CNS to be the acquisition, coding, storage, manipulation and retrieval of information. All sensory information – both external and internal in origin – is assimilated and accommodated to provide the cognitive organisation of overt behaviors. A knowledge of the extent to which psychoactive drugs, and benzodiazepines in particular, affect information processing is clearly of interest at clinical, psychopharmacological and theoretical levels.

The purpose of this essay is threefold: first, to examine the extent to which critical flicker fusion threshold (CFFT) can be regarded as a measure of information processing (ability, capacity and efficiency); second, to report the effects of benzodiazepine receptor ligands on CFFT; and third, to speculate on the relationship between information processing, CFFT and benzodiazepines.

2 Critical Flicker Fusion Threshold

Historians would remind CFFT researchers that Ptolemy, in the second century B.C., was among the first to report a visual flicker or "stroboscopic" phenome-

[1] Human Psychopharmacology Research Unit, Department of Psychology, University of Leeds, Leeds LS2 9JT, UK.

non when he observed that the spokes on the wheels of a moving carriage appeared stationary. However, it was probably the physiological research of Plateau in Belgium and Talbot in England during the 1830s that began the study of visual functions using the CFFT. Much of the development and refinement of CFFT techniques and methods was in the hands of ophthalmologists and sensory physiologists concerned with the mechanisms of visual perception and retinal function. These researchers, primarily interested in physiology, studied the various influences – age, physique, physiology, neurophysiology, personality, as well as somatic, endocrine and psychiatric states – that could affect the CFFT.

It was not until the early 1950s that psychopharmacologists turned their attention to CFFT as a tool for measuring "vigilance"; it is this use of the CFFT to appraise the effects of psychotropics on CNS function in a reliable, sensitive and valid way that is now our main concern. SMITH and MISIAK (1976), reviewing the early studies (1951–1976) of CFFT and psychoactive drugs, were able to present data on 19 different drugs. HINDMARCH (1982), reviewing work published between 1976 and 1981, found controlled studies on 42 different drugs, with many compounds being investigated on a dose-related basis. A contemporary review of CNS-active drugs and CFFT would doubtless show a similar increase in the use of CFFT as a measure in psychopharmacological research.

The popularity of CFFT is doubtless due to the ease with which measurements can be taken and to the non-invasive nature of the task. At the most formal level, it is necessary to show that the measure is both reliable and valid.

2.1 Reliability

The reliability of a test measure is usually indicated by the extent to which the results obtained in one situation are obtainable in another situation using similar techniques and assessments. HINDMARCH (1982) showed the effects of clobazam, a 1,5-benzodiazepine, on CFFT in a range of experiments with different treatment regimens but utilising the same apparatus for the assessment of CFFT. The changes of CFFT produced by different dosages of clobazam in 13 discrete assays are presented in Table 1. As can be seen, clobazam reduced CFFT in only two instances, and neither was statistically significant. In the remaining 11 instances, seven dose regimens produced a significant rise in CFFT and four instances were found where clobazam produced a non-significant rise in CFFT. A closer examination of the two instances where clobazam reduced CFFT reveal that an overnight (as opposed to daytime) assessment schedule was involved and that this could be the experimental variable which accounted for the lowered CFFT. However, it would appear that the CFFT technique used in these studies was sensitive to a rise in CFFT in 85% of the cases, and significantly so 54% of the time. These findings, as well as suggesting the reliability of CFFT as an index of CNS activity, demonstrate the importance of replication of results in psychopharmacological research. Consistent reductions in CFFT after a range of doses of ethanol (0.32–1.29 g absolute alcohol/kg) and following different doses of various 1,4-benzodiazepine sedatives, tranquillisers and hypnotics also show the reliability of the assessment measure (HINDMARCH 1982).

Table 1. Changes produced by clobazam with respect to placebo controls on CFFT (after HINDMARCH 1982)

Dose regimen: clobazam	CFFT change
1 × 20 mg	+
1 × 20 mg	+*
1 × 30 mg	−
1 × 30 mg	+*
1 × 40 mg	+
3 × 10 mg	+
5 × 10 mg	+*
12 × 10 mg	+*
15 × 10 mg	+*
5 × 20 mg	+*
3 × 30 mg	+*
5 × 30 mg	−
5 × 40 mg	+

+, increase with respect to placebo; +*, significant ($p < 0.05$) increase with respect to placebo; −, decrease with respect to placebo.

2.2 Validity

Any interrelationships between CFFT and other measures of CNS function would be taken as evidence of the validity of CFFT as an index of some aspects of psychological state and/or behavior. BOBON et al. (1982), reviewing the published work then available, showed that CFFT thresholds correlated with, amongst other things, alpha and beta activity on the electroencephalogram, sub-

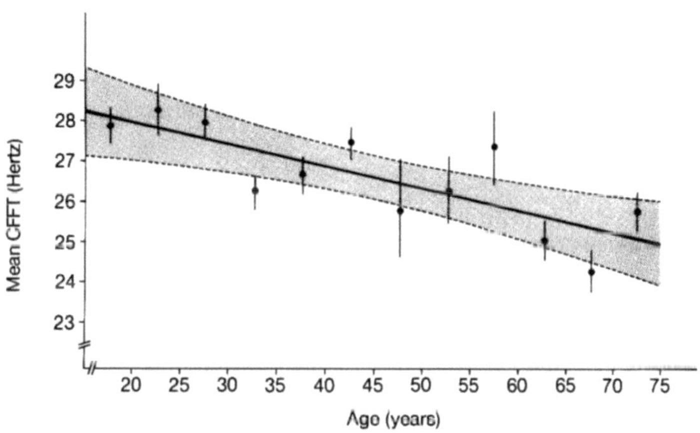

Fig. 1. Regression of mean CFFT on age in 5-year counts. *Shaded area* indicates 95% confidance limits. From FREWER (1986)

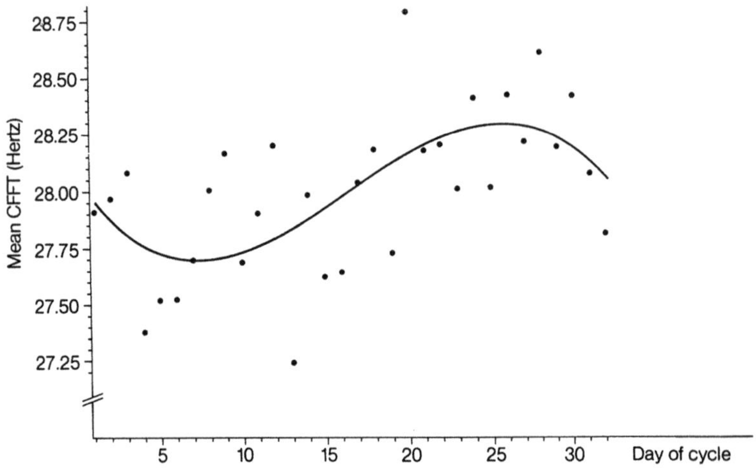

Fig. 2. Mean CFFT changes over the menstrual cycle; the curve is fitted from a cubic regression, $p<0.003$ ($n=47$)

jective ratings of arousal, fine motor movement, choice reaction time and duration of the spiral aftereffect. PARROTT (1982) also demonstrated the close concordance between CFFT and other measures of CNS alertness and psychological test performance. The applicability of CFFT is extended and its validity, as a measure of the efficiency of information processing, is increased when the relationship between CFFT and age is considered; the reduction in cognitive functions associated with senescence is reflected in the fall of CFFT. CFFTs also reflect changes in information processing due to anxiety, circadian rhythms, menstrual cycles, endocrine activity, cardiovascular variables, metabolic functions and neurological disorders (BOBON et al. 1982; FREWER 1986; L. DYE, unpublished observations). Of particular interest is the correlation reported by BOBON et al. (1982) between CFFT and intelligence. Although there are some inconsistencies and negative findings, the bulk of the evidence shows that an increase in CFFT is associated with a more efficient processing of information as evidenced by better performance on various verbal and non-verbal subscales of the Wechsler Intelligence Test.

2.3 Measurement

BOBON et al. (1982) preface their text on CFFT in psychopathology and psychopharmacology with an insightful caveat, pertinent to all researchers, "When CFFT falls into disgrace it is … due to differences in apparatus and design of the trials as well as the great number of … variables which modify CFF thresholds". In order to minimise interstudy differences due to apparatus we have employed the same technique for measuring CFFT and a constant mode of presentation since 1972. The results and discussion that follow are all based on work conducted on the CFFT equipment embodied in the Leeds Psychomotor Tester (LPT; Leeds Psychomotor Services, Front Street, Acomb, York YO2 3BJ, UK); furthermore, the studies use a similar methodology and experimental protocol (subjects acting as their own controls in a balanced crossover design) with placebos and/or verum controls.

The CFFT as presented on the LPT requires the subject to discriminate the flicker in a set of four light-emitting diodes fixed at the corners of a 1-cm square. When viewed binocularly at 1-m the square of diodes is in foveal fixation. Subjects are required to press a button to indicate if the diodes are "flickering" or not. In the majority of circumstances the method of limits for at least three ascending and three descending scales (WOODWORTH and SCHLOSBERG 1958) is used to determine the CFFT. In some instances, a computer-assisted presentation has used "frequency" and "forced choice" methods to find the mean CFFT. The CFFT is taken as the mean of the individual readings which is the point halfway between the threshold for flicker/fusion (ascending scales), and the threshold for fusion/flicker (descending scales). In psychopharmacological terms, this mean is the "point of subjective equality" and the frequency at which the diodes may have an equal chance of being seen as either "flickering" or "not flickering".

2.4 Psychological Disorders and CFFT

We have shown that CFFT measures what might be called "information processing", since changes in CFFT are reflected in changes in other measures of intellectual performance and psychological functioning. Furthermore, we have demonstrated a correlation between CFFT changes and those changes in intellectual and cognitive function associated with aging.

BOBON et al. (1982) point out that at the turn of the century Pierre JANET in the Salpêtrière showed a relationship between a reduced CFFT and certain cases of hysteria and depression. More recently SIEGFRIED and O'CONNOLLY (1986) have shown that CFFT changes monitor the cognitive effects of antidepressants in elderly patient populations and SIEGFRIED (1988) has demonstrated the utility of CFFT both in discriminating between antidepressants and in profiling a drug's particular action in the clinical situation.

Both high-anxiety subjects and anxious patients have lower CFFTs than age-matched "normals" (KRUGMAN 1947; BUHLER 1954; GOLDSTONE 1955; WAGONER and COHEN 1956; JONES 1958; WAGONER 1960; HINDMARCH 1979; CLYDE 1981; FREWER 1986; FREWER and HINDMARCH 1988). As information processing is assigned a primary role in activating affective, anxiety and behavioral functions (BECK 1985), it is to be expected that changes in cognitive function should reflect changes in anxiety state. BECK (1970) found a meaningful correlation between fluctuations in patients' free-floating anxiety and certain cognitive functions. As CFFT reflects changes in anxiety level brought about by antianxiety drugs or psychotherapy (HINDMARCH 1979; PAES DE SOUSA et al. 1981; HILL et al. 1981; GRINGRASS and BEAUMONT 1985; THOMPSON 1985), it seems reasonable to assume that CFFT is reflecting the change in cognitive function (information processing capacity) underlying the observed change in anxiety levels.

A similar argument can be developed with regard to changes in CFFT and age (Fig. 1). Several researchers have demonstrated an obvious and often significant regression of CFFT with age (SIMONSEN et al. 1941; BROZEK and KEYS 1945; WEEKERS and ROUSSEL 1946; MISIAK 1947; COLGAN 1954; COPPINGER 1955; MCFARLAND et al. 1958; HINDMARCH 1981; BOBON et al. 1982; FREWER 1986; FREWER and HINDMARCH 1988). Changes in the function of the CNS have been observed to accompany the aging process (BRIZLE et al. 1975), and others (THOMPSON and MARSH 1973) have shown a slowing of the alpha rhythm to be commensurate with aging. A decrease in alpha activity with age has been taken as an index of decline in the functional efficiency of the CNS (BIRREN et al. 1980). As GRUNBERGER et al. (1982) have reported a correlation between CFFT and alpha activity, it can be postulated that the age-related decline of CFFT is indicative of a decline in the efficiency of the CNS – particularly with regard to information processing. The notion that information processing efficiency decreases with age and is reflected in decreased CFFT measures is well supported by the work of DI LOLLO et al. (1982), who used two flash thresholds (a measure similar to CFFT) and tests of visual information processing. FREWER (1986), reviewing the published work on CFFT, age and anxiety, concluded that CFFT represents a measure of the efficiency of the CNS in processing information.

3 Benzodiazepine Receptor Ligands and CFFT

Due to the different treatment regimens used and the variation in the times of measuring CFFT, it is difficult to compare directly the various studies on the effects of benzodiazepines on CFFT as measured by the LPT. The studies and data

Table 2. Effects of single doses benzodiazepine receptor ligands on CFFT

	Dose (mg) producing:	
	Significant decrement with respect to placebo	No Significant change with respect to placebo
Alprazolam	0.5, 1, 1.5	0.25
Bromazepam	6	
Clobazam	30, 40, 60	5, 10, 15, 20
Clorazepate	15	
Diazepam	10, 20, 30	5
Flunitrazepam	1	0.25, 0.5
Flurazepam	15, 30	15
Lorazepam	1, 2, 3	
Lormetazepam	1.5, 2	0.5, 1
Metaclazepam	20	5, 10
Midazolam	5, 10, 15, 20	
Nitrazepam	5, 10	2.5
Oxazepam	12.5, 20, 25, 30, 35, 37.5, 50	
Temazepam	20, 30, 40	10, 15, 20
Triazolam	0.25, 0.5	0.125, 0.25
Zopiclone		7.5

Table 3. Effects of repeated doses of benzodiazepines on CFFT

	Dose (mg) producing:		
	Significant impairment	No significant change	Significant improvement
Alprazolam		0.5 b.i.d. ×3	
Bromazepam		3 t.i.d. ×28	
Chlordiazepoxide	10 t.i.d. ×5		
Clobazam		10 t.i.d. ×4	10 t.i.d. ×5
		15 b.i.d. ×3	20×7
		20×3	20×14
		40×14	20×4
Clorazepate		15×4	
		15×14	
Diazepam	5 t.i.d. ×1	5 t.i.d. ×5	
Flunitrazepam		0.5 nocte ×7	
		1 nocte ×7	
Flurazepam	15 nocte ×7	15 nocte ×4	
	15 nocte ×14		
Ketazolam		30 nocte ×3	
Lorazepam		1 b.i.d. ×5	
		2 mane ×5	
Lormetazepam		0.5 nocte ×7	
		1 nocte ×7	
		2 nocte ×7	
Nitrazepam	10 nocte ×2	5 nocte ×4	
	10 nocte ×6	5 nocte ×5	
Oxazepam		40 b.i.d. ×1	
Oxazolam		30 nocte ×3	
Temazepam	30 nocte ×4	10 nocte ×4	
		20 nocte ×4	
		40 nocte ×7	
		60 nocte ×7	
Triazolam		0.5 nocte ×4	

All changes are with respect to placebo or baseline controls.

used to draw up the following tables are summarised in Hindmarch (1980, 1982, 1984), Bhatti, Alford and Hindmarch (this volume), and Frewer (1986). Table 2 shows the effects of single doses of benzodiazepines on CFFT and Table 3 the effects of repeated doses. These tables are by no means exhaustive in their coverage of the published data but are intended to illustrate the fact that different benzodiazepines have different effects on CFFT. It is well accepted that dose, treatment and experimental and methodological considerations confound any watertight generalisations from these tables, or indeed from the published papers in their entirety. However, it is possible to arrive at a parsimonious division and classification of benzodiazepine receptor ligands with respect to the effects they produce on CFFT under controlled experimental conditions.

Many of the drugs listed in Tables 2 and 3 show considerable evidence of reducing CFFT – especially following single doses. Some become less impairing with repeated doses. For example, 30 mg clobazam is needed to reduce CFFT in a single dose, whereas 10 mg t.i.d. has no effect on CFFT in repeated dose regimens. The single-dose reduction of CFFT is also ameliorated when repeated doses of diazepam, bromazepam, nitrazepam, temazepam and triazolam are given. This could be a reflection of the rapid development of tolerance to the effects of the drugs or a reflection of the lack of data from large samples of controlled studies with both volunteers and patients.

4 Discussion

Should CFFT be an index of information processing, we would expect those drugs with a great effect on CFFT to have a similarly potent action on tests of "working" or "short-term" memory. Table 4 illustrates the effects of various dose regimens of benzodiazepines on "short-term memory" measured in a variety of ways (Curran 1986). The whole picture is more complex, as there are discrete but definite differences between studies in their use of patients (or volunteers), the timing of the "memory" tests and the type of memory test used. It is, however, evident that alprazolam, diazepam, flunitrazepam, flurazepam, lorazepam, midazolam and triazolam are more potent amnesic agents in their disruption of a variety of tests of memory and information processing, and they are certainly worse than bromazepam, clorazepate, loprazolam, metaclazepam, nitrazepam and temazepam. However, clobazam (especially following repeated doses), lormetazepam and zopiclone have even less effect on memory – within clinical dose ranges – than the other benzodiazepine receptor ligands listed. In general, a consideration of Tables 2–4 will show an overall concordance between the magnitude of the disruptive effect of the individual drugs on CFFT and their amnesic activity.

Needless to say, more controlled and large-scale studies need to be done in order to evaluate these tentative conclusions, but it would appear that CFFT is sensitive to the action of benzodiazepine receptor ligands in that it represents and measures an aspect of information processing capacity and efficiency related to short-term or working memory. The extent to which CFFT will prove useful in future psychopharmacological research will depend on the extent to which the present assertions are confirmed in practice. It can only be hoped that researchers will provide more information from studies on patient populations and other large studies to elucidate and extend some interesting possibilities regarding CFFT, information processing and benzodiazepine receptor ligands.

Table 4. Effects of oral benzodiazepine receptor ligands on human memory

	Dose (mg)	Results
Alprazolam	1×1	Immediate recall ✗
Bromazepam	6 t.i.d. ×14	Nonsense syllables ✗, digit span =
Clobazam	10×1, 20×1	Telephone numbers =, trigrams =, city map =,
	30×1	Word recall ✗
	40×1, 60×1	Digits backwards ✗
	10 t.i.d. ×28	Object recall =
Clorazepate	7.5×1	Delayed recall words =
	15×1	Immediate recall =
Diazepam	5×1	Telephone numbers ✗, letter recall ✗, digit recall ✗, picture recall ✗, digits backwards ✗
	10×1	Digits backwards ✗, paired associates ✗, telephone numbers ✗
	20×1	Cued recall ✗, picture recognition ✗, telephone numbers ✗
	30×1	Telephone numbers ✗, digit backwards ✗
	10×3	Paired associates ✗, geometric figures ✗, face recognition =, paired associates ✗ (after 1 day), paired associates = (after 14 days)
Flunitrazepam	0.5×1	Picture recall ✗
	1×1	Picture recognition ✗, Sternberg ✗
	2 nocte ×3	Morning recall ✗
	15 nocte ×1	Digits backwards ✗, word/number pairs ✗
	30 nocte ×2	Task recall ✗, delayed recall ✗
	30 nocte ×6	Task recall ✗
Loprazolam	1×1	Address/name recall ✗
Lorazepam	1×1	Sternberg ✗, word recall ✗
	2×1	Delayed recall ✗
	2.5×1	Paired associates ✗, digit span ✗, recognition
	3×1	Word recall ✗, geometric pattern ✗
	4×1	Picture recall ✗
	1 t.i.d. ×28	Object recognition ✗
Lormetazepam	1.5 nocte ×2	Immediate recall =, delayed recall ✗
	1 nocte ×1	Sternberg =
	1 nocte ×1	Telephone numbers =
Metaclazepam	5×1	Telephone numbers =, digit recall =
	10×1	Telephone numbers ✗, digit recall ✗
	20×1	Telephone numbers ✗, digit recall ✗
Midazolam	15×1	Immediate recall ✗, delayed recall ✗
Nitrazepam	5×1	Digit span ✗, word recall =
	5 nocte ×1	Digits backwards ✗, telephone numbers ✗
Temazepam	10×1	Digit recall =, paired associates =
	20×1	Paired associates ✗, backwards digits ✗, telephone numbers ✗
	30 nocte ×2	task recall ✗
Triazolam	0.25 nocte ×1	Sternberg ✗, digits backwards ✗, telephone numbers ✗
	0.5 nocte ×1	Task recall ✗, word recall ✗, word recognition ✗
	0.25 nocte ×6	Task recall ✗
	0.5 nocte ×6	Task recall ✗
Zopiclone	7.5 nocte ×1	Sternberg =, telephone numbers =, digits backwards =

✗, significant (with respect to control and/or baseline conditions) impairment of memory task; =, no difference from control condition.

References

Beck AT (1970) Role of fantasies in psychotherapy and psychopathology. J Nerv Ment Dis 150:3–17
Beck AT (1985) Theoretical perspectives on clinical anxiety. In: Turman AH, Maser JD (eds) Anxiety and the anxiety disorders. Erlbaum, London, pp 183–196
Birren JE, Woods AM, Williams MV (1980) Behavioural slowing with age. Causes, organisation and consequences. In: Pook LW (ed) Aging in the 1980's: Psychological issues. APA, Washington DC
Bobon DP, Lecoq A, Von Frenkell R, Mormont I, Laverque G, Lottin T (1982) La frequence critique de fusion visuelle en psychopathologie et en psychopharmacologie. Imprimerie des Sciences, Brussels, pp 112 (Acta medica belgica)
Brizle KR, Harkin JC, Ordy JM, Kaack B (1975) Accumulation of senile plaques in the aging nervous system. In: Brody H (eds) Clinical, morphological and neurochemical aspects of the aging central nervous system. Raven, New York
Brozek J, Keys A (1945) Changes in flicker fusion frequency with age. J Consult Psychol 9:87–90
Buhler RA (1954) Flicker fusion threshold and anxiety level. Dissertation Abstracts 14:1255
Colgan CM (1954) Critical flicker fusion, age and intelligence. Am J Psychol 67:711–713
Coppinger NW (1955) The relationship between critical flicker fusion and chronological age for varying levels of stimulus brightness. J Gerontol 10:48–53
Clyde CA (1981) The influence of personality on response to low doses of benzodiazepines. In: Hindmarch I, Stonier PD (eds) Clobazam. Academic, London, pp 75–86 (Royal Society of Medicine international congress and symposium series, no 43)
Curran HV (1986) Tranquillising memories: a review of the effects of benzodiazepines on human memory. Biol Psychiatry 23:179–213
Di Lollo V, Arnett JL, Kruk RV (1982) Age related changes in rate of visual information processing. J Exp Psychol 8:225–237
Frewer LJ (1986) Some psychopharmacological variables affecting the critical flicker fusion threshold. PhD Thesis, University of Leeds
Frewer LJ, Hindmarch I (1988) Critical flicker fusion frequency (CFF) in psychopharmacology. In: Hindmarch I, Stonier PD (eds) Human psychopharmacology, vol II. Wiley, London (in press)
Goldstone S (1955) Flicker fusion measurement and anxiety level. J Exp Psychol 49:200–202
Gringras M, Beaumont G (1985) The effectiveness of repeated nocturnal doses of clobazam and dipotassium clorazepate on clinical measures of anxiety, patient ratings of mood and sleep and objective assessments of CNS activity. In: Hindmarch I, Stonier PD, Trimble MR (eds) Clobazam: human psychopharmacology and clinical applications. Oxford University Press, Oxford, pp 73–79
Grunberger J, Saletu B, Berner P, Stohr H (1982) CFF and assessments of pharmacodynamics. Pharmacopsychiatria 15 (Suppl 1):29–36
Hill AJ, Walsh RD, Hindmarch I (1981) Tolerability of nocturnal doses of clobazam in anxious patients in general practice. In: Hindmarch I, Stonier PD (eds) Clobazam. Academic, London, pp 133–140 (Royal Society of Medicine international congress and symposium series, no 43)
Hindmarch I (1979) A preliminary study of the effects of repeated doses of clobazam on aspects of performance, arousal and behaviour in a group of anxiety rated volunteers. Eur J Clin Pharmacol 16:17–21
Hindmarch I (1980) Psychomotor function and psychoactive drugs. Br J Clin Pharmacol 10:189–209
Hindmarch I (1981) Measuring the effects of psychoactive drugs on higher brain function. In: Burrows GD, Werry JS (eds) Advances in Human Psychopharmacology, vol 2. JAI Press, Connecticut, pp 99–127
Hindmarch I (1982) Critical flicker fusion frequency (CFFF): the effects of psychotropic compounds. Pharmacopsychiatria 15 (Suppl 1):44–48

Hindmarch I (1984) Psychological performance models as indicators of the effects of hypnotics on sleep. In: Hindmarch I, Ott H, Roth T (eds) Sleep, benzodiazepines and performance. Springer, Berlin Heidelberg New York, pp 58–68

Jones O (1958) Relationship between visual and auditory discrimination and anxiety level. J Gen Psychol 59:111–118

Krugman H (1947) Flicker fusion frequency as a function of anxiety reaction. Psychosom Med 4:269–272

McFarland RA, Warren AB, Karis C (1958) Alterations in critical flicker fusion as a function of age. J Exp Psychol 56:26–32

Misiak H (1947) Age and sex differences in the critical flicker fusion. J Exp Psychol 31:318–322

Paes De Sousa M, Figuiera M-L, Loureiro F, Hindmarch I (1981) Lorazepam and clobazam in anxious elderly patients. In: Hindmarch I, Stonier PD (eds) Clobazam. Academic, London, pp 120–123 (Royal Society of Medicine international congress and symposium series no 43)

Parrott AC (1982) Critical flicker fusion thresholds and their relationship to other measures of alertness. Pharmacopsychiatria 15 (Suppl 1):39–43

Siegfried K (1988) Towards a clinical classification of antidepressant profiles. In: Hindmarch I, Stonier PD (eds) Human psychopharmacology vol II. Wiley, London (in press)

Siegfried K, O'Connolly M (1986) Cognitive and psychomotor effects of different antidepressants in the treatment of old age depression. Int Clin Psychopharmacol 1:221–230

Simonsen E, Enzer N, Blankstein SS (1941) The influence of age on the fusion frequency of flicker. J Exp Psychol 29:252–255

Smith JM, Misiak H (1976) Critical flicker frequency (CFF) and psychotropic drugs in normal human subjects – a review. Psychopharmacology 47:175–182

Thompson DJ (1985) Comparison of drug treatment (clobazam) and relaxation therapy in the management of anxiety, preliminary communication. In: Hindmarch I, Stonier PD, Trimble MR (eds) Clobazam: human psychopharmacology and clinical applications. Oxford University Press, pp 101–103

Thompson LW, Marsh GR (1973) Psychophysiological studies of aging. In: Eisdorfer K (ed) The psychology of adult development and aging. APA, Washington DC, pp 19–32

Wagoner RA (1960) Differences in response latency and response variability between high and low anxiety subjects in a flicker fusion task. J Abnorm Soc Psychol 61:355–359

Wagoner RA, Cohen LD (1956) Analysis of patterns of response of anxious and non-anxious subjects to flickering light. Percept Mot Skills 6:167–170

Weekers R, Roussel F (1946) Introduction a l'etude de la frequence de fusion en clinique. Ophthalmologica 112:305–319

Woodworth RS, Schlosberg H (1958) Experimental psychology. Methuen, London

Individual Differences in Benzodiazepine-Induced Changes of Memory

P. Netter[1]

Abstract

Antero- and retrograde amnesia are observed as side effects of most types of benzodiazepines. They have rarely been investigated with respect to physical and personality factors, or to prior experiences, present expectations, and emotional states of the subjects, all of which are well known to modify drug response. By reviewing research on benzodiazepine-induced changes of memory in preoperative, anxious, and depressed patients as well as in healthy subjects, it is demonstrated that differences in benzodiazepine-induced amnesic effects may depend on:

1. subject variables like predrug level of anxiety, depressive symptomatology, memory capacity, experiences with benzodiazepine-type drugs, and expectations of treatment outcome, and secondary factors like social environment and treatment setting
2. interactions between these subject variables and type of schedule (times of acquisition, treatment, and testing), type of learning material, and dose of drug
3. the extent of benzodiazepine-induced changes in anxiety or depression, cortical and emotional arousal (alertness and activity) as well as physiological effects of benzodiazepines

Special emphasis should be placed on the investigation of drug-induced changes of covariation between psychological measures which may provide valuable information for differential prediction and on mechanisms of drug action.

1 Introduction

Searching the literature shows that there have been very few studies of individual differences in the effects of benzodiazepines on memory. For one thing, most studies on benzodiazepines and amnesia are performed for clinical purposes and aim to determine drug- and dose-related differences rather than patient-related factors. Of 129 studies conducted between 1978 and 1981 on benzodiazepines and memory, 42% were performed on preoperative patients and only 27% on healthy volunteers. Strikingly, even the studies on healthy subjects did not seem to pay much attention to the influence of subject-related factors.

When trying to elucidate interindividual differences in drug-induced changes of memory, one should be aware that there are several factors which may contribute to these differences. The main sources of variance in memory research are those related to (1) the drug, (2) the learning material, (3) the subject, (4) interactions between two or more of these factors, (5) changes brought about by the drug in psychological functions other than memory.

[1] Department of Psychology, University of Giessen, Otto-Behaghel-Str. 10, 6300 Giessen, FRG.

(1) Drug-related variables influencing pharmacological results are type of drug, time, dosage, duration of treatment, and mode of application.
(2) The learning material can vary in complexity, meaningfulness, mode of presentation, mode of testing recall, time of presentation, and time between testing recall and medication.
(3) Subject-related variables represent the full range of state and trait variables on psychological and physiological or morphological bases. These comprise: type of personality and type of disease, drug experience, drug expectations, attitudes towards the experiment, and levels of cortical, emotional, and autonomic arousal prior to drug intake and prior to learning.
(4) It is evident from psychopharmacological studies that interactions between these factors will emerge, indicating, for example, that amnesic effects may only be pronounced in introverts who do not smoke and have no previous experience with these drugs, whereas smoking, previous drug exposure, and introversion by themselves might not influence amnesic effects.
(5) A question implicitly involved in individual differences in drug effects is the one of covariation between psychological processes affected by the drug, since analysis of differences for instance between subjects exhibiting high and low drug-induced changes in vigilance or sleeping behavior is equivalent to computing correlations between amnesic and vigilance-related effects of drugs.

Only a very limited subset of the variables mentioned in points 1–5 can be touched upon here. Examples of external experimental factors (points 1 and 2) will only be dealt with briefly insofar as they represent possible sources of interaction with individual differences in pharmacokinetics and basic intellectual capacity.

A major part of this article will be devoted to subject-related factors and their interactions (points 3 and 4), and finally the matter of concomitance between benzodiazepine-induced changes of memory and other psychological functions will be elucidated.

2 External Experimental Factors

One of the drug factors relevant for producing individual differences in responsiveness is the type of benzodiazepine applied. Although it has been observed by several authors (Healy et al. 1983; Paes de Sousa et al. 1981; Roth et al. 1981; Siegfried et al. 1981; Silbernagel and Netter 1980) that lorazepam produces higher mean anterograde amnesic effects than clorazepate, diazepam, flunitrazepam, triazolam, clobazam, or brotizolam when given in doses with equivalent sedative effect, there is always a certain percentage of individuals who deviate from the observed rank order of amnesic drug effects when drugs are applied in crossover experiments. Different levels of susceptibility become evident with respect to drug doses and the time course of drug actions, as can be seen from Figs. 1, 2. The figures indicate that with both doses of diazepam and flurazepam, there are always a number of subjects resistant to anterograde amnesia.

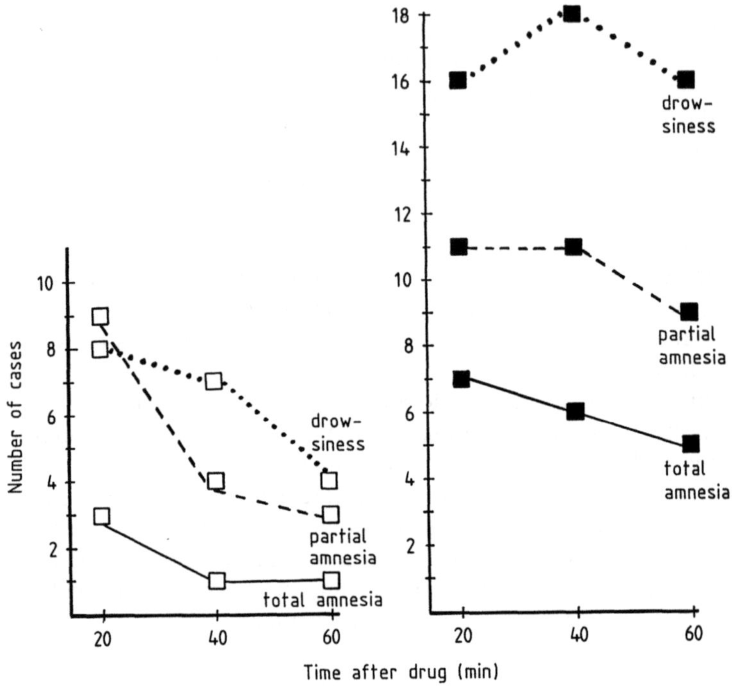

Fig. 1. Relationship between number of cases experiencing drowsiness and amnesia after diazepam (□, 10 mg; ■, 20 mg) and time after administration, in 25 subjects (data from GEORGE and DUNDEE 1977)

Comparisons between 10 and 20 mg diazepam on the one hand and between the high doses of diazepam and flurazepam on the other indicate that the response curves, given as the number of subjects in whom a particular effect occurs, differ with dose and type of drug. This suggests that responsiveness of individual subjects may not remain constant across ranges of drugs, doses, and times, particularly since dose-response relationships for memory decrements may be valid for some drugs (e.g., lorazepam) and not for others (e.g., diazepam, clorazepate), as HEALEY et al. (1983) have shown. Data on such a lack of correlation between dose and impairment are given for midazolam and diazepam in Fig. 3 (MAGNI et al. 1983). Furthermore, the lack of correspondence between plasma level of benzodiazepines (in particular those with long half-lives) and response has been observed in clinical studies (CURRY 1974; LIN and FRIEDEL 1979; SMITH et al. 1976). It also seems to occur for amnesic effects if performance is compared to the concomitant decline of serum levels as shown in a study by HILLESTAD et al. (1974) (Fig. 4). Thus, individual differences of absorption, distribution, degradation, and elimination of drugs will only explain part of the variance in amnesic effects, a major part probably being due to differences in the pharmacodynamics.

Learning material and time of learning and recall in relation to drug application can be shown to be further relevant variables likely to interact with individual performance capacity.

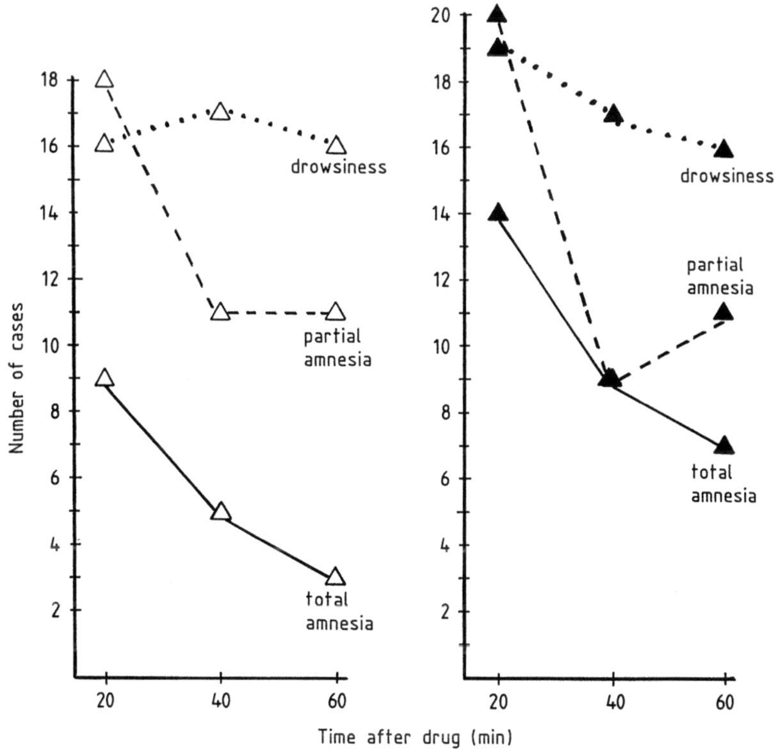

Fig. 2. Relationship between number of cases experiencing drowsiness and amnesia after flurazepam (△, 1 mg; ▲, 2 mg) and time after administration in 25 subjects (data from GEORGE and DUNDEE 1977)

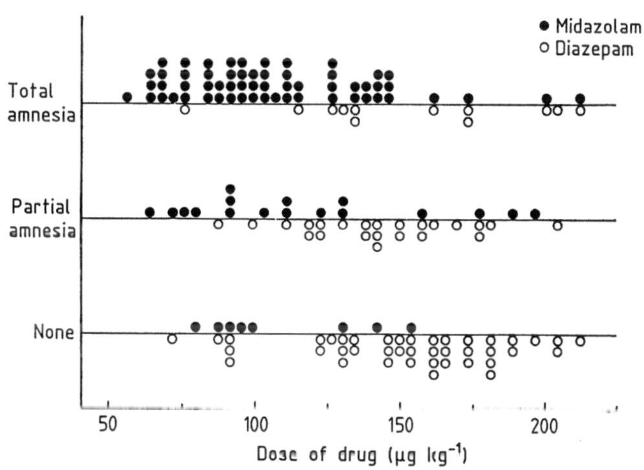

Fig. 3. Lack of correlation between dose and amnesia for midazolam (●) and diazepam (○) (MAGNI et al. 1983)

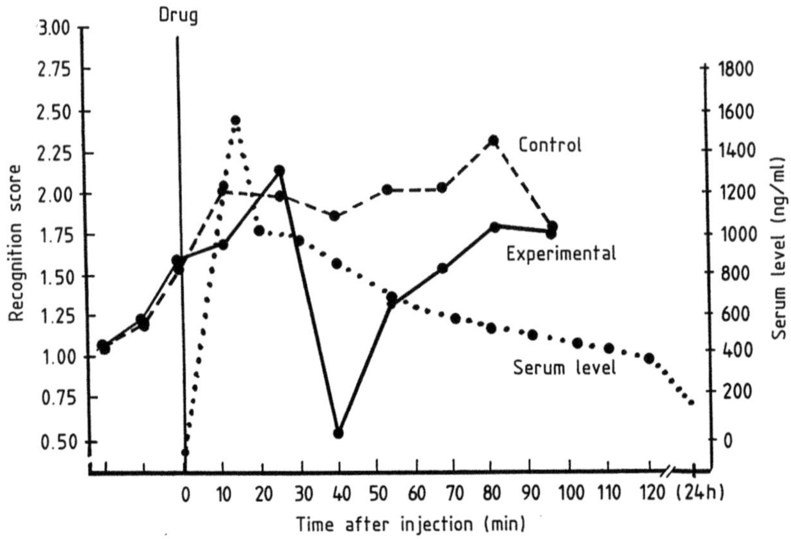

Fig. 4. Mean recognition scores for experimental and control groups over time. Serum concentration resulting from intravenous administration of 20 mg diazepam is also plotted (from HILLESTAD et al. 1974)

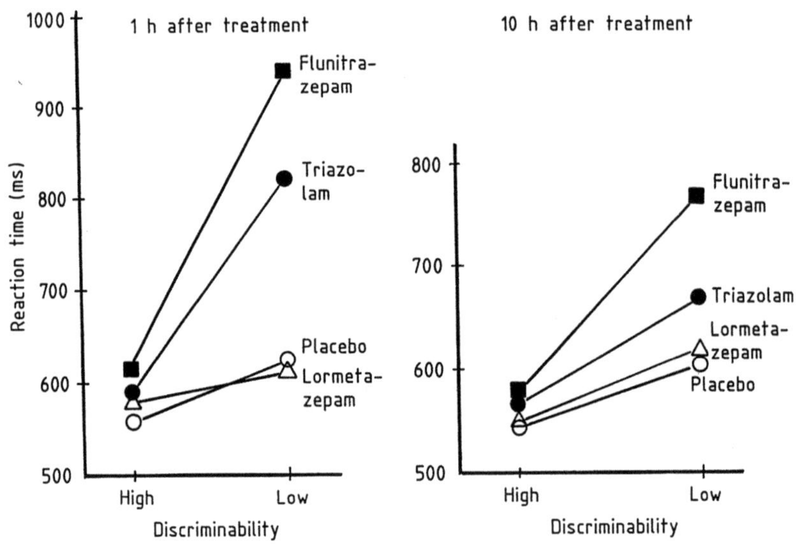

Fig. 5. Joint effects of drug and discriminability on reaction time (RT) 1 and 10 h following treatment with placebo (○), lormetazepam (△), triazolam (●), or flunitrazepam (■) (after SUBHAN 1984)

There exists a consensus that material learned prior to drug intake is not affected, but that with immediate and delayed recall of material learned after drug intake, there is considerable variation depending on methods of assessment and type of items. Thus, according to DARLEY et al. (1973) recall is more impaired than recognition; for diazepam impairment of recall is more pronounced if items are not categorized, whereas type of item has no effect with hyoscine. Similarly, BROWN et al. (1977) have reported smaller recall-recognition differences for lorazepam than for diazepam.

The relevance of type of material learned is demonstrated by the findings that benzodiazepine impairs detailed but not general recall (BIXLER et al. 1979) and that, for instance, visual amnesia is more pronounced with lorazepam than with diazepam but that recall of acoustic or pain stimuli is equally impaired with both drugs (STUDD and ELTRINGHAM 1980).

If the difficulty of learning material varies, this variable may interact with the type of drug and time of assessment, as may be seen from Fig. 5 which shows data obtained by SUBHAN (1984). These demonstrate that low item discriminability (difficult material) considerably increases the response time in memory scanning with flunitrazepam and triazolam but not lormetazepam 1 h after drug application, whereas 10 h after medication this interaction between drug and type of item is barely visible. These influences tend to interact with personality and other subject variables, as the following section shows.

3 Subject-Related Factors

Studies using psychometric measures in healthy volunteers usually do not consider personality differences, although in some studies age and sex have been recorded and could have been tested for influences on the effects of drugs. When personality traits have been assessed, either the assessment has been included to ensure high anxiety levels to start with in order to simulate anxious patients (NAKANO et al. 1978) or if both high and low anxious subjects have been tested performance measures other than memory have been used (CLYDE 1981; MÜNTE et al. 1984; JANKE et al. 1979; DEBUS and JANKE 1986).

Psychological subject-related factors, which are of primary interest here, are more or less based on or influenced and modified by somatic factors; their interactions are shown in Fig. 6. All of these somatic factors may be more or less closely related to some personality dimension, such as extraversion-introversion, neuroticism and psychoticism, or impulsivity. These relations are indicated by the arrows pointing from the personality factors to the intervening variables as well as to the primary level of memory function, the latter being based on the level of physiological brain activity and information processing capacity. In addition, more temporary factors like smoking habits, attitudes, and expectations will also be considered in this section.

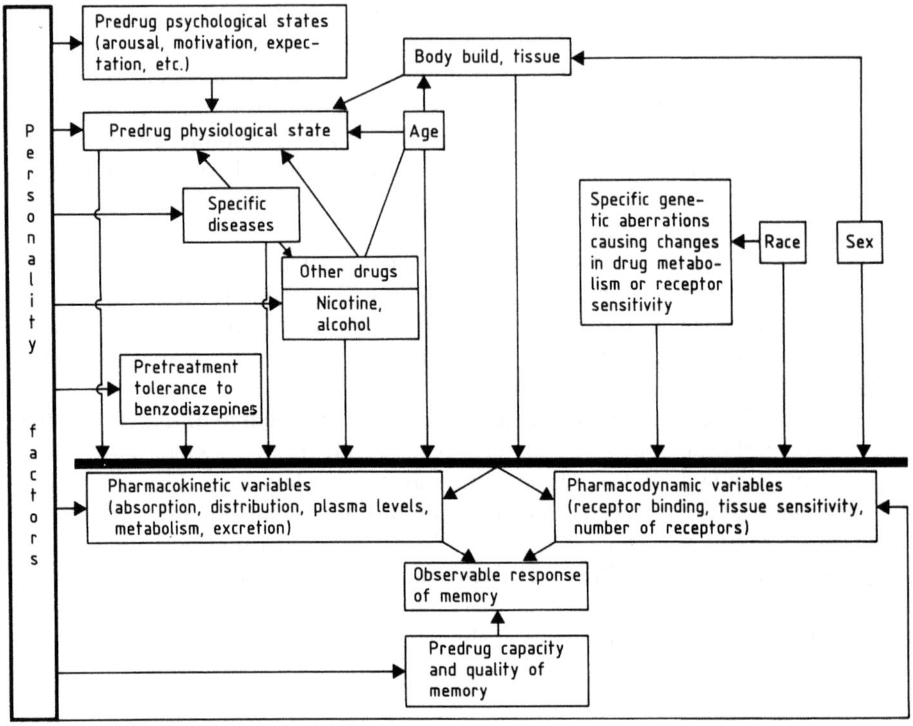

Fig. 6. Somatic factors serving as mediators between personality factors and drug-induced changes of memory (after NETTER 1983)

3.1 Cigarette Smoking

Smoking could modify drug-induced changes of memory in two ways: (a) via the vigilance-reducing effect of smoking deprivation which, in combination with the benzodiazepine, might reduce retention and recall, or (b) via induction of drug-metabolizing enzymes which would result in accelerated metabolism of the benzodiazepine, thus leading to a decrease of the amnesic effect. We will probably be confronted with a confounding of both effects, which could explain why in some smokers we observe increased memory performance with a benzodiazepine, whereas in others it is decreased or unchanged.

In a study in my laboratory, the effect of a single dose of 30 mg clobazam on retro- and anterograde amnesia was investigated in 24 male students according to the procedure given in Table 1 (FISCHER 1984). At t_2 retrograde amnesia was measured and at t_3, anterograde amnesia.

Interactions of smoking habits, sex of subjects, and drug effects were tested by analysis of variance with respect to recall of numerical, figural, and verbal material. The results for retrograde amnesia are given in the upper and lower parts of Fig. 7 for verbal and numerical memory respectively (triplets and telephone numbers). It is evident that for both numerical and verbal memory under placebo

Table 1. Procedure of the experiment on clobazam and memory (Fischer 1984)

Time	Phase of experiment	Tests applied	Functions measured	Phase of memory assessment	Length of interval
0800	Predrug baseline t_1	EWL 1	Emotional state		
		CFF 1	Critical flicker fusion frequency		
		CRT 1	Choice reaction time		
0820 ↓		Tel. nos.	Numerical LTM	Learning of series A	2 h 50 min
		City map	Figural LTM		
		Triplets	Verbal LTM		
0827		Digit span 1	STM	Acquisition + recall	
0835	Drug applic. t_2				2 h 35 min
1100	Peak serum level of drug	EWL 2			
1110 ↓		Tel. nos. 2	LTM (= retrograde amnesic effect)	Reproduction of series A	2 h 50 min
		City map 2			
		Triplets 2			
1117		CFF 2	Cortical arousal		
		CRT 2			
1125 ↓		Tel. nos.		Learning of series B	2 h 20 min
		City map			
		Triplets			
1132		Digit span 2	STM	Acquisition + Recall	
1245 ↓	Decline of serum level t_3	EWL 3	Emotional state	Reproduction of series B	
		Tel. nos. 3	LTM (= anterograde amnesic effect)		
		City map. 3			
1252		Triplets 3			
1300		Questionnaires about experiences during experiment	Attitudes		
			Motivations		
			Habits		

conditions, female nonsmokers performed better than female smokers while there is no difference in males; these relationships were completely reversed after clobazam (the two female groups were about equal, but smoking males performed worse than nonsmoking males; the interactions drug × sex × smoking habit were significant at $p = 0.023$ and $p = 0.006$ for verbal and numerical material respectively). This could be explained by higher motivation in females stimulating an effort to increase performance under placebo when no drug interferes. With clobazam, detrimental anterograde amnesic effects seem to affect high initial levels of performance more than low levels, and nicotine deprivation and clobazam may have been acting additively in females. The hypothesis of nicotine-induced induction of liver enzymes may be true only for males (who possibly smoke more), which may have led to faster degradation and less effectiveness of clobazam with respect to memory impairment.

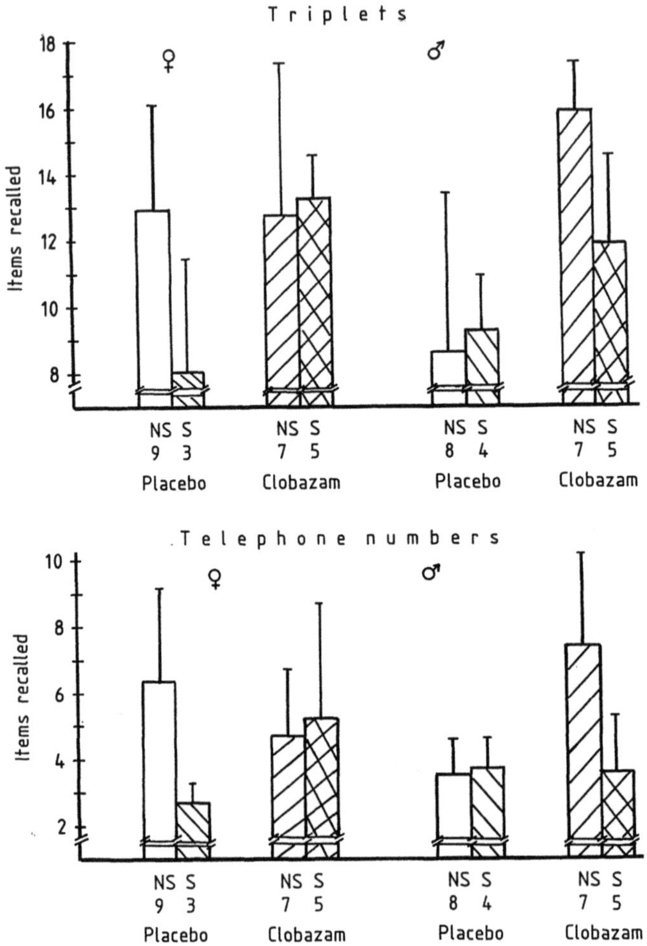

Fig. 7. The effect of clobazam on verbal (triplets) and numerical (telephone numbers) memory in male and female smokers (*S*) and nonsmokers (*NS*), tested at t_2. Interaction drug × smoking × sex: triplets, $p=0.023$; telephone numbers, $p=0.006$

For anterograde amnesia (t_3), interactions only became significant for the triplet test ($p=0.023$). To make intraindividual comparisons, results for t_2 and t_3, though measuring different aspects of drug action, are presented as within-group response curves in Fig. 8. The figure demonstrates that both smoking and nonsmoking females improved their performance for triplets from the first (t_2) to the second memory test (t_3) under placebo and showed a decrease under clobazam, the effect in each condition being more pronounced with smokers than with nonsmokers. In males on the other hand, there was no change in either condition for smokers, but a marked deterioration from t_2 to t_3 under clobazam in nonsmokers. Unfortunately, in this design it was not possible to decide whether the higher values observed at t_2 under clobazam were drug induced or were due to initial intergroup differences in memory capacity.

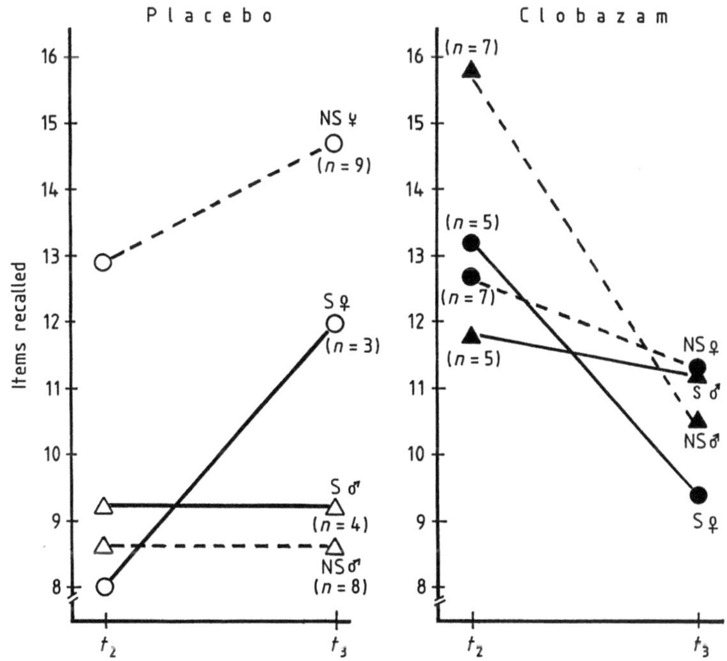

Fig. 8. The effect of clobazam on verbal memory (triplets) in male and female smokers (*S*) and nonsmokers (*NS*). t_2, 2 h 50 min after drug intake (retrograde amnesia), interaction drug × sex × smoking habit, $p = 0.019$. t_3, 3 h 15 min after drug intake (anterograde amnesia), interaction drug × sex × smoking habit, $p = 0.023$

Since no predrug data on long-term memory were available to answer the question of how drug-free memory capacity might serve to explain the drug × sex × smoking interaction, we examined initial scores on tests of short-term memory (STM), and found a significantly lower mean immediate recall in Digit Span Backwards in smokers than in nonsmokers (means, 5.0 vs 4.1; $p = 0.034$). These results correspond to the results of testing long-term memory (LTM) at t_2; shown in Fig. 8, which may be due either to smoking deprivation or to the higher extraversion scores of smokers. These results for STM could not, however, explain the higher levels at t_2 in the clobazam group, which therefore must not be attributed to a higher memory capacity in this group but may be due to a positive retrograde effect of the drug on retention.

3.2 Attitudes and Expectations

Several studies have shown that subjects' expectations of particular effects of a drug may influence drug-induced changes of behavior. This has also been demonstrated in studies on drugs and memory, such as an experiment by Nash and Zimring (1969) which showed that subjectively experienced as well as objectively as-

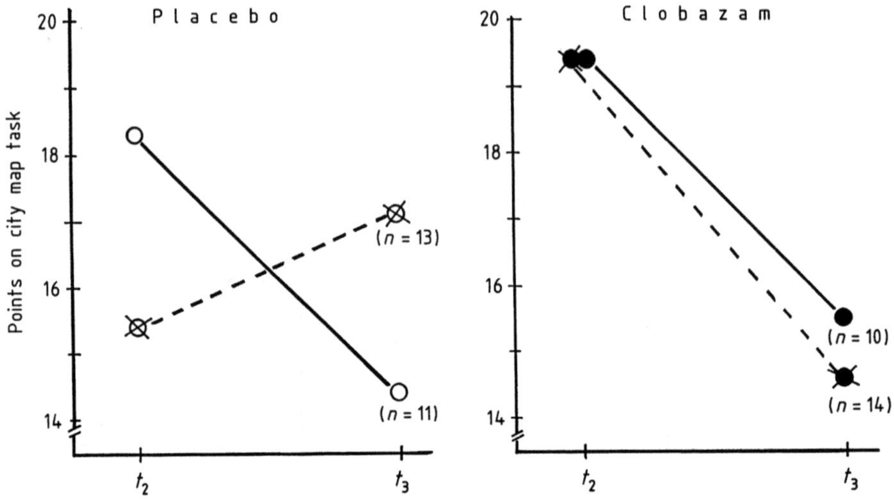

Fig. 9. The effect of clobazam on figural memory (city map task) in subjects with a positive (●, ○) or negative (✖, ⊠) attitude towards tranquilizers. t_2, reproduction of material learned 10 min prior to drug intake (interaction drug × attitude, $p > 0.10$). t_3, reproduction of material learned 2 h 50 min after drug intake (interaction drug × attitude, $p = 0.06$)

sessed STM was better in subjects who expected high therapeutic efficacy from a nootropic drug claimed to increase cerebral blood flow.

Therefore, in a study on the influence of a single oral dose of 30 mg clobazam (FISCHER 1984) the subjects' attitudes to tranquilizers were included as an independent factor in a three-factor analysis of variance (drug, sex, and attitude – positive vs negative according to a questionnaire on attitudes to drugs; JANKE and FRANK 1970). The results shown in Fig. 9 demonstrate that a positive attitude to tranquilizers only acts on memory under placebo conditions by causing a deterioration of LTM (city map task) performance; this is probably in line with the expectation of a subject (even one performing fairly well) that a powerful tranquilizer will cause an increase in experienced drowsiness after a short time. Subjects who do not believe in tranquilizers display less suggestibility in their performance under placebo but instead show a practice related increase at t_3. The influence of attitude to drugs is suppressed by clobazam, the anterograde amnesic effect of which is evidently more dominant than the effect of the memory-modulating attitude to tranquilizers (Fig. 9).

3.3 Motivation

The influence of a motivational factor related to coping with experienced drug effects was also investigated in the study mentioned above (FISCHER 1984). A question asking how strongly subjects had tried to counteract the drowsiness attributed to drug effects yielded a group factor of high and low motivation which interacted significantly ($p = 0.032$) with the factors drug and sex of subjects for re-

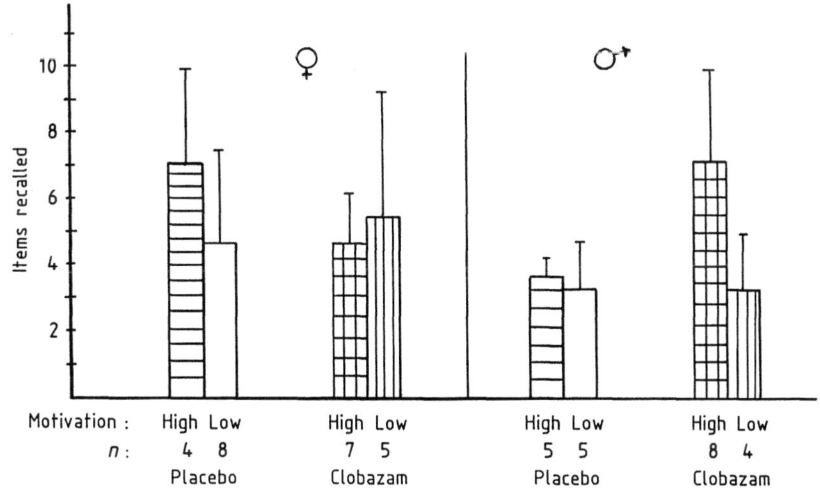

Fig. 10. The influence of sex and motivation to counteract drowsiness attributed to drug effects on recall of telephone numbers at t_2 (2 h 50 min after drug intake, retrograde amnesic effect). Interaction drug × sex × motivation, $p=0.032$

call of telephone numbers at t_2 (Fig. 10). In females given placebo, the effort to counteract fatigue yielded better scores in recall of telephone numbers, whereas this endeavor did not seem to be very successful under the influence of the benzodiazepine. Male subjects, however, tended to obtain better results if their effort was to counteract a true instead of an imagined effect. These findings could be interpreted by the different positions of the two sexes on the inverted U-shaped curve relating performance to level of arousal. In many studies females are more severely affected by sedatives and the effort required to overcome the effects shifts them beyond their optimal level of arousal, whereas in males only the effort required to overcome a true obstacle – the clobazam effect – seems to shift them to their optimum arousal level.

3.4 Anxiety and Neuroticism

Although, according to EYSENCK's (1967) drug postulate, extraversion would be predicted to show more interaction with benzodiazepine-induced changes of performance, few results have been reported on this matter. In our own study, no significant interaction between the effects of extraversion and clobazam on memory could be observed (FISCHER 1984). More studies were interested in the effect of anxiety, the prediction being that anxiety-reduced performance would be restored with a mild dose of a benzodiazepine.

Frequently, however, interactions observed with anxiety are higher order, as in the experiment reported by DESAI et al. (1983). Stepwise analysis of a fivefold interaction between speed of item presentation, the condition of articulatory suppression during acquisition of memory material, serial position of items, drug,

Table 2. Procedure of evaluation and results obtained for interactions in successively performed analyses of variance in a 5-factor design involving drug (diazepam – placebo) and anxiety state (high – low STAI score) by DESAI et al. (1983)

5-way ANOVA	$F=2.79$	Rate of presentation ×		Articulatory suppression ×	Serial position ×	Anxiety state × Drug
4-way ANOVA	Fast $F<1$ n.s.	Slow $F=2.16$ $p<0.05$		Articulatory suppression ×	Serial position ×	Anxiety state × Drug
3-way ANOVA				No $F=1.2$ n.s.	Yes $F=2.26$ $p<0.05$	Serial position × Anxiety state × Drug

The rate of presentation branches into Fast and Slow. Under Slow, articulatory suppression branches into No and Yes.

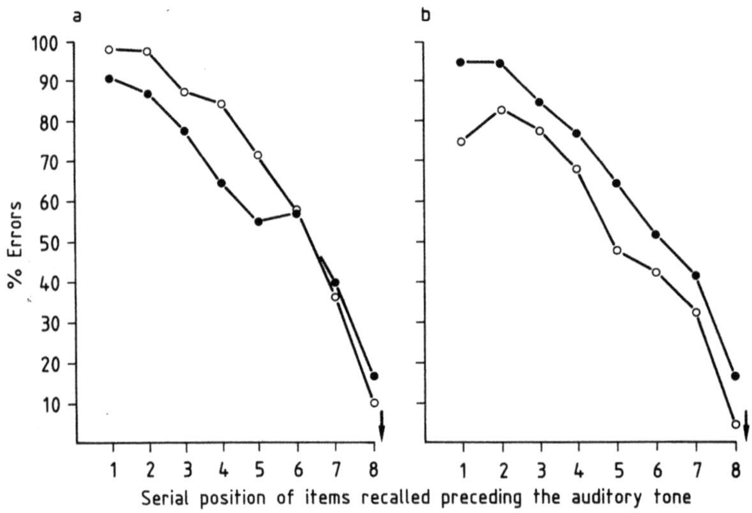

Fig. 11. Effect of diazepam (●) or placebo (○) in high (*a*) and low anxiety (*b*) subjects on short term memory test performed under the conditions of slow item presentation and articulary suppression (see text). ↓, auditory tone; thus, serial position 8 identifies the item most recently presented (DESAI et al. 1983)

and anxiety state (Table 2) revealed the results presented in Fig. 11. Under the task-facilitating condition of slow item presentation and the more difficult condition of not being permitted to articulate the material in the process of learning, subjects scoring high on state anxiety (determined according to the State Anxiety Inventory; SPIELBERGER et al. 1970) performed better with diazepam than with placebo if items were presented at positions 1–5, i.e., long before the tone signalling interruption, while the reverse held for low-anxiety subjects with all item po-

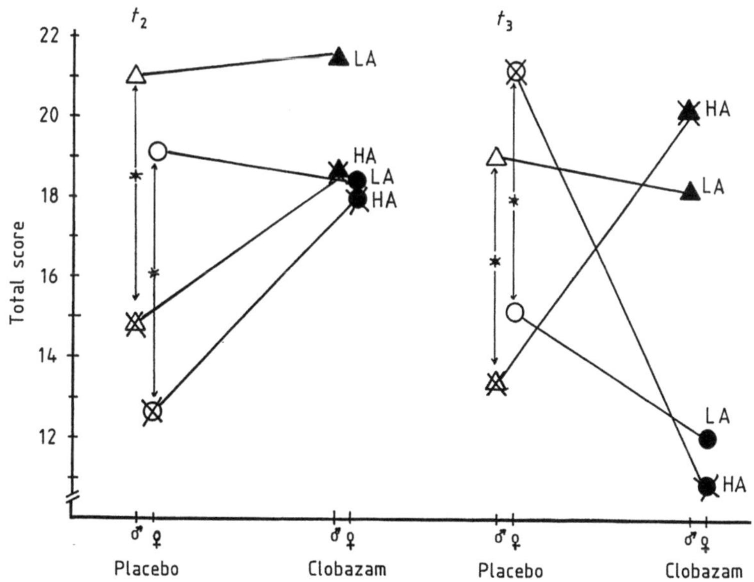

Fig. 12. Tests for retrograde (t_2) and anterograde (t_3) amnesic effects of clobazam (city map) in subjects divided according to sex and trait anxiety level. ▲△, low-anxiety males; ●○, low anxiety females; high anxiety males; high anxiety females t_2: Main effects: drug, $p=0.10$; anxiety, $p=0.02$ t_3: Interaction drug × sex × anxiety, $p=0.06$. (after KOEPPEN et al. 1985)

sitions. This would support the theory that arousal reduction increases performance in high-anxiety subjects. Similar results were found for changes in psychomotor performance induced by clobazam (CLYDE 1981). With memory functions, the conditions evidently have to be favorable for high-anxiety subjects, since, in the study of DESAI et al. (1983) with fast presentation of items and for the items presented at the end of a series, the benefit of diazepam disappeared. Similarly, in our study the beneficial effect of clobazam on memory for material learned before drug intake (Fig. 12, left) was primarily evident in high-anxiety subjects, although the interaction drug × anxiety did not reach statistical significance but was obscured by the main effect of anxiety (lower overall performance levels in high-anxiety than low-anxiety subjects). In the test for anterograde amnesic effects clobazam-induced improvement was only observed in anxious males, whereas high-anxiety females performed considerably worse under clobazam than under placebo (Fig. 12, right; interaction drug × sex × anxiety, $p=0.06$). This result could be explained in two ways:

1. The same dose of benzodiazepine given to males and females may mean a considerably higher concentration of drug in females, causing a greater shift on the inverted U-shaped arousal-performance curve with females than with males.
2. Since for anterograde amnesia the drug-performance-personality relationship was similar to the U-shaped function plotted by HINDMARCH (1979) with dose of drug held constant for varying degrees of neuroticism (Fig. 13), the reason

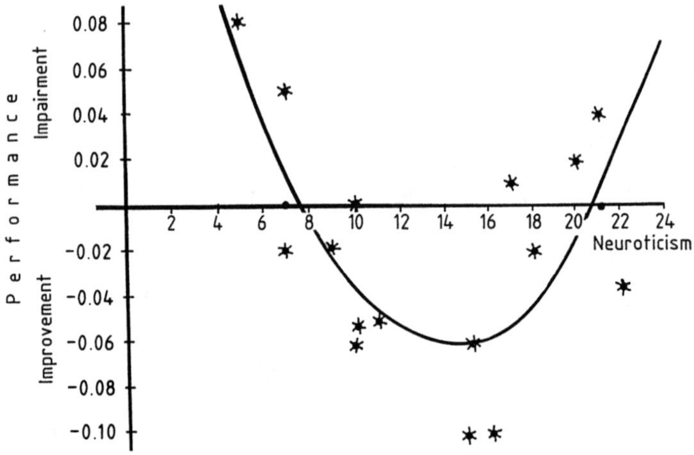

Fig. 13. Effect of an acute dose of clobazam (20 mg) compared with placebo on complex psychomotor function plotted against individual neuroticism scores (HINDMARCH 1979)

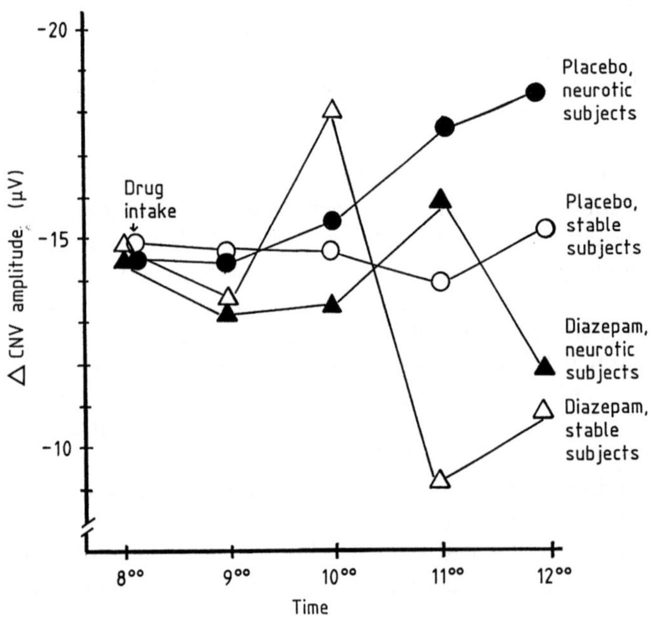

Fig. 14. CNV amplitude changes at several times after intake of diazepam (▲, △) and placebo (●, ○) in subjects scoring high (▲, ●) and low (△, ○) on the neuroticism scale (from MÜNTE et al. 1984)

for the higher susceptibility of anxious females than males could result from their higher mean scores on the anxiety and neuroticism scales, which would predict more pronounced decrements of performance.

A possible underlying neurological basis is provided by data on benzodiazepine-induced shifts in contingent negative variation (CNV) from MÜNTE et al. (1984), some of which are presented in Fig. 14. In the contingent negative variation (CNV) measure presented, the stable subjects produced higher baseline amplitude negativities than the highly neurotic subjects (15.6 μV as opposed to −13.7 μV; $p < 0.001$).

Diazepam induced an increase of CNV amplitude 2 h after drug intake and a marked decrease 3 h after intake in stable subjects, whereas there was a delayed and much lower increase in the neurotic group, whose amplitudes stayed below the respective placebo values throughout the experiment. This is interpreted by the authors as indicating higher initial emotional arousal in neurotic subjects; in these subjects arousal reduction by diazepam occurs later and to a lesser extent than in stable subjects, as indicated by their lower amplitude negativities. These results can be interpreted in terms of the distraction-arousal model outlined by TECCE et al. (1978), which assumes a linear relationship between CNV amplitude and attention and an inverted U-shaped relationship between CNV amplitude and arousal. In neurotics, the distracting arousal level would tend be reached before its attention-increasing property can become effective. Although the authors themselves do not provide data on memory, it is conceivable that arousal – which

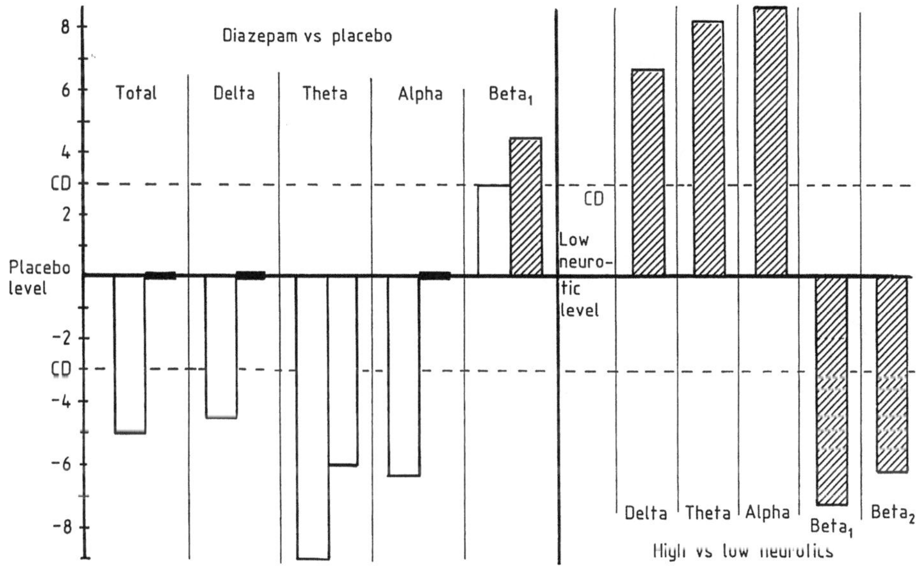

Fig. 15. *Left*, Differences in placebo-corrected changes of EEG criteria (absolute power) induced by 5 mg diazepam in healthy male subjects divided according to the FPI neuroticism scale into high (■) and low (□) scorers. *Right*, Differences between high and low scorers under placebo conditions (absolute power). *CD*, critical difference ($p = 0.05$). After HEINZE and KÜNKEL (1979)

in neurotics is claimed to be particularly high in the limbic system and hippocampus – will interfere with retrieval processes in which the hippocampus area is also involved.

Differences between high and low scorers on neuroticism in terms of absolute power of EEG wave types under diazepam and with placebo are given in Fig. 15, which also shows differences in standard units between high and low scorers on neuroticism under placebo conditions. It is characteristic of neurotics that they seem to display a lower total response to diazepam and lower changes in delta, theta, and alpha power in particular, with a slightly greater $beta_1$ diazepam-placebo difference, than stable subjects. By contrast their placebo values in respect of absolute power are higher for delta, theta, and alpha but lower for $Beta_1$ and $Beta_2$ than those of stable subjects (Fig. 15, right).

4 Covariations Between Drug-Induced Changes in Memory and Changes in Other Cerebral Functions

In Sect. 3.4 the question was raised of whether underlying neurophysiological differences could help to explain personality-related differences in susceptibility to drug-induced memory disturbances. This leads to a closer inspection of those psychological processes that share functions with memory processes and that therefore may be concomitantly affected by the drugs. Functions which can be disturbed as concomitants of memory deficits in old age (LAUTER 1973) are depicted (Fig. 16) in a model similar to the one developed by KANOWSKI and COPER (1982) for components of age-related cerebral insufficiency.

The overlap between functions in this model would suggest that drug-induced deficits covary in a similar way to concomitants of age-related memory deterioration. Therefore, the matter of individual differences is extended to the question of whether the memory functions of subjects highly susceptible to sedative or hyp-

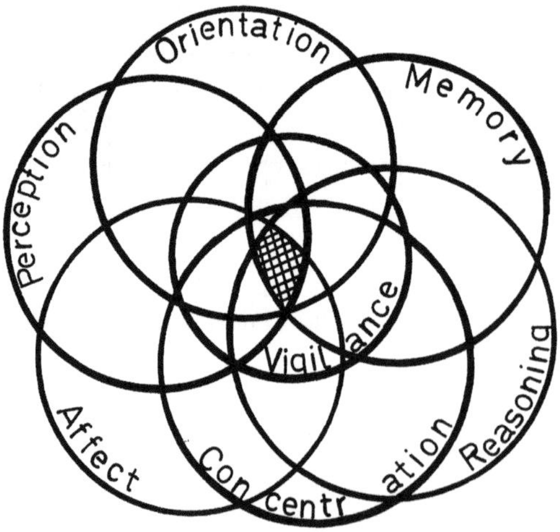

Fig. 16. Model of interrelations between memory-related mental processes

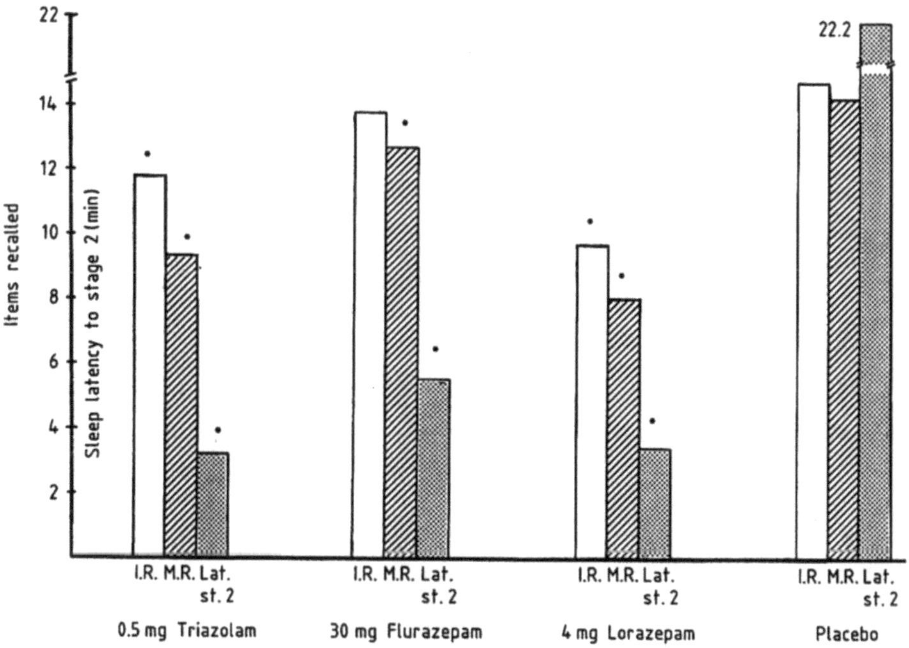

Fig. 17. Effects of 0.5 mg triazolam, 30 mg flurazepam, 4 mg lorazepam, and placebo on memory and sleep latency in 11 male volunteers. *I. R.*, Immediate recall; *M. R.*, morning recall; *Lat. st. 2*, latency to stage 2 sleep; $^*p \leqq 0.05$ in comparison with placebo (after ROTH et al. 1981)

notic effects of benzodiazepines are also more affected. Unfortunately, the majority of papers present group means obtained for different psychological aspects of sedation but – due to low sample sizes – few of them report data on intraindividual covariation between changes of memory and changes of other parameters.

One of the variables unanimously claimed to be related to degree and duration of benzodiazepine-induced amnesia in sleep and learning studies is the hypnotic potency of the drug. This is shown by the relations between latency to stage 2 sleep after awakening at night and immediate and morning recall of learned material (ROTH et al. 1981). Figure 17 shows that there is a relationship between memory parameters and sleep latency for three benzodiazepines, but that this is not observed under placebo. The relationship between hypnotic potency and amnesia can be confirmed by plotting sleep latency group means against mean recall in eight different groups, as was done by ROEHRS et al. (1984; Fig. 18). This indicated that benzodiazepine-impaired consolidation is further disrupted by sleep onset (ROTH et al. 1981).

Less agreement exists with respect to correlations between degree of sedation and memory loss. BIXLER et al. (1979) reported data on alertness, coordination of gait, and recall in a study with three drug and two placebo nights. This indicated that the occurrence of the sharpest impairment after the first drug night and the recovery of function partly on the following drug nights and clearly after the last placebo, ran in parallel in the three functions (Fig. 19). However, doubts have

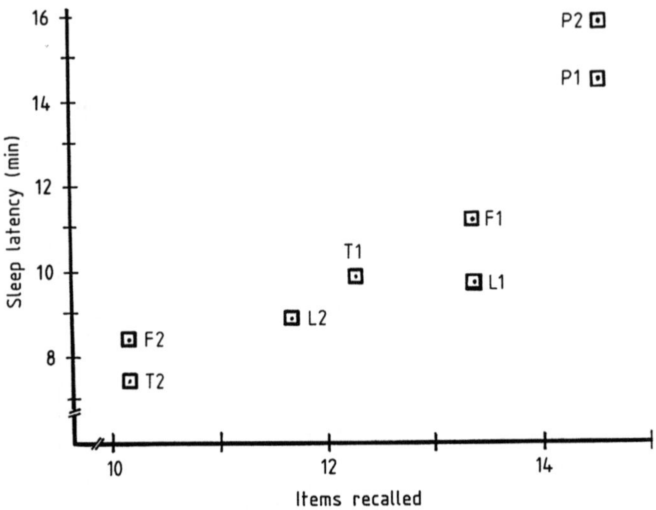

Fig. 18. Relation of sleep latency and number of items recalled in the morning. *F*, Flunitrazepam; *L*, lorazepam; *P*, placebo; *T*, triazolam. (from ROEHRS et al. 1984)

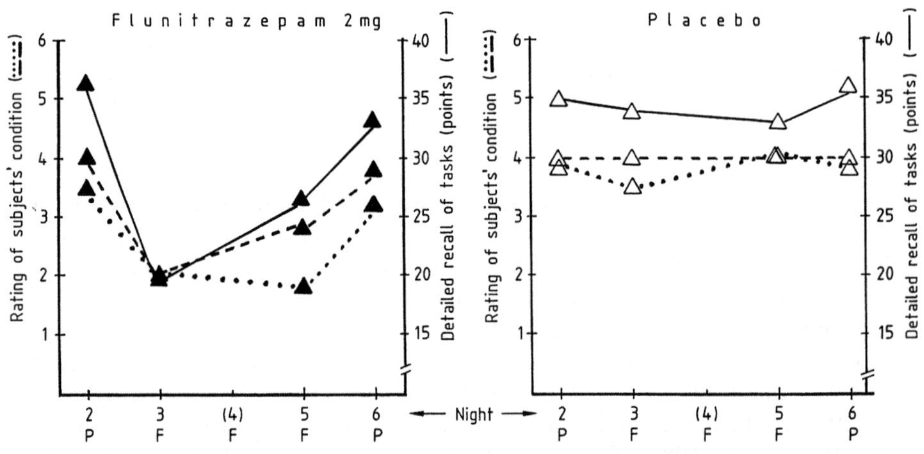

Fig. 19. Concomitance between impairment of memory, alertness, and coordination of gait in a group of four subjects after three nights with 2 mg flunitrazepam following and preceding a placebo night. ···, Alertness; ---, coordination of gait; —, detailed recall; *P*, placebo; *F*, flunitrazepam. Alertness and coordination rated by experimenter (after BIXLER et al. 1979)

been cast on the value of sedative effects for predicting degree of amnesia: (a) on the basis of clinical observations (DUNDEE and PANDIT 1972), (b) since sometimes neither of the effects is dose related (MAGNI et al. 1983), and (c) since in long-term studies tolerance is developed for some of the sedation-related functions (for instance motor response time) but not for others (critical flicker fusion, CFF; SEPPÄLÄ et al. 1980), and therefore dissociation of impairment is to be expected at least in long-term treatment.

Table 3. Correlations between changes of critical flicker fusion frequency (CFF $t_1 - t_2$) and sensory choice reaction time (CRT $t_1 - t_2$) with figural memory (city map) at t_2 and t_3 under placebo and clobazam (from FISCHER 1984)

	City map t_2		City map t_3	
	Placebo	Clobazam	Placebo	Clobazam
CFF $t_1 - t_2$	0.36*	−0.27 (*)	0.36*	−0.52**
CRT $t_1 - t_2$	−0.12	0.58**	0.13	0.38*

(*) $p \leqq 0.10$, *$p \leqq 0.05$, **$p \leqq 0.01$.

We tried to elucidate relationships between impairment of memory and vigilance in two studies. The only interesting correlations in the clobazam study (FISCHER 1984) are shown in Table 3. Decrease in cortical vigilance as assessed by CFF from predrug to 2.5 h after application is related to good figural memory performance under placebo but to bad performance under clobazam with respect to antero- and retrograde memory deficits. The latter relationship also holds for the sensory part of a choice reaction time task. This implies that those individuals who perceive less rapidly under the placebo condition have better memory consolidation (and perhaps retrieval) capacity than the ones whose processing of sensory input remains constant with time. Instead, with benzodiazepine a drug-induced decline in visual vigilance is a predictor for degree of retro- as well as anterograde amnesia. It could be speculated that suppression of the input information processing channel might normally facilitate consolidation of previously acquired information, but that the benzodiazepine-induced sedation overrides the optimal level of input reduction.

The second analysis is based on data obtained in a study comparing 30 mg flurazepam with 0.2 and 0.5 mg brotizolam and placebo in four separate groups, each of 22 male subjects. Anterograde amnesia of incidentally and intentionally learned material reproduced 24 h later was determined under nondrug conditions, and objective and subjective indicators of arousal and activation were assessed in the actual drug condition (SILBERNAGEL 1979; SILBERNAGEL and NETTER 1980). Correlations between these measures computed separately for each drug condition are shown for flurazepam and placebo in Table 4.

Framed pairs of correlations are significantly different ($p < 0.05$) for the two drug conditions.

It is evident that covariations between different measures of incidental learning ($r_{1 \times 3}$, $r_{2 \times 4}$; see Table 4) or between incidental and intentional learning ($r_{3 \times 5}$) present in the placebo condition are reduced to zero with the benzodiazepine. This may be due to the greater impairment of recall than recognition reported to be characteristic of other benzodiazepines (BROWN et al. 1977), which might remove the associations. The explanation is probably also valid for the lack of correlation between recognition upon simultaneous presentation (which was not significantly impaired compared to placebo) and recall of intentionally learned material (which was much worse than under placebo).

Table 4. Correlations between measures of memory and activation in the placebo (*P*) and in the flurazepam (*F*) group

					Drug	(1)	(2)	(3)	(4)	(5)	(6)	(7)	(8)	
Learning memory	Incidental/ learning	Slide	Spontaneous reproduction	(1)	P F									
			Successive presentation	(2)	P F	0.59** 0.15	Spontaneous presentation							
			Simultaneous presentation	(3)	P F	0.66** 0.05	0.75 0.91**	Successive presentation						
			Questionnaire about events	(4)	P F	0.39* 0.33	0.52* −0.13	0.27 −0.02	Simultaneous presentation					
			Memory (IST) (intentional learning)	(5)	P F	0.02 0.18	0.32 −0.17	0.42* −0.19	−0.01 0.14	Questionnaire about events				
Activation	Motor activity		Tapping	(6)	P F	0.15 −0.22	0.06 0.43	−0.14 0.32	−0.23 0.55*	0.09 0.07	Memory (1st)			
	Motivation		Preference for mental activity	(7)	P F	−0.58** −0.10	−0.45* 0.39*	−0.50* 0.51*	−0.17 0.08	−0.18 0.06	−0.28 0.03	Motor activation (tapping)		
	Emotional state (EWL)		Deactivation (inverted)	(8)	P F	−0.41* −0.13	−0.26 0.31	−0.29 0.45	−0.28 0.14	−0.17 0.11	−0.12 −0.05	0.74** 0.64**	Motivation (preferance for mental activity) TPF	
			Extraversion	(9)	P F	0.46 0.14	0.24 0.23	0.27 −0.21	0.36 0.11	−0.13 0.40*	0.09 0.06	−0.06 −0.20	0.01 −0.18	Deactivation (inverted): EWL

Slides: (2)–(4) diagonal labels (Spontaneous presentation, Successive presentation, Simultaneous presentation)

□, Significant difference between correlation coefficients $p \leqq 0.05$, (*) $p \leqq 0.10$, * $p \leqq 0.05$, ** $p \leqq 0.01$.

With indicators of motor, motivational and emotional activation, the shift of signs of correlation coefficients from placebo to flurazepam was in the opposite direction. Under placebo, negative associations between all three parameters of drive or energy on the one hand and measures of LTM on the other indicated better memory capacity for incidentally learned material in subjects who were lower in tapping speed, who were not particularly motivated for creative mental activity, and who felt more deactivated under placebo. Thus, a certain amount of passivity under normal conditions seems to facilitate storage of incidentally perceived information. Under flurazepam the relationship is inverted: less motor, motivational, and emotional deactivation by the drug yields less anterograde amnesia for incidentally learned material, indicating that the relaxing or deactivating property of the drug does impair consolidation of automatic information processing but seems to be irrelevant for retention of intentionally stored information (as indicated by about zero correlations between the IST memory test and the measures of activation).

From these evaluations, it can be seen that consideration of drug-induced changes in covariations may not only yield information on individual differences but also reveal insights into possible mechanisms of drug action upon subunits of memory processes.

5 Summary

Factors representing sources of variance for individual differences in benzodiazepine-induced changes of memory have been discussed from examples in the literature. The following were discussed:

1. External experimental factors, such as type of drug, dosage, plasma levels, time of learning and reproduction with respect to time of drug application, type of material, its variation in meaningfulness and method of reproduction could be shown to be basic sources of variance liable to interact with personality factors.
2. Among subject-related factors evidence was provided for influences of the following variables and their interactions upon benzodiazepine-induced impairment of memory:
   a) Smoking: In smokers, who display worse memory performance and vigilance prior to drug intake, benzodiazepines may improve recall in males and impair it in females.
   b) Attitude to sedatives: A positive attitude to sedatives shown to deteriorate memory performance with placebo may no longer be relevant in drug conditions.
   c) Motivation: The motivation to counteract suspected drug-induced sedation, responsible for increased performance with placebo, is more pronounced when true drug effects are experienced in male subjects and cannot overcome benzodiazepine-induced amnesic effects in females.
   d) Anxiety and neuroticism: There is evidence that anxiety and neuroticism, observed to be predictors of beneficial effects of benzodiazepines, are prob-

ably related by a U-shaped function to benzodiazepine-induced impairment of memory. Very highly neurotic subjects, in particular females, are adversely affected and moderately neurotic ones (anxious male subjects) are positively affected by the drug. These findings could be corroborated by EEG differences between subjects with high and low neuroticism scores.

3. Individual differences in concomitant drug-induced changes of arousal and activation may be responsible for the effects of drugs on memory consolidation, as can be shown by shifts in correlational signs obtained under placebo and a benzodiazepine. Thus, figural LTM consolidation under drug conditions could be shown to be better if little impairment of vigilance occurs, whereas with placebo decreased visual vigilance predicts better memory performance. Furthermore, performance in incidental learning facilitated by passivity under placebo is decreased if passivity is further enhanced by a benzodiazepine, a relationship which is not true for memory of intentionally learned material.

References

Bixler EO, Scharf MB, Soldatos CR (1979) Effects of hypnotic drugs on memory. Life Sci 25:1379–1388

Brown J, Lewis V, Brown MW, Hon G, Bowes JB (1977) Amnesic effects of intravenous diazepam and lorazepam. Experientia 34:501–502

Clyde CA (1981) The influence of personality on response to low doses of benzodiazepines. In: Hindmarch J, Stonier PD (eds) Clobazam. Academic, London, pp 75–86 (Royal Society of Medicine, International congress and symposium series no 43)

Curry SH (1974) Concentration-effect relationships with major and minor tranquilizers. Clin Pharmacol Ther 16:192–197

Debus G, Janke W (1986) Allgemeine und differentielle Wirkungen von Tranquillantien bei gesunden Personen im Hinblick auf Angstreduktion. In: Janke W, Netter P (eds) Angst und Psychopharmaka. Kohlhammer, Stuttgart, pp 135–150

Desai N, Taylor-Davies A, Barnett DB (1983) The effects of diazepam and oxprenolol on short-term memory in individuals of high- and low-state anxiety. Br J Clin Pharmacol 15:197–202

Darley CK, Tinklenberg JR, Hollister TE, Atkinson RC (1973) Marijuana and retrieval from short-term memory. Psychopharmacologia 29:231–238

Dundee JW, Pandit SK (1972) Anterograde amnesic effects of pethidine, hyoscine and diazepam in adults. Br J Pharmacol 44:140–144

Eysenck HJ (1967) The biological basis of personality. Thomas, Springfield, IL

Fischer C (1984) Zur Auswirkung von Clobazam (Frisium[R]) auf längerfristiges und kurzfristiges Behalten sowie auf verschiedene Stimmungsmaße unter Berücksichtigung differentieller Aspekte. Unpublished MA thesis, University of Giessen

George KA, Dundee JW (1977) Relative amnesic actions of diazepam, flunitrazepam and lorazepam in man. Br J Clin Pharmacol 4:45–50

Healy M, Pickens R, Meisch R, McKenna T (1983) Effects of clorazepate, diazepam, lorazepam, and placebo on human memory. J Clin Psychiatry 44:436–439

Heinze HJ, Künkel H (1979) The significance of personality traits in EEG evaluation of drug effects. Pharmacopsychiatry 12:155–164

Hillestad L, Hansen T, Nelson H, Drivines A (1974) Diazepam metabolism in normal man: serum concentration and clinical effects after intravenous, intramuscular and oral administration. Clin Pharmacol Ther 16:479–484

Hindmarch I (1979) Some aspects of the effects of clobazam on human motor performance. Br J Clin Pharmacol 7 (Suppl 1):77–82

Janke W, Frank R (1970) Arzneimitteleinstellungsfragebogen. University of Giessen and University of Düsseldorf, unpublished questionnaire
Janke W, Debus G, Longo N (1979) Differential psychopharmacology of tranquilizing and sedating drugs. In: Bossier JR (ed) Differential psychopharmacology of anxiolytics and sedatives. Karger, Basel, pp 13–98
Kanowski S, Coper H (1982) Das hirnorganische Psychosyndrom als Ziel pharmakologischer Beeinflussung. In: Bente D, Coper H, Kanowski S (eds) Hirnorganische Psychosyndrome im Alter. Springer, Berlin Heidelberg New York, pp 3–21
Koeppen D, Netter P, Fischer C (1985) Individual differences in the effect of clobazam on memory functions. Pharmacopsychiatry 18:12–14
Lauter H (1973) Psychosyndrom, organisches. In: Müller C (ed) Lexikon der Psychiatrie. Springer, Berlin Heidelberg New York, pp 380–382
Lin KM, Friedel RO (1979) Relationship of plasma levels of chlordiazepoxide and metabolites to clinical response. Am J Psychiatry 131:18–23
Magni VC, Frost RA, Leung WC, Cotton PB (1983) A randomized comparison of midazolam and diazepam for sedation in upper gastrointestinal endoscopy. Br J Anaesthesiol 55:1095–1100
Münte TF, Heinze HJ, Künkel H, Scholz M (1984) Personality traits influence the effects of diazepam and coffeine on CNV magnitude. Neuropsychobiology 12:60–67
Nakano S, Gillespie HK, Hollister LE (1978) A model for evaluation of antianxiety drugs with the use of experimentally induced stress. Comparison of nabilone and diazepam. Clin Pharmacol Ther 23:54–62
Nash MM, Zimring FM (1969) Prediction of reaction to placebo. J Abn Psychol 74:568–573
Netter P (1983) Somatic factors as predictors of psychotropic drug response. In: Janke W (ed) Response variability to psychotropic drugs. Pergamon, Oxford, pp 67–96
Paes de Sousa M, Figuira M-L, Loureiro F, Hindmarch I (1981) Lorazepam and clobazam in anxious elderly patients. In: Hindmarch I, Stones PD (eds) Clobazam. Academic, London, pp 119–123 (Royal Society of Medicine international congress symposium series, no 43)
Roehrs T, McLenaghan A, Korhorek G, Zorick F, Roth T (1984) Amnesic effects of lormetazepam. In: Hindmarch I, Ott H, Roth T (eds) Sleep, benzodiazepines and performance. Springer, Berlin Heidelberg New York, pp 165–172
Roth T, Zorick F, Sickelsteel J, Stepanski E (1981) Effects on benzodiazepines on sleep and wakefulness. Br J Clin Pharmacol 11 (Suppl 1):31–35
Seppälä T, Palva E, Mattila MJ, Kortila K, Shrotiya RD (1980) Benzodiazepines: multiple-dose effects on psychomotor skills and memory; comparison with diazepam and interactions with ethanol. Psychopharmacology 69:209–218
Siegfried K, Koeppen D, Malerczyk V, Sittig W, Taeuber K, Badian M (1981) A double-blind comparison of the acute effects of clobazam and lorazepam on memory and psychomotor function. In: Hindmarch I, Stonier PD (eds) Clobazam. Academic, London, pp 13–21 (Royal Society of Medicine international congress symposium series, no 43)
Silbernagel W (1979) Eine experimentelle Untersuchung zur Beeinflussung der anterograden Amnesie nach Gabe von We 941 (Brotizolam) und Flurazepam. Unpublished MA thesis, University of Mainz
Silbernagel W, Netter P (1980) Ein Verfahren zur Erfassung der anterograden Amnesie bei verschiedenen Benzodiazepin-Derivaten. Arzneimittelforschung 30:12–32
Smith RC, Dekirmenjian H, Davis J, Casper R, Gosenfeld C, Tsai C (1976) Blood level, mood and MHPG responses to diazepam in man. In: Gottschalk LA, Merli S (eds) Pharmacokinetics of psychoactive drugs. Spectrum, New York
Spielberger CD, Gorsuch RL, Lushene RE (1970) Manual for the state-trait anxiety inventory. Consulting Psychological Press, Palo Alto
Studd C, Eltringham RJ (1980) Lorazepam as night sedation and premedication: a comparison with diazepam. Anaesthesia 35:60–64
Subhan Z (1984) The effects of benzodiazepine on short-term memory and information processes. In: Hindmarch I, Ott H, Roth T (eds) Sleep, benzodiazepines and performance. Springer, Berlin Heidelberg New York, pp 173–181
Tecce JJ, Savignano-Bowman J, Cole JO (1978) Drug effects on contingent negative variation and eyeblinks: the distraction-arousal hypothesis. In: Lipton MA, DiMascio A, Killan KR (eds) Psychopharmacology. A generation of progress. Raven, New York, pp 745–758

Clinical and Psychological Aspects of Memory Dysfunction

Clinical Relevance of Effects of Benzodiazepines on Learning and Memory

R. G. LISTER[1], H. WEINGARTNER[2], M. J. ECKARDT[1] and M. LINNOILA[1]

Abstract

The effects of benzodiazepines on learning and memory are examined in the various clinical situations in which these drugs are used. Alterations in performance arising from the conditions for which benzodiazepines are prescribed are also considered. Current evidence indicates that, in anxious patients, as in normal volunteers, benzodiazepines impair the acquisition of new information (episodic memory). Although some tolerance may develop to these impairments, deficits are observed even after patients have been taking their medication chronically. Like amnesic patients, benzodiazepine-treated subjects may be unaware of their impaired ability to learn. The effects of the impairments on behavioral psychotherapies are considered.

1 Introduction

Benzodiazepines remain the most widely prescribed class of psychotropic drugs, not only because of their efficacy in treating anxiety and sleeping disorders, seizures, and producing muscle relaxation, but also because they have been considered very safe. However, recent evidence suggests that long-term use of benzodiazepines may lead to dependence, and a large number of reports document psychomotor impairment in benzodiazepine-treated subjects, both in normal volunteers and in anxious patients (SAARIO et al. 1976; DE GIER et al. 1981; LINNOILA et al. 1983). It is also widely accepted that benzodiazepines can produce cognitive impairments when given to normal volunteers. This paper focuses on these cognitive impairments and examines their clinical relevance in various patient populations that receive benzodiazepine medication. We begin by considering, briefly, the nature of the cognitive impairments in normal volunteers. We then examine, in turn, the clinical situations in which benzodiazepines are administered and discuss whether cognitive impairments are likely to be observed, and their clinical implications. Finally, we discuss evidence that benzodiazepines differ from one another in their tendencies to produce learning impairments.

[1] Laboratory of Clinical Studies, DICBR, NIAAA, Building 10 Room 3C218, 9000 Rockville Pike, Bethesda, MD 20892, USA.
[2] Department of Psychology, George Washington University, Washington DC, USA.

2 Benzodiazepine-Induced Amnesia in Normal Volunteers

The amnesia caused by benzodiazepines in normal volunteers was the subject of a recent review (Lister 1985). The evidence from a large number of studies indicated that benzodiazepines do not impair all aspects of cognitive functioning. Their greatest effect is observed in tests of long-term episodic memory in which impairment results from deficits in the acquisition of new information (see also Clarke et al. 1970; Brown et al. 1982; Hinrichs et al. 1982; Lister and File 1984; Ghoneim et al. 1984). Retrieval of information from episodic memory (Clarke et al. 1970; Liljequist et al. 1978; Brown et al. 1982) or of previously acquired knowledge (Brown et al. 1983; Ghoneim et al. 1984) does not appear to be impaired. It also appears that benzodiazepine-treated subjects may be unimpaired in their ability to learn new procedures and skills (Lister and File 1984; Ghoneim et al. 1986).

2.1 Benzodiazepines as Preanesthesia Medications

Benzodiazepines are frequently used as premedicants in anesthetic practice (see Dundee and Kawar 1982). In fact, the first clinical reports that benzodiazepines caused amnesia came primarily from anesthesiologists (Brandt and Oakes 1965; McClish 1966; Brown and Dundee 1968; Fox et al. 1968; O'Neil and Verril 1969). In these circumstances, the anterograde amnesic properties of benzodiazepines constitute a desirable side-effect. McKay and Dundee (1980) noted that amnesia for "neutral" experimental stimuli (such as picture postcards) is greater than that for emotionally significant stimuli (such as being taken to an operating room). Several comparisons between benzodiazepines have been made (George and Dundee 1977; McKay and Dundee 1980; Kothary et al. 1981). Korttila and Linnoila (1976) reported that a dose of flunitrazepam 0.01 mg/kg caused more marked amnesia for abdominal pinching than an equivalent dose of diazepam (Korttila and Linnoila 1975). McKay and Dundee (1980), however, suggested that these two drugs produced comparable degrees of amnesia. The need for further studies comparing the amnesic effects of clinically equivalent doses of different benzodiazepines is discussed in more detail below.

3 Anxiety, Learning, and Memory

Before considering whether benzodiazepines, when used as anxiolytics, are likely to alter a patient's ability to learn and remember, it is important to determine whether anxiety alone alters these cognitive functions.

There have been a large number of studies investigating the effects of anxiety on learning and memory. Two features of these studies should be noted in reading the following discussion. First, the majority have used student volunteers and few have examined clinically anxious patient populations. The relationships between high levels of anxiety in student volunteers and anxiety observed in clinical populations are unclear. Secondly, features of laboratory testing are not always applicable to everyday life. For example, the test anxiety that plays a role in the performance of a student in the laboratory may be irrelevant for an individual trying to learn and remember a shopping list. Test anxiety is, however, likely to be important for a student studying for examinations. For discussions of the learning and memory of everyday life the reader is referred elsewhere (BADDELEY 1981; NEISSER 1982; BRUCE 1985).

Based on their studies of test anxiety, LIEBERT and MORRIS (1967) proposed a distinction between worry, which they considered to be the cognitive component of anxiety, and emotionality, which involves changes in physiological functioning. EYSENCK (1979) has argued that worry and other task-irrelevant cognitive activities associated with anxiety always impair the quality of performance, by competing with task-relevant information for space in the processing system. Further, citing KAHNEMAN (1973), he suggests that highly anxious subjects attempt to compensate for these impairments by increasing effort expenditure. The overall effect of anxiety on performance depends on the balance between these two factors. Consistent with this analysis is the observation that the effect of anxiety depends on the difficulty of the task. EYSENCK (1979) commented on 13 studies in which high anxiety improved performance on easy learning tasks. In 11 of the 13 studies, high anxiety impaired performance on difficult tasks which required maximum working memory capacity. Whether a similar pattern of results would be obtained with clinically anxious patients is worthy of investigation.

Various cognitive views of anxiety disorders have been proposed (e.g., BUTLER and MATHEWS 1983; LANG 1985; SARASON 1985; SPIELBERGER 1985). Little attempt has been made, however, to interpret the anxiolytic effects of the benzodiazepines within a cognitive framework. Recent investigations suggest that mood state-dependent learning, or more correctly mood state-dependent retrieval, may be important. These studies view mood as a context that biases the encoding strategies that subjects use in interpreting events, or as context that alters retrieval processes. Most studies on mood state-dependent retrieval have focused on the effects of happy and sad moods (e.g., WEINGARTNER et al. 1977; BOWER 1981). Memories associated with a happy mood are more easily retrieved than depressing memories when a subject is in a happy mood (TEASDALE et al. 1980; BOWER 1981). This increased availability of memories congruent with one's mood is likely to increase the intensity of the mood state (TEASDALE 1983). CLARK et al. (1983) elegantly demonstrated the importance of arousal as a cue for state-dependent retrieval. They showed that material learned in a high state of arousal was best retrieved in a high state of arousal. Further, the method of increasing arousal did not have to be the same before retrieval as it was before acquisition. A similar mechanism may be important in anxiety states. Thus, a subject in an anxious state may have ready access to memories associated with anxiety, and thereby perceive situations as more threatening than in a nonanxious state (BUTLER and MATHEWS 1983). For example, an anxious individual may be afraid to open a letter, because of readily available memories of opening letters containing anxiogenic stimuli such as bad news or large bills. The same subject in a tranquil state may have more difficulty accessing these anxious memories and, therefore, not be threatened by the same unopened letter. Perhaps the benzodiazepines reduce anxiety, at least in part, by selectively reducing the availability of memories associated with high states of anxiety.

In view of the complexity of the effect of anxiety on performance in tests of learning, it is not surprising that there are few studies in which benzodiazepine treatment is also investigated. In one such study, DESAI et al. (1983) found that a group of students with high state anxiety performed worse in their test of learning than students with low state anxiety. A 5-mg dose of diazepam improved the performance of the high-anxiety group but had no such effect in the students with low anxiety. There was no mention of the effects of diazepam on subjects' self-ratings of anxiety in this study. In contrast to this study, HARTLEY et al. (1982) reported a tendency for diazepam to slow the retrieval of information from semantic memory to a greater extent in high- than in low-anxiety subjects.

Although the vast majority of studies investigating the effects of benzodiazepines on learning and memory have administered only a single dose of the compound under investigation, in the clinical management of anxiety benzodiazepines are usually taken chronically. Since tolerance develops to many of the behavioral effects of benzodiazepines (FILE 1985), tolerance to their amnesic properties might also develop. GHONEIM et al. (1981) showed that although there was some attenuation of diazepam-induced learning impairments after 3 weeks of treatment, some impairment still remained. More recently, they showed that oxazepam also impaired learning after 3 weeks of chronic treatment (GHONEIM et al. 1986). The study of LUCKI et al. (1986; LUCKI and RICKELS 1986), discussed in more detail below, suggests that the impairments remain after much longer periods of treatment.

Several studies demonstrate that clinically anxious patients taking benzodiazepines can exhibit impairments in learning and memory (ANGUS and ROMNEY 1984; LUCKI et al. 1986; LUCKI and RICKELS 1986). LUCKI et al. (1986) and LUCKI and RICKELS (1986) compared the performance of patients who were taking benzodiazepines for anxiety-related disorders with a group of clinically anxious patients who were not receiving medication. It is worth noting that in this study the benzodiazepine-treated group performed (nonsignificantly) worse than the drug-free group in all performance tests reported (digit-symbol substitution, symbol copying, letter cancellation, and verbal learning). These authors also examined the effect of acute treatment with a benzodiazepine on the performance of the long-term users by comparing predrug and postdrug performance. Delayed recall of verbal material was significantly impaired following benzodiazepine administration, clearly demonstrating that benzodiazepines impair learning in patient populations. Since these patients had been taking their medication daily for an average of 60 months, any tolerance to the impairments should already have developed.

The limited clinical data are consistent with the studies in normal volunteers suggesting that benzodiazepines primarily impair acquisition processes. Benzodiazepines are, therefore, relatively contraindicated in situations in which patients have to acquire information. In particular, caution should be exercised in prescribing benzodiazepines to anxious students preparing for examinations. Drug-induced amnesia is likely to exacerbate rather than relieve preexam nervousness and may have adverse effects on performance.

4 Benzodiazepines and the Effects of Behavioral Therapies

Another important clinical issue is whether treating a patient with a benzodiazepine might interfere with behavioral psychotherapies, i.e., treatments that rely on various forms of retraining or conditioning. Such treatments, if successful, must involve learning and later recall. A number of authors have suggested that this may be the case, but for different reasons. Some suggest that benzodiazepine-induced amnesia will reduce the impact of behavioral psychotherapy (ANGUS and ROMNEY 1984), while others suggest that the anxiolytic effect will prevent patients from developing tolerance to the stressful situations they encounter (GRAY et al. 1982).

Several factors need to be considered. First, the effects may depend on the type of behavioral therapy being used. Consider, for example, the use of flooding to treat phobic anxiety. Patients are given prolonged exposure to the phobic stimulus. The therapeutic effect of this manipulation is considered to reflect simple extinction. Several studies show that the use of diazepam in conjunction with this method actually improves the therapeutic outcome (MARKS et al. 1972; JOHNSTON and GATH 1973). It has been suggested that the beneficial effect of diazepam in these circumstances may have resulted from diazepam-treated subjects tolerating a greater exposure to the phobic stimulus (HAFNER and MARKS 1976). In these circumstances, the beneficial effect of diazepam clearly outweighed any amnesic effect.

It is likely that several different types of learning are involved in behavioral therapies and benzodiazepines may not impair all of these. Alterations in semantic memory and the learning of new procedures to cope with stress play a role in some therapies. The mechanisms involved in altering semantic memory (TULVING 1983, 1984) and the effects of benzodiazepines on such mechanisms require further investigation. Benzodiazepines may not impair certain forms of procedural learning (LISTER and FILE 1984; GHONEIM et al. 1986).

5 Other Clinical Uses

Several groups have examined the effects of benzodiazepines on learning and memory in a setting in which they are used as nighttime hypnotic medications. These studies have consistently shown that benzodiazepines impair learning during nighttime awakenings (BIXLER et al. 1979; ROTH et al. 1980; SPINWEBER and JOHNSON 1982; ROEHRS et al. 1983).

This effect should also be noted in the light of recent studies suggesting that some benzodiazepines may reduce the adverse effects of traveling across time zones (SEIDEL et al. 1984; TUREK and LOSEE-OLSEN 1986). Possible undesirable amnesic effects in the context of international travel need to be considered.

Benzodiazepines (primarily clonazepam) are in some circumstances used in the treatment of epilepsy. The difficulties in dissociating impairments in performance arising from clinical levels of anxiety from those resulting from drug treatment have already been mentioned. The problems are compounded in epilepsy

(THOMPSON and TRIMBLE 1981) since once a subject is diagnosed as epileptic, he or she will invariably receive some form of medication. A drug-free control population is, therefore, unlikely to be found. Further, epileptics may receive head injuries during grand mal seizures and some performance impairments may result from these, although it should be noted that clonazepam is used most frequently in epilepsies other than grand mal.

The question of whether the sedative and amnesic effects of benzodiazepines are related was discussed in some detail in a previous review (LISTER 1985) in which it was concluded that the relationship between sedation and amnesia remained unresolved. Some studies suggested a link between sedation and amnesia (e.g., ROTH et al. 1980; FILE and LISTER 1982; ROEHRS et al. 1983) and others suggested a dissociation (PANDIT et al. 1976; GHONEIM et al. 1981). There have been several recent studies that suggest amnesia and sedation can be dissociated. LUCKI et al. (1986) and LUCKI and RICKELS (1986) showed learning impairments shortly after benzodiazepine administration in patients who had been taking benzodiazepines for 60 months, and who appeared tolerant to the drugs' sedative action as assessed by subjects' self-ratings and a symbol copying test (although critical flicker fusion threshold was reduced). A study by ROACHE and GRIFFITHS (1985) suggests that the amnesia caused by triazolam cannot be totally attributed to sedation. They used a within-subject design to compare the effects of four doses of triazolam with four doses of sodium pentobarbital. A comparison of the effects of the lowest dose of triazolam (0.5 mg) with pentobarbital 400 mg showed that the triazolam caused a greater learning impairment than the barbiturate, although the barbiturate caused a greater degree of sedation as assessed by staff ratings and several other behavioral parameters. MCKAY and DUNDEE (1980) suggested that it may be more difficult to dissociate amnesia and sedation if benzodiazepines are administered orally, than if they are administered intravenously. They found that the amnesia induced by oral diazepam, flunitrazepam, and lorazepam paralleled, both in extent and duration, the soporific action of these drugs. No similar correlation was found when these drugs were given intravenously (PANDIT et al. 1976; GEORGE and DUNDEE 1977). Recently completed studies by the National Institute for Mental Health have also shown that the effects of diazepam on memory are highly correlated ($r > 0.90$) with its sedative effects when it is administered intravenously in increasing doses (WOLKOWITZ et al. 1987). However, in a second study in which subjects were pretreated with either placebo or the benzodiazepine antagonist RO 15–1788, the memory acquisition impairments of diazepam persisted despite the blockade of the sedative effect (HOMMER et al., in preparation). Other studies, however, have reported that RO 15–1788 will reverse, at least partly, the amnesic effects of benzodiazepines (O'BOYLE et al. 1983; GENTIL et al. 1985).

Since sedatives generally impair learning (e.g., BIRNBAUM and PARKER 1977; WILLIAMS and RUNDELL 1983) the sedative properties of the benzodiazepines probably contribute to their amnesic effects even if they are not underlying cause. SHAGASS and coworkers (SHAGASS 1954; SHAGASS and NAIMAN 1956) found that the sedation threshold of anxious subjects is higher than that of controls, and sedation thresholds correlate significantly with degree of anxiety. This reduced sensitivity of anxious patients to sedatives may lead to a reduced susceptibility to benzodiazepine-induced amnesia.

6 Differences between Benzodiazepines

The question of whether benzodiazepines differ in their tendency to produce amnesia is important for several reasons. If benzodiazepines were found to differ from one another, then this would affect the choice of the benzodiazepine used, both in situations where amnesia is desirable (e.g., in anesthesia) and where it is highly undesirable (e.g., in anxious outpatients). Investigation of the characteristics of the benzodiazepines predictive of high amnesic potency would assist in determining the mechanisms underlying benzodiazepine-induced amnesia and would be of benefit in the search for drugs such as β-carbolines that might improve learning (see below). Unfortunately there are problems with most of the studies suggesting that benzodiazepines differ from one another in their abilities to cause amnesia when used in comparable clinical doses. Many studies have either failed to examine an adequate number of doses or failed to determine whether the doses of drugs used were, in fact, equivalent. SCHARF et al. (1984) compared the effects of 1 and 2 mg lorazepam with 7.5 and 15 mg clorazepate. Only the 2 mg dose of lorazepam produced significant learning impairments and on this basis they suggested that short- and long-acting benzodiazepines may differ in their amnesic properties. However, the authors failed to report any other behavioral measures to suggest that the 2 mg dose of lorazepam was equivalent to the highest dose of clorazepate. Further, the implication that any difference between the two drugs could be attributed to a difference in half-life is hard to justify on the basis of the data presented. Clorazepate and lorazepam differ in many of their properties (e.g., lipophilicity, receptor affinity), any one of which might underly the suggested differences in the behavioral effects of these drugs. HEALEY et al. (1985) also suggested that lorazepam may have more pronounced amnesic properties than either diazepam or clorazepate. This study also failed to demonstrate the equivalence of the doses used, and a ceiling effect may have masked possible amnesic effects of diazepam and clorazepate.

Alprazolam and lorazepam have tended to feature predominantly in reports of amnesic side effects to the Food and Drug Administration (ROBERT NELSON, personal communication). There have also been suggestions that triazolam is particularly likely to induce amnesia (SHADER and GREENBLATT 1983). There is a clear need for studies comparing these compounds with other benzodiazepines in an experimental setting to determine whether these reports do in fact reflect real differences in amnesic effects.

7 Conclusions

Current data suggest that it may be possible to dissociate the amnesic effect of the benzodiazepines from their therapeutic effects. Evidence suggesting a dissociation between the sedative and amnesic effects has already been mentioned. It has been proposed that the anxiolytic effects of some benzodiazepines are observed at doses lower than those needed to induce amnesia (MCKAY and DUNDEE 1980).

However, very few well-controlled investigations into the effects of benzodiazepines on learning and memory in patient populations have been performed. The most probable reason for this is the difficulty in designing and carrying out such experiments. There is a great need for placebo-controlled studies examining the effects of both acute and chronic benzodiazepine administration on learning and memory function in clinically anxious populations. When possible, these studies should attempt to compare the amnesic effects of different benzodiazepines, and examine the relationship between the therapeutic and amnesic effects of these drugs.

A further problem has been that in the clinical studies that have been performed memory has been considered a unitary phenomenon, and a single measure has been used to quantify it. The recent developments in cognitive psychology suggest that this assumption is a gross oversimplification, and if we are to understand the effects of benzodiazepines (or for that matter any drug) on learning and memory and the clinical relevance of these effects, further studies are needed comparing drug effects on putatively different learning and memory processes (see Lister and Weingartner 1987). The typical test of episodic learning involved in memorizing lists of words represents only one domain.

Another important point concerns the effects of benzodiazepines on metacognition. Clinicians should note that if a patient does not complain of memory impairments this does not necessarily imply that such impairments are not present. The ability to judge the accuracy of memory performance appears to be disrupted in amnesic Korsakoff's disease patients. Benzodiazepine-treated subjects may be impaired in their ability to learn but be unaware of the impairment (e.g., Malpas 1972; Hinrichs et al. 1982; Roache and Griffiths 1985, 1986). Roache and Griffiths (1985) compared the effects of pentobarbital and triazolam. They found that staff ratings of drug effects were in agreement with the subjects' ratings for pentobarbital-treated subjects but not for triazolam-treated subjects. Subjects that received triazolam underestimated their impairments. A similar pattern of results was obtained in a study comparing the effects of lorazepam and meprobamate (Roache and Griffiths 1986). Lorazepam-treated subjects underestimated their degree of impairment whereas meprobamate-treated subjects did not. It is difficult to determine at what stage a learning impairment becomes noticeable, and what factors mediate an individual's awareness of his or her impairment. The results of Roache and Griffiths suggest that the impaired ability to perceive deficits may be more important for benzodiazepines than for other sedative agents. The effects of benzodiazepines on metacognitive (self-monitoring) processes are worthy of further systematic exploration and are clearly as important as the well-documented memory impairing effects of these drugs.

Finally, it may be noted that the drugs discussed in this article are all benzodiazepine agonists. In recent years a large number of different compounds have been synthesized which have high affinities for benzodiazepine receptors and which have actions opposite to those of the benzodiazepines. For example, these agents, which have been called "inverse agonists," cause anxiety and are proconvulsant or induce seizures. A logical possibility is that rather than impairing acquisition processes, these compounds may facilitate a subject's ability to learn. Preliminary evidence from animal experiments suggests that this may be the case

(VENAULT et al. 1986). There is a clear need for experiments investigating the specificity of these effects, examining the relationships between the facilitation of learning and increases in arousal. It is too early to say whether inverse agonists would have a clinical use either in learning-impaired or normal subjects, and the potentially hazardous effects of these compounds such as increased anxiety and (kindled) seizures will need to be considered carefully.

References

Angus WR, Romney DM (1984) The effect of diazepam on patients' memory. J Clin Psychopharmacol 4:203–206

Baddeley AD (1981) The cognitive psychology of everyday life. Br J Psychol 72:257–269

Birnbaum IM, Parker ES (1977) Alcohol and human memory. Hillsdale NJ, Erlbaum

Bixler EO, Scharf MB, Soldatos CR, Mitsky DJ, Kales A (1979) Effects of hypnotic drugs on memory. Life Sci 25:1379–1388

Bower GH (1981) Mood and memory. Am Psychol 36:129–148

Brandt AL, Oakes FD (1965) Preanesthesia medication: double blind study of a new drug, diazepam. Anesth Analg 44:125

Brown J, Lewis V, Brown M, Horn G, Bowes JB (1982) A comparison between transient amnesias induced by two drugs (diazepam or lorazepam) and amnesia of organic origin. Neuropsychologia 20:55–70

Brown J, Brown M, Bowes JB (1983) Effects of lorazepam on forgetting, on retrieval from semantic memory and on manual dexterity. Neuropsychologia 21:501–512

Brown SS, Dundee JW (1968) Clinical studies of induction agents XXV: diazepam. Br J Anaesth 40:108–112

Bruce D (1985) The how and why of ecological memory. J Exp Psychol (Gen) 114:78–90

Butler G, Mathews A (1983) Cognitive processes in anxiety. Adv Behav Res Ther 5:51–62

Clark MS, Milberg S, Ross J (1983) Arousal cues arousal-related material in memory: implications for understanding effects of mood on memory. J Verb Learn Verb Behav 22:633–649

Clarke PRF, Eccersley PS, Frisby JP, Thornton JA (1970) The amnesic effect of diazepam (Valium). Br J Anaesth 42:690–697

de Gier JJ, 't Hart BJ, Nelemans FA, Bergman H (1981) Psychomotor performance and real driving performance of outpatients receiving diazepam. Psychopharmacology 73:340–344

Desai N, Taylor-Davies A, Barnett DB (1983) The effects of diazepam and oxprenolol on short-term memory in individuals of high and low state anxiety. Br J Clin Pharmacol 15:197–202

Dundee JW, Kawar P (1982) Benzodiazepines in anesthesia. In: Usdin E, Skolnick P, Tallman JF, Greenblatt D, Paul SM (eds) Pharmacology of benzodiazepines. Macmillan, London, pp 313–328

Eysenck MW (1979) Anxiety, learning and memory: a reconceptualization. J Res Personality 13:363–385

File SE (1985) Tolerance to the behavioral actions of benzodiazepines. Neurosci Biobehav Rev 9:113–121

File SE, Lister RG (1982) Do lorazepam-induced deficits in learning result from impaired rehearsal, reduced motivation or increased sedation? Br J Clin Pharmacol 14:545–550

Fox GS, Wynands JE, Bhambhami M (1968) A clinical comparison of diazepam and thiopentone as induction agents to general anaesthesia. Can Anaesth Soc J 15:281–290

Gentil V, Nogueira RP, Gorenstein C, Moreno RA, Draggan TG, Singer J (1985) Effects of flunitrazepam on memory and their reversal by two antagonists. Paper presented at IVth world congress of biological psychiatry, Philadelphia

George KA, Dundee JW (1977) Relative amnesic actions of diazepam, flunitrazepam and lorazepam in man. Br J Clin Pharmacol 4:45–50

Ghoneim MM, Mewaldt SP, Berie JL, Hinrichs JV (1981) Memory and performance effects of single and 3-week administration of diazepam. Psychopharmacology 73:147–151

Ghoneim MM, Mewaldt SP, Hinrichs JV (1984) Dose response analysis of the behavioral effects of diazepam: I Learning and memory. Psychopharmacology 82:291–295
Ghoneim MM, Hinrichs JV, Mewaldt SP (1986) Comparison of two benzodiazepines with differing accumulation: behavioral changes during and after 3 weeks of dosing. Clin Pharmacol Ther 39:491–500
Gray JA, McNaughton N, Holt L, Tsaltas E, Feldon J, Shemer A (1982) The effects of anti-anxiety drugs on tolerance for stress. In: Levy A, Spiegelstein MY (eds) Behavioral models and the analysis of drug action. Amsterdam, Elsevier, pp 175–194
Hafner J, Marks I (1976) Exposure in vivo of agoraphobics: contributions of diazepam, group exposure and anxiety evocation. Psychol Med 6:71–88
Hartley LR, Spencer J, Williamson J (1982) Anxiety, diazepam and retrieval from semantic memory. Psychopharmacology 76:291–293
Healey M, Pickens R, Meisch R, McKenna T (1985) Effects of clorazepate, diazepam, lorazepam, and placebo on human memory. J Clin Psychiatry 44:436–439
Hinrichs JV, Mewaldt SP, Ghoneim MM, Berie JL (1982) Diazepam and learning: assessment of acquisition deficits. Pharmacol Biochem Behav 17:165–170
Johnston D, Gath D (1973) Arousal levels and attribution effects in diazepam-assisted flooding. Br J Psychiatry 123:463–466
Kahneman D (1973) Attention and effort. Prentice Hall, Englewood Cliffs NJ
Korttila K, Linnoila M (1975) Skills related to driving after intravenous sedation: dose-response relation with diazepam. Br J Anaesth 47:457–463
Korttila K, Linnoila M (1976) Amnesic actions of and skills related to driving after intravenous flunitrazepam. Acta Anaesth Scand 20:160–168
Kothary SP, Brown ACD, Pandit UA, Samra SK, Pandit SK (1981) Time course of antirecall effect of diazepam and lorazepam following oral administration. Anesthesiology 55:641–644
Lang PJ (1985) The cognitive psychophysiology of emotion: fear and anxiety. In: Tuma AH, Maser JD (eds) Anxiety and the anxiety disorders. Erlbaum, Hillsdale NJ, pp 131–170
Liebert RM, Morris LW (1967) Cognitive and emotional components of test anxiety: a distinction and some initial data. Psychol Rep 20:975–978
Liljequist R, Linnoila M, Mattila MJ (1978) Effect of diazepam and chlorpromazine on memory functions in man. Eur J Clin Pharmacol 13:339–343
Linnoila M, Erwin CW, Brendle A, Simpson D (1983) Psychomotor effects of diazepam in anxious patients and healthy volunteers. J Clin Psychopharmacol 3:88–96
Lister RG (1985) The amnesic action of benzodiazepines in man. Neurosci Biobehav Rev 9:87–94
Lister RG, File SE (1984) The nature of lorazepam-induced amnesia. Psychopharmacology 83:183–187
Lister RG, Weingartner H (1987) Neuropharmacological strategies for understanding psychobiological determinants of cognition. Hum Neurobiol 6:119–127
Lucki I, Rickels K (1986) The behavioral effects of benzodiazepines following long-term use. Psychopharmacol Bull 22:424–433
Lucki I, Rickels K, Geller AM (1986) Chronic use of benzodiazepines and psychomotor and cognitive test performance. Psychopharmacology 88:426–433
Malpas A (1972) Subjective and objective effects of nitrazepam and amylobarbitone sodium in normal human beings. Psychopharmacology 27:373–378
Marks IM, Viswanathan R, Lipsedge MS, Gardner R (1972) Enhanced relief of phobias by flooding during waning diazepam effect. Br J Psychiatry 121:493–505
McClish A (1966) Diazepam as an intravenous induction agent for general anaesthesia. Can Anaesth Soc 13:562–575
McKay AC, Dundee JW (1980) Effect of oral benzodiazepines on memory. Br J Anaesth 52:1247–1257
Neisser U (ed) (1982) Memory observed: remembering in natural contexts. Freeman, San Francisco
O'Boyle C, Lambe R, Darragh A, Taffe W, Brick I, Kenny M (1983) RO 15–1788 antagonises the effects of diazepam in man without affecting its bioavailability. Br J Anaesth 55:349–355
O'Neil R, Verrill PJ (1969) Intravenous diazepam in minor oral surgery. Br J Oral Surg 7:12–14

Pandit SK, Heisterkamp DV, Cohen PJ (1976) Further studies of the anti-recall effect of lorazepam: a dose-time-effect relationship. Anesthesiology 45:495–500
Roache JD, Griffiths RR (1985) Comparison of triazolam and pentobarbital: performance impairment, subjective effects and abuse liability. J Pharmacol Exp Ther 234:120–133
Roache JD, Griffiths RR (1986) Behavioral effects and abuse liability of lorazepam and meprobamate in humans. Pharmacologist 28:202
Roehrs T, Zorick FJ, Sicklesteel JM, Wittig RM, Hartse KM, Roth T (1983) Effects of hypnotics on memory. J Clin Psychopharmacol 5:310–313
Roth T, Hartse KM, Saab PG, Piccione PM, Kramer M (1980) The effects of flurazepam, lorazepam and triazolam on sleep and memory. Psychopharmacology 70:231–237
Saario I, Linnoila M, Mattila MJ (1976) Modification of diazepam or thioridazine of the psychomotor skills related to driving: a sub-acute trial in neurotic out-patients. Br J Clin Pharmacol 3:843–848
Sarason IG (1985) Cognitive processes, anxiety and the treatment of anxiety disorders. In: Tuma AH, Maser JD (eds) Anxiety and the anxiety disorders. Erlbaum, Hillsdale NJ, pp 87–107
Scharf MB, Khosla N, Brocker N, Goff P (1984) Differential amnestic properties of short- and long-acting benzodiazepines. J Clin Psychiatry 45:51–53
Seidel WF, Roth T, Roehrs T, Zorick F, Dement WC (1984) Treatment of a 12-hour shift of sleep schedule with benzodiazepines. Science 224:1262–1264
Shader RI, Greenblatt DJ (1983) Triazolam and anterograde amnesia: all is not well in the Z-zone. J Clin Psychopharmacol 3:273
Shagass C (1954) The sedation threshold. A method for estimating tension in psychiatric patients. EEG Clin Neurophysiol 6:221–233
Shagass C, Naiman J (1956) The sedation threshold as an objective index of manifest anxiety in psychoneurosis. J Psychosom Res 1:49–57
Spielberger CD (1985) Anxiety, cognition and affect: a state-trait perspective. In: Tuma AH, Maser JD (eds) Anxiety and the anxiety disorders. Erlbaum, Hillsdale NJ, pp 171–182
Spinweber CL, Johnson LC (1982) Effects of triazolam (0.5 mg) on sleep, performance, memory, and arousal threshold. Psychopharmacology 76:5–12
Teasdale JD (1983) Negative thinking in depression: cause, effect or reciprocal relationship? Adv Behav Res Ther 5:3–25
Teasdale JD, Taylor R, Fogarty SJ (1980) Effects of induced elation-depression on the accessibility of memories of happy and unhappy experiences. Behav Res Ther 18:339–346
Thompson PJ, Trimble MR (1981) Further studies on anticonvulsant drugs and seizures. Acta Neurol Scand Suppl 89:51–58
Tulving E (1983) Elements of episodic memory. Clarendon, Oxford
Tulving E (1984) Precis of elements of episodic memory. Behav Brain Sci 7:223–268
Turek FW, Losee-Olsen S (1986) A benzodiazepine used in the treatment of insomnia phase-shifts the mammalian circadian clock. Nature 321:167–168
Venault P, Chapouthier G, Prado de Carvalho L, Simiand J, Morre M, Dodd RH, Rossier J (1986) Benzodiazepine impairs and β-carboline enhances performance in learning and memory tasks. Nature 321:864–866
Weingartner H, Miller H, Murphy DL (1977) Mood-state-dependent retrieval of verbal associations. J Abnorm Psychol 86:276–284
Williams HL, Rundell OH (1983) Secobarbital effects on recall and recognition in a levels-of-processing paradigm. Psychopharmacology 80:221–225
Wolkowitz O, Weingartner H, Thompson K, Pickar D, Paul SM, Hommer DW (1987) Diazepam-induced amnesia. Am J Psychiatry 144:25–29

The Effect of Anxiolytic Drugs on Memory in Anxious Subjects

I. LUCKI[1] and K. RICKELS

Abstract

The benzodiazepines (BZs), represented by diazepam, are the class of drugs used most frequently to treat clinical anxiety disorders. Since it is known that acute BZ intake impairs memory function, the effects of BZs on memory were evaluated in chronic users of BZ medications. In addition, the acute effects of diazepam were compared with those of the non-BZ anxiolytic buspirone on memory function in anxious subjects. Memory function was evaluated by a free verbal recall procedure where subjects recalled a list of 16 noncategorized nouns immediately after the word list was read (immediate recall) and again 20 min later (delayed recall). When the chronic BZ users were tested for free verbal recall during their first visit, 4–14 h after their last dose, they did not differ in immediate or delayed recall from an age- and sex-matched group of unmedicated anxious subjects. At a subsequent visit, the acute effects of BZ medications were studied 60–90 min after the subjects took their usual dose. Although acute BZ administration did not alter immediate recall, delayed recall was significantly impaired in the chronic BZ users. Thus, complete tolerance does not develop to the acute memory-impairing effects of BZs after long-term use. Acute administration of the anxiolytic drugs diazepam (5 mg) or buspirone (5 or 10 mg) did not alter immediate recall in another group of unmedicated anxious subjects. Diazepam selectively impaired delayed recall of the word list when compared with placebo. In contrast, neither dose of buspirone altered delayed recall. To the extent that such effects on verbal recall tests are reflected in a patient's daily activities, the failure of buspirone to adversely affect memory function could contribute to its usefulness as an alternative antianxiety therapy.

1 Introduction

The benzodiazepines (BZs) remain one of the most widely used families of drugs in medicine. The BZs have been found helpful in the treatment of anxiety and insomnia, as muscle relaxants, in the control of seizures, or as adjuncts in treatment of a variety of other medical problems. Among the side effects of BZs, it has long been known that intravenous administration of BZs such as diazepam and lorazepam impairs memory function in normal subjects (CLARKE et al. 1970; BROWN et al. 1978). This effect was viewed as contributing to the usefulness of these drugs as presurgical anesthetic agents. Subsequently, it has become appreciated that oral administration of BZs also causes memory-impairing effects at doses that might be expected to be used in medical practice. Memory-impairing effects have been demonstrated for BZs used primarily in the treatment of anxiety, such as diazepam (GHONEIM et al. 1984 a), lorazepam (LISTER and FILE 1984), alprazolam

[1] Department of Psychiatry, University of Pennsylvania, Philadelphia, PA 19104, USA.

(BLOCK and BERCHOU 1984), and oxazepam (MEWALDT et al. 1986). Furthermore, other BZs that are used primarily as hypnotics also impair memory functions (BIXLER et al. 1979). Although the initial administration of BZs impairs a variety of motor functions and causes subjective feelings of sedation and drowsiness, higher doses of BZs may be required to impair psychomotor performance consistently than to produce cognitive impairment (GHONEIM et al. 1984b).

The conditions under which BZs cause memory impairment have been extensively studied with normal volunteers (for review, see LISTER 1985). Most studies have employed verbal learning techniques for measuring memory. It has consistently been found that BZs produce an anterograde-type amnesia, i.e., memory is impaired for information presented after, but not before, ingestion of the drug (GHONEIM and MEWALDT 1975; HINRICHS et al. 1982). Since recall ability remains intact after diazepam, it has been suggested that BZs impair the consolidation of new information into long-term memory storage. The effects of BZs on recall of word lists are more easily detected following delayed rather than immediate recall (GHONEIM and MEWALDT 1975; MEWALDT et al. 1983), with hard rather than easy words, and with longer rather than shorter word lists (MEWALDT et al. 1983). Even words that are learned under BZs seem to be recalled less well, since recognition memory is impaired and confused by synonyms or by similar-sounding words (RICHARDSON et al. 1984). BZs' amnesic effects are also reflected in poorer learning of tasks involving repeated presentations of word lists or by impaired learning of paired associates (HINRICHS et al. 1982). Although increasing arousal by presenting noise increases the retention of verbally presented material, this could not overcome the memory-impairing effect of lorazepam (LISTER and FILE 1984). Memory for pictures or nonverbal material is similarly disrupted by BZs (LISTER and FILE 1984). However, performance on certain memory tests, such as digit span or learning nonsense syllables, may be more resistant to the effects of BZs (JONES et al. 1978). In the case of digit span, this may reflect a greater contribution of immediate memory than long-term retention in the manner that this test is usually administered. It is unlikely that BZs' memory-impairing effects are caused by state-dependent learning since verbal material learned prior to drug administration can be recalled adequately, whereas posttrial administration of diazepam will not reinstate verbal material forgotten under the influence of diazepam (PETERSEN and GHONEIM 1980).

Despite the predominant use of BZ medications in the treatment of anxiety disorders, very few studies have been done on BZs' memory-impairing effects using anxious subjects. In one study, diazepam (5 mg) impaired the short-term memory of low-anxiety student volunteers but improved the memory of high-anxiety subjects (BARNETT et al. 1981). However, HARTLEY et al. (1982) reported that diazepam impaired recall in both anxious and nonanxious student volunteers. In a diverse group of patients being treated for anxiety or insomnia, ANGUS and ROMNEY (1984) found that short-term diazepam treatment impaired paired-associate learning on the Wechsler Memory Test, and this effect was reversed following cessation of drug treatment. GHONEIM et al. (1984c) reported that acute administration of 10 mg diazepam impaired immediate and delayed recall in a group of patients suffering from agoraphobia or panic disorder. After 2 weeks of treatment (30 mg/day), the patients' memories were not affected when they were tested 5–

8 h after their last dose. The remainder of this paper will review some of our own work concerning the measurement of memory function in clinically anxious subjects taking BZ medications.

2 Acute Effects of Antianxiety Medications in Anxious Subjects

In view of the few studies that have examined BZs' behavioral effects in anxious subjects, we studied the acute effects of diazepam in patients diagnosed as having a generalized anxiety disorder. A dose of 5 mg diazepam was administered to subjects, as representing a dose usually prescribed for clinical treatment, and the effects on verbal recall ability were measured. Buspirone is one of several non-BZ medications currently being studied for potential alternative or adjunctive treatment of anxiety-related disorders (GOLDBERG and FINNERTY 1979; RICKELS et al. 1982). Buspirone is not thought to produce its anxiolytic effects at brain BZ receptors (SKOLNICK et al. 1984). Since buspirone had not been previously evaluated on tests of memory, the effect of diazepam on verbal recall ability was compared with that of buspirone in anxious subjects. Some of these results were reported previously (LUCKI et al. 1987).

2.1 Subjects and Design

The subjects were 14 males and 25 females with an average age of 34.3 ± 1.6 years (± 1 SEM; range 19–60 years). Nearly all of the subjects (35, 90%) had completed high school and nearly half of them (18, 46%) had graduated from college. After a psychiatric interview, each subject was diagnosed as having a generalized anxiety disorder (DSM III criteria) of moderate intensity or greater for at least 3 months' duration. The severity of each subject's symptoms had to be rated at 18 or greater on the Hamilton Anxiety Scale and be scored 9 or greater on the Covi Anxiety Scale, but 8 or less on the Raskin Depression Scale, in order for them to be included in the study. The subjects were drug free for at least 2 weeks preceding the behavioral tests.

The tests of free verbal recall ability involved reading a list of 16 noncategorized nouns to the subjects at a rate of 1 word/2 s. Immediately after the word list was read, the subjects wrote as many of the words as they could recall on a blank sheet of paper. Delayed recall was measured similarly 20 min later when the subjects again recorded as many words that they could recall from the same list. Rehearsal of the word list was hindered by presenting the subjects with two psychomotor tests that did not involve memory between the immediate and delayed recall tests. The digit span test was also administered as described previously by WECHSLER (1955).

Immediate and delayed recall were measured as the absolute number of words remembered after either no delay or after the 20-min delay, respectively. In addition, the relative memory decay was measured as the percentage of words that were remembered immediately that could no longer be recalled after the 20-min delay, i.e., (1 − delayed recall/immediate recall) × 100. The memory decay measured the amount of information from the word list that each individual forgot during 20 min. Separate word lists were used composed of nouns matched for their relative frequency of use and imagery (THORNDIKE and LORGE 1944; PAIVIO et al. 1968).

Each subject received his or her first memory test at the initial interview. The memory test following drug ingestion occurred during a second session 4–10 days later. The subjects were assigned randomly to one of four treatment groups: (1) placebo ($n=10$); (2) 5 mg diazepam ($n=10$); (3) 5 mg buspirone ($n=10$); or (4) 10 mg buspirone ($n=9$). Each subject was tested for immediate and delayed verbal recall beginning approximately 70 min after ingesting the drug capsules. The memory tests were conducted under double-blind conditions; i.e., neither the subjects nor the test administrator were aware of the drug condition. The experiment code was broken only after completion of the entire study.

The subjects continued taking their medication for the next 7 days. The dosage schedules for the three drug groups were: (1) 5 mg diazepam; days 1–3, 15 mg/day; days 4–7, 20 mg/day; (2) 5 mg buspirone, days 1–3, 15 mg/day; days 4–7, 20 mg/day; (3) 10 mg buspirone, days 1–3, 20 mg/day; days 4–7, 30 mg/day. Subjects who received placebo initially continued to take placebo capsules three times daily for 7 days. On day 8, the subjects reported to the laboratory for testing approximately 3 h after taking their last dose of medication.

2.2 Results

When tested at baseline, the groups did not differ in immediate recall, delayed recall, or digit span performance, according to analysis of variance ($p>0.05$). On day 1, approximately 70 min after the first ingestion of drug, neither immediate recall ability nor digit span performance were altered by diazepam or buspirone. However, as shown in Table 1, diazepam significantly impaired verbal recall after the 20-min delay when compared with placebo. Similarly, the percentage memory decay over the 20-min period was increased significantly by diazepam. Neither dose of buspirone altered delayed verbal recall when compared with placebo.

The effects of diazepam on immediate and delayed recall were studied in detail as a function of the list serial position. As shown in Fig. 1, the serial position curves for immediate recall did not differ significantly between subjects that received placebo or diazepam, according to analysis of variance ($p>0.05$). However, the impairing effect of diazepam on delayed recall can be seen clearly in Fig. 2. The serial position curves for delayed recall differed significantly between the placebo and diazepam groups, according to analysis of variance ($p<0.01$). It appeared that diazepam more severely impaired the delayed recall of words at the end than at the beginning of the word list. The serial-position curves generated by subjects treated with buspirone did not differ from the curves for subjects given placebo (data not shown).

The subjects were tested again on day 8 approximately 3 h after the last ingestion of their medication. Although diazepam still appeared to reduce delayed recall and to increase the memory decay, none of the groups differed significantly

Table 1. Effects of antianxiety medications on verbal recall by anxious subjects

	Day 1	Day 8
Immediate recall		
Placebo	9.2±0.8	8.7±0.7
Diazepam 5 mg	8.4±0.4	7.5±0.5
Buspirone 5 mg	9.4±0.5	8.2±0.4
Buspirone 10 mg	9.0±0.8	9.4±0.9
Delayed recall		
Placebo	6.8±0.8	6.2±0.8
Diazepam 5 mg	4.2±0.4*	4.1±0.5
Buspirone 5 mg	7.0±0.3	5.3±0.6
Buspirone 10 mg	6.3±0.8	6.6±1.4
Memory decay (%)		
Placebo	27.0±5.4	28.1±6.5
Diazepam 5 mg	49.8±5.1*	44.8±5.5
Buspirone 5 mg	24.6±3.3	36.5±4.9
Buspirone 10 mg	31.0±4.5	34.6±8.6
Digit span		
Placebo	13.4±0.5	13.7±0.6
Diazepam 5 mg	13.0±0.9	14.7±0.6
Buspirone 5 mg	13.5±0.6	14.2±0.4
Buspirone 10 mg	13.9±0.4	15.4±0.4

All values represent the mean ±1 SEM.
* Value differs significantly from the corresponding placebo group, according to Dunnett's test ($p<0.01$).
Drug effects on day 1 were measured 70 min after drug ingestion, whereas drug effects on day 8 were measured approximately 180 min after drug ingestion.

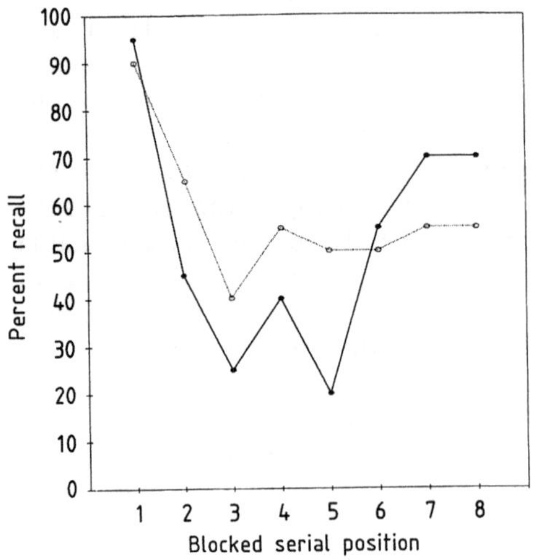

Fig. 1. Serial position curves for immediate recall by anxious subjects. Subjects ($n=10$) were given either placebo (○) or diazepam (●) 70 min prior to being read a 16-word list of noncategorized nouns. Values represent the mean percentage recall frequency of words by subjects as a function of the position within the list. The serial position curves of the groups did not differ significantly according to analysis of variance

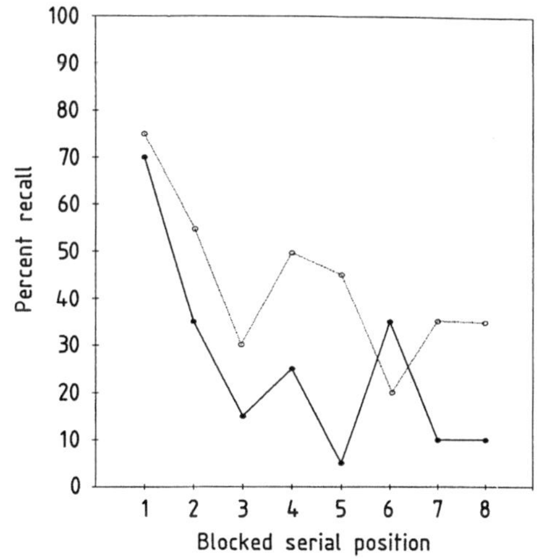

Fig. 2. Serial position curves for delayed recall by anxious subjects given either placebo (○) or 5 mg diazepam (●). Values represent the mean percentage recall frequency of words by subjects as a function of the position within the list. Delayed recall was assessed approximately 20 min after the word list was read to the subjects. The overall curves for delayed recall differed significantly after medication (●) when compared with baseline values (○), according to analysis of variance ($p<0.01$)

in their verbal recall ability or digit span performance at this time, according to analysis of variance ($p>0.05$). This difference in diazepam's ability to impair memory after subchronic treatment could be related, however, to the 3-h delay after drug ingestion on day 8 compared with the 70-min delay on day 1.

3 Verbal Recall Ability in Chronic Users of BZ Medications

An important issue concerning the clinical significance of BZs' ability to impair memory is whether this effect persists following prolonged administration of the drugs. Many of the behavior-impairing effects caused by the initial administration of BZs, including the subjective feelings of drowsiness and sedation or impairment of psychomotor performance, demonstrate tolerance following continued use of BZ medications (ARANKO et al. 1983; LILJEQUIST et al. 1979). In a previous study using normal subjects, GHONEIM et al. (1981) reported that diazepam impaired verbal recall after 3 weeks of daily administration but noted some development of tolerance to this effect.

We attempted to address this issue by studying the psychomotor, cognitive, and subjective effects of BZs in a group of chronic users of these medications (LUCKI et al. 1986; LUCKI and RICKELS 1986). A total of 54 subjects (31 males and 23 females) were examined who had taken BZ medications on a daily basis for at least 1 year. The subjects were usually taking BZs for treatment of an anxiety-related disorder: diazepam ($n=17$), lorazepam (17), alprazolam (10), clorazepate (7), chlordiazepoxide (1), oxazepam (1), or flurazepam (1). The average duration of use was 58 months. In nearly every case, the daily intake of BZ medications was within the range acceptable for normal therapeutic use.

The subjects were first tested at their initial interview. They reported to the clinic without any special instructions regarding their drug intake, except not to take their drug immediately before their appointment. In most cases, then, behavioral testing occurred between 4 and 14 h after the subject's last dose of medication, with the last dose generally being taken either the night before or earlier in the day of the interview. Since most of the subjects were being treated for anxiety-related disorders, their performance was compared with that of a group of drug-free anxious subjects who were matched for sex, age, and educational level with the group of chronic BZ users.

When tested at their initial interview, the chronic BZ users did not differ from the drug-free anxious subjects in verbal recall ability. Immediate recall, delayed recall, and memory decay were not significantly altered compared with the control group (LUCKI et al. 1986; LUCKI and RICKELS 1986). Digit span was also not altered in chronic BZ users (LUCKI and RICKELS 1986). In addition, the chronic BZ users did not show altered performance on a variety of psychomotor tests, including digit-symbol substitution performance, letter cancellation, choice reaction time, or tapping. Chronic BZ users did show reduced critical flicker fusion thresholds and reported greater subjective feelings of tranquilization than the drug-free anxious subjects, when tested at their initial interview.

Although verbal recall ability appeared unaltered in chronic BZ users, the tests were administered a considerable time after most subjects had taken their previous dose of medication. Of the chronic BZ users, 20 were able to return to the clinic so that verbal recall ability could be examined shortly after administration of their usual dose of medication. Some of these results have been reported previously (LUCKI et al. 1986; LUCKI and RICKELS 1986).

3.1 Subjects and Procedure

Of the chronic BZ users, 20 returned to the clinic about 3 weeks after their initial visit to assess the acute effect of their BZ medications. Of this subset of subjects, 8 were taking diazepam, 7 lorazepam, 3 clorazepate, 1 alprazolam, and 1 chlordiazepoxide. They were instructed to eat only a light breakfast (fruit juice and no caffeinated beverages) and to refrain from taking their medication immediately prior to the appointment. Upon their arrival, the subjects were tested on a battery of tests, including free verbal recall as previously described. The subjects then took an assigned dose of their medication and were retested 60–90 min later. The dose administration to these subjects was assigned according to their regular medication: 5 mg diazepam, 1 mg lorazepam, 7.5 mg clorazepate, 0.5 mg alprazolam, or 10 mg chlordiazepoxide. These are generally considered therapeutically equivalent doses.

3.2 Results

The effect of acute administration of BZs on immediate and delayed recall in chronic BZ users is shown in Figs. 3, 4, respectively. Acute administration of BZs did not alter significantly immediate recall of the 16-word list of unrelated nouns (baseline $= 8.7 \pm 0.5$ words (mean $\pm$ SEM); postdrug $= 8.1 \pm 0.5$ words; $p > 0.05$).

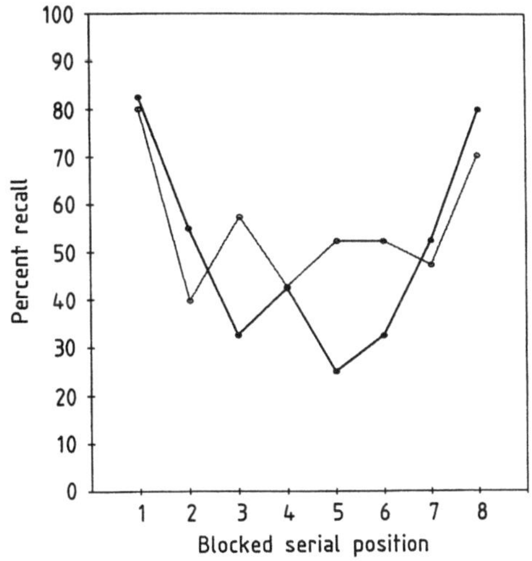

Fig. 3. Serial position curves for immediate recall by chronic BZ users ($n=20$) after acute administration of their medication. The chronic users were first tested for recall ability at their baseline interview, 4–14 h after last taking their medication (○). These values were then compared with values obtained at a second interview 60–90 min after the subjects took their medication (●). Values represent the mean percentage recall frequency of words for subjects as a function of the position within the list. The two serial position curves for immediate recall did not differ significantly according to analysis of variance

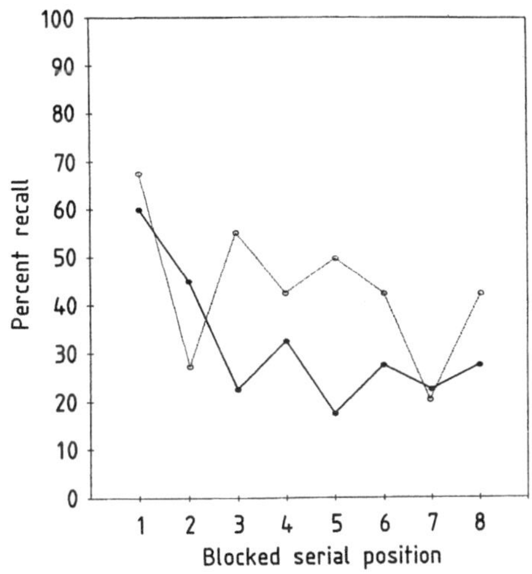

Fig. 4. Serial position curves for delayed recall by chronic BZ users after acute administration of their medication. Delayed recall was assessed approximately 20 min after the word list was read to the subjects. Values represent the mean percentage recall frequency of words for subjects as a function of the position within the list. The overall serial position curve for delayed recall differed significantly after medication (●) when compared with baseline values (○), according to analysis of variance ($p<0.01$)

In agreement, the serial position curve for immediate recall did not change significantly for chronic BZ users shortly after taking their medication, according to analysis of variance (Fig. 3). In contrast, acute administration of BZs significantly impaired delayed recall of the word list (baseline $=6.6\pm0.5$ words; postdrug $=5.0\pm0.5$ words; $p<0.05$ according to Student's t test for related comparisons). The serial position curve for delayed recall differed significantly after medication, according to analysis of variance with repeated measures for both drug and serial position ($p<0.01$). Figure 4 shows that acute BZ administration tended to reduce the recall of words that occurred at the middle and at the end of the word list (except for position 7) when compared with the performance at baseline. In addition,

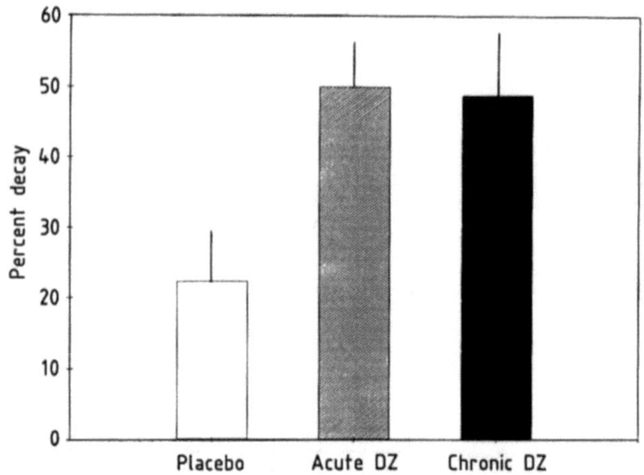

Fig. 5. The effect of acute administration of 5 mg diazepam on the delayed recall ability of chronic diazepam users ($n=8$). Chronic diazepam users (*Chronic DZ*) were examined for both immediate and delayed verbal recall 70 min after administration of their medication. These values were compared with drug-free anxious subjects who were given either placebo or 5 mg diazepam 70 min prior to testing (*Acute DZ*). The percentage memory decay is defined as the percentage of words that were remembered immediately that could no longer be recalled after the 20 min delay (see Sect. 2.1). The vertical bars represent the mean percentage memory decay with the vertical lines indicating 1 SEM. Diazepam (5 mg) significantly increased the memory decay for both drug-free and chronic diazepam users when compared with placebo ($p<0.01$)

acute BZ administration significantly increased the memory decay of the chronic BZ users (baseline $=26.1 \pm 3.6\%$; postdrug $=41.1 \pm 4.4\%$; $p<0.05$).

Figure 5 presents the effect on memory decay of acute administration of 5 mg diazepam to eight chronic diazepam users. The intake of the eight diazepam users averaged 11 mg/day, and the average duration of use was 103 months. These results are compared with the acute effect of 5 mg diazepam or placebo given to drug-free anxious subjects that were described in an earlier section. Acute administration of 5 mg diazepam increased memory decay significantly in both drug-free and chronic diazepam users when compared with placebo, according to analysis of variance ($p<0.01$). The memory decay did not differ significantly between the two groups of subjects given diazepam ($p>0.05$).

4 Discussion

The present results confirmed the amnesic action of diazepam when administered initially to anxious subjects. Impairment of delayed recall and increased memory decay were caused by a dose of 5 mg diazepam that is likely to be used in medical treatment. The effects of this low dose of diazepam appeared to be somewhat selective. Diazepam did not alter immediate recall or digit span in anxious subjects. Similarly, Ghoneim et al. (1984a) reported that comparable doses of diazepam impaired verbal recall in normal subjects. This dose of diazepam was lower than that required to impair psychomotor performance (GHONEIM et al. 1984b), making diazepam's cognitive effects among the most sensitive of its actions.

Diazepam (5 mg) particularly appeared to impair the recall of words presented at the end of the list. Words positioned at the beginning of the list represent recall from long-term memory storage, whereas words at the end of the list are thought to be recalled from recent memory (ATKINSON and SCHIFFRIN 1968). Low doses of diazepam may accelerate the loss of information from recent memory, and this may account for at least some of the amnesic effects of BZ medications. These results are not entirely consistent with a previous study by MEWALDT et al. (1983), who showed higher doses of diazepam impaired the recall of words from all positions of the list. These investigators suggested that BZs impair memory by inhibiting the transfer of information from short-term to long-term storage. However, several differences in procedure, subjects, and dose of diazepam used between the studies could have contributed to the different effects found. Different mechanisms could contribute to diazepam's amnesic effects at various doses. In contrast to diazepam, neither 5 nor 10 mg of the non-BZ anxiolytic buspirone altered immediate recall, delayed recall, or digit span in anxious subjects. Thus, this non-BZ anxiolytic drug lacks the memory-impairing side effects demonstrated by most BZ medications.

In chronic BZ users, verbal recall was not impaired at their initial examination, which occurred many hours following their last dose, despite persistently high plasma BZ concentrations. However, delayed recall, but not immediate recall, was impaired when measured shortly after the chronic users took their medication. Thus, memory appeared to be impaired in chronic users for only a brief time following ingestion of their medication. This may also explain why anxious subjects who took diazepam for 7 days did not show significant memory impairment when tested 3 h (instead of 1 h) after their last dose. In general, it appears that the memory-impairing effect of 5 mg diazepam only lasts for a few hours following oral administration (GHONEIM et al. 1984a; KOTHARY et al. 1981). The lack of correlation between the time course of amnesic effects and the pharmacokinetic disposition of BZ medications supports previous suggestions that pharmacodynamic interactions at central BZ receptors may be involved in determining the time course of their behavioral effects (ELLINWOOD et al. 1985). Importantly, clinical studies involving patients taking BZ medications should attend to the time following the last dose as a critical experimental variable.

GHONEIM et al. (1981) suggested that some tolerance developed to diazepam's memory impairment in normal subjects after 3 weeks of continuous administration. However, the ability of acute BZ administration to impair memory in chronic users indicates that complete tolerance to this effect does not occur. Moreover, in the present study, eight chronic (for nearly 9 years) users of diazepam did not show evidence of tolerance to diazepam's amnesic effects when compared with anxious subjects receiving diazepam for the first time.

The effect of BZ medications on memory has implications for the safety of their long-term clinical use, especially since complete tolerance failed to develop in chronic users. Because the memory-impairing effects of BZs would be limited only to new information presented during critical periods of susceptibility shortly after drug ingestion, it is not clear whether patients would suffer from or be aware of any changes in their memory. Moreover, BZs sometimes improve retention of verbal material that was presented prior to drug administration (HINRICHS et al. 1984). On the other hand, many patients take BZ medications several times daily. Some may be more susceptible to BZs' amnesic effects when working at certain occupations. Certain kinds of patients, such as cardiac patients or the elderly, appear to be more susceptible to BZs' amnesic effects (FRAZURE-SMITH and ROLICZ-WOLOSZYK 1982; SALZMAN et al. 1975). It is important to evaluate the *clinical* significance of BZs' amnesic effects by verifying how taking BZ medications alters functioning in real-life settings. One could then properly evaluate the risks of BZs' amnesic effects within the perspective of the need for medical treatment.

Acknowledgements. This research was supported by USPHS grants MH 08957 and DA 05186 and by funds from Mead Johnson Pharmaceutical Division, Bristol Myers Co.

References

Angus WR, Romney DM (1984) The effect of diazepam on patient's memory. J Clin Psychopharmacol 4:203–206

Aranko K, Mattila MJ, Seppala T (1983) Development of tolerance and cross-tolerance to the psychomotor actions of lorazepam and diazepam in man. Br J Clin Pharmacol 15:545–552

Atkinson RC, Shiffrin RM (1968) The control of short-term memory. Sci Am 225:82–90

Barnett DB, Taylor-Davies A, Desai N (1981) Differential effect of diazepam on short term memory in subjects with high or low level anxiety. Br J Clin Pharmacol 11:411–412

Bixler EO, Scharf MB, Soldatos CR, Mitsky DJ, Kales A (1979) Effects of hypnotic drugs on memory. Life Sci 25:1379–1385

Block RI, Berchou R (1984) Alprazolam and lorazepam effects on memory acquisition and retrieval processes. Pharmacol Biochem Behav 20:233–241

Brown J, Lewis V, Brown MW, Horn G, Bowes JB (1978) Amnesic effects of intravenous diazepam and lorazepam. Experientia 34:501–502

Clarke PRF, Eccersley PS, Frisby JP, Thornton JA (1970) The amnesic effect of diazepam (Valium). Br J Anaesthesia 42:690–697

Ellinwood EH Jr, Heatherly DG, Nikaido AM (1985) Comparative pharmacokinetics and pharmacodynamics of lorazepam, alprazolam and diazepam. Psychopharmacology 86:392–399

Frazure-Smith N, Rolicz-Woloszyk E (1982) Memory problems after ischemic heart disease episodes: effects of stress, benzodiazepines and smoking. J Psychosom Res 26:613–622

Ghoneim MM, Mewaldt SP (1975) Studies on human memory: the interactions of diazepam, scopolamine, and physostigmine. Psychopharmacology 52:1–6

Ghoneim MM, Mewaldt SP, Berie JL, Hinrichs JV (1981) Memory and performance effects of single and 3-week administration of diazepam. Psychopharmacology 73:147–151

Ghoneim MM, Hinrichs JV, Mewaldt SP (1984a) Dose-response analysis of the behavioral effects of diazepam: I. Learning and memory. Psychopharmacology 82:291–295

Ghoneim MM, Mewaldt SP, Hinrichs JV (1984b) Dose-response analysis of the behavioral effects of diazepam: II. Psychomotor performance, cognition and mood. Psychopharmacology 82:296–300

Ghoneim MM, Hinrichs JV, Noyes R, Anderson DJ (1984c) Behavioral effects of diazepam and propranolol in patients with panic disorder and agoraphobia. Neuropsychobiology 11:229–235

Goldberg HL, Finnerty RJ (1979) The comparative efficacy of buspirone and diazepam in the treatment of anxiety. Am J Psychiatry 136:1184–1187

Hartley LR, Spencer J, Williamson J (1982) Anxiety, diazepam and retrieval from semantic memory. Psychopharmacology 76:291–295

Hinrichs JV, Mewaldt SP, Ghoneim MM, Berie JL (1982) Diazepam and learning: Assessment of acquisition deficits. Pharmacol Biochem Behav 17:165–170
Hinrichs JV, Ghoneim MM, Mewaldt SP (1984) Diazepam and memory: retrograde facilitation produced by interference reduction. Psychopharmacology 84:158–162
Jones DM, Lewis MJ, Spriggs TLB (1978) The effects of low doses of diazepam on human performance in group administered tasks. Br J Clin Pharmacol 6:333–337
Kothary SP, Brown ACD, Pandit UA, Samara SK, Pandit SK (1981) Time course of antirecall effect of diazepam and lorazepam following oral administration. Anesthesiology 55:641–644
Liljequist R, Palva E, Linnoila M (1979) Effects of learning and memory of 2-week treatments with chlordiazepoxide lactam, *N*-desmethyldiazepam, oxazepam and methyloxazepam, along or in combination with alcohol. Int Pharmacopsychiatry 14:190–198
Lister RG (1985) The amnesic action of benzodiazepines in man. Neurosci Biobehav Rev 9:87–94
Lister RG, File SE (1984) The nature of lorazepam-induced amnesia. Psychopharmacology 83:183–187
Lucki I, Rickels K (1986) The behavioral effects of benzodiazepines following long-term use. Psychopharmacol Bull 22:424–433
Lucki I, Rickels K, Geller AM (1986) Chronic use of benzodiazepines and psychomotor and cognitive test performance. Psychopharmacology 88:426–433
Lucki I, Rickels K, Giesecke MA, Geller AM (1987) Differential effects of the anxiolytic drugs diazepam and buspirone on memory function. Br J Clin Pharmacol 23:207–211
Mewaldt SP, Hinrichs JV, Ghoneim MM (1983) Diazepam and memory: support for a duplex model of memory. Memory Cognition 11:557–564
Mewaldt SP, Ghoneim MM, Hinrichs JV (1986) The behavioral actions of diazepam and oxazepam are similar. Psychopharmacology 88:165–171
Paivio A, Yuille JC, Madigan SA (1968) Concreteness, imagery and meaningfulness values for 925 nouns. J Exp Psychol (Suppl) 76:1–25
Petersen RC, Ghoneim MM (1980) Diazepam and human memory: influence on acquisition, retrieval, and state-dependent learning. Prog Neuropsychopharmacol 4:81–89
Richardson JTE, Frith CD, Scott E, Crow TJ, Cunningham-Owens D (1984) The effects of intravenous diazepam and hyoscine upon recognition memory. Behav Brain Res 14:193–199
Rickels K, Wiseman K, Norstad N, Singer M, Stoltz D, Brown A, Danton J (1982) Buspirone and diazepam in anxiety: a controlled study. J Clin Psychiatry 43:81–86
Salzman C, Shader RI, Harmantz J, Robertson L (1975) Psychopharmacologic investigations in elderly volunteers: effect of diazepam in males. J Am Geriatr Soc 23:451–457
Skolnick P, Paul SM, Weissman BA (1984) Preclinical pharmacology of buspirone hydrochloride. Pharmacotherapy 4:308–314
Thorndike EL, Lorge L (1944) The teachers word book of 30,000 words. Columbia University Press, New York
Wechsler D (1955) Manual for the Wechsler Adult Intelligence Scale. Psychological Corp., New York

Sleep and Memory

T. Roth[1], T. Roehrs, A. Zwyghuizen-Doorenbos,
E. Stepanski and R. Wittig

Abstract

Generally sleep is considered a time of amnesia. It is not uncommon for an individual to experience 8 h of sleep and have no memory for events during that time. Similarly, a substantial proportion of the population has no memory for dreams that occurred during the night, despite the fact that the literature on awakening during rapid eye movement (REM) sleep clearly shows that individuals normally have four to six "dream experiences" a night. Research on this issue seems to indicate that the lack of memory cannot be explained by the organisms' inability to perceive stimuli. The data indicate that although perceptual thresholds are elevated, organisms can clearly perceive stimuli, and, in fact, can discriminate between them during sleep. The amnesia also cannot be explained by a defect in long-term memory, as studies have indicated that stimuli put into the memory during wakefulness are more efficiently retrieved after a sleep period than after a comparable period of wakefulness. The most likely explanation for the amnestic property of sleep seems to be the inability of organisms to transfer information from short-term memory to long-term memory during sleep. There are several sources of evidence to support this hypothesis. First, the probability of remembering a stimulus given during wakefulness is related to the proximity of sleep onset to the stimulus. Generally, information put into the system within 5 min of sleep onset is lost from memory. Secondly, disorders of excessive daytime somnolence which cause individuals to have frequent microsleeps are often associated with complaints of memory problems. When these patients are kept fully alert, or if their condition is reversed, they show perfectly normal memory functioning. Finally, anterograde amnesia is often associated with drugs that produce sleepiness. A good example of such a class of drugs is the benzodiazepines. The effect of benzodiazepines on anterograde amnesia is dose dependent and related to their hypnotic activity. Generally, when the drug produces sleep onset within 5 min of the stimulus it is likely to produce amnesia for the event. In summary, the process of memory consolidation is inhibited during sleep. This factor must be considered in all studies evaluating the *direct* effect of benzodiazepines on memory processes.

It is generally recognized that sleep is a time of amnesia. It is not uncommon for an individual to experience 8 h of sleep and have no memory of any events which occurred during that time. It is similarly not uncommon for an individual to actually awaken during the middle of the night, perform a brief activity, immediately go back to sleep and have no memory of awakening. Two important discoveries of modern sleep research attest to the amnesic properties of sleep. The first of these phenomena is the high rate of dream recall when subjects are woken during rapid eye movement (REM) sleep relative to morning recall of dreams. With the advent of sleep laboratory based dream research it became clear that most individuals experience four to six REM periods a night and awakenings

[1] Henry Ford Hospital, Sleep Disorders and Research Center, W Grand Boulevard, Detroit, MI 48202, USA.

from these REM periods result in an 80% rate of dream recall (GOODENOUGH 1978). In addition, it has been shown that non-REM (NREM) awakenings also result in dream recall, albeit less frequently (FOULKES and VOGEL 1974). This high rate of nightly dream experiences contrasts strikingly with the observation that the average individual recalls only a single dream about once every couple of days (WEBB and KERSEY 1967). The second discovery which attests to the amnesic properties of sleep relates to the subjective evaluation of sleep in insomnia patients in comparison with objective laboratory measures. It is a well established fact that patients with insomnia complaints overestimate the degree of sleep disturbance they experience. These patients generally overestimate their sleep latency and underestimate their total sleep time (CARSKADON et al. 1976). However, these same patients also underestimate the number of nocturnal awakenings. Underestimation of nocturnal awakening can be explained by the fact that these individuals have no memory of brief awakenings. Only when they are awake for an extended period of time (greater than 3 min) do they have a memory of the fact that they woke up (HAURI, personal communication).

The observation of an inherent inability to acquire new information during sleep was studied directly by individuals interested in sleep learning. In the classic study of sleep learning, subjects were presented with information throughout the night. They recalled items presented while awake but could not recall items presented while there were EEG signs of sleep (SIMON and EMMONS 1956). In the same study it was found that some recall did occur for items presented during the transition from wakefulness to sleep. However, recall for these items was poorer than for those presented while the subject was awake. Thus, it can be concluded that learning is not possible during sleep and is impaired during the transition to sleep.

The question arises as to what aspect of memory is inhibited during sleep. Using a simplistic model of memory, three possibilities arise. First, memory is impaired because there is a failure of stimulus registration during sleep. Secondly, memory is not possible because of an inhibition of memory consolidation. Finally, there is the possibility that long-term memory is somehow impaired during sleep.

Failure of stimulus registration does not seem to be a possible explanation for the memory impairment. Research studies have clearly shown that although auditory thresholds are elevated during sleep compared to wakefulness, subjects do perceive stimuli (WILLIAMS et al. 1966). In fact, auditory stimuli presented during sleep can evoke responses which were previously learned when subjects were awake. It also seems clear that individuals can discriminate between (i.e., differentially respond to) stimuli presented during sleep (WILSON and ZUNG 1966). Finally, sedative hypnotics administered before sleep further raise the arousal threshold but they do not prevent stimulus registration.

The amnesic properties of sleep also cannot be explained by defects in long-term memory. In fact, early research on this issue indicated that memory is better over periods of sleep than over comparable periods of waking activity (JENKINS and DALLENBACH 1924). More recent research has confirmed these findings and has shown that recall is best after periods of NREM sleep, worst after periods of wakefulness, and intermediate after REM sleep (YAROUSH et al. 1971). Although there is controversy regarding the explanation of these findings, there is no doubt that long-term memory is not impaired during sleep.

The most likely explanation for memory dysfunction during sleep is that sleep per se inhibits memory consolidation. This view argues that memory consolidation is an active process and the process is inhibited during the sleep state. There are several lines of evidence which support this position.

The dream recall literature provides support for the memory consolidation hypothesis. The recall of REM-period dreams is greatly dependent on the proximity of the awakening to the dream experience (DEMENT and KLEITMAN 1957). While recall of a dream is highly probable if the awakening is made in REM sleep, delaying the awakening by several minutes into NREM sleep greatly reduces the probability of dream recall (BAKELAND and LASKY 1968). Generally, it is concluded that dreams are never recalled unless the dream experience is interrupted by an awakening. Further, the morning recall of the nighttime dream reports is dependent on the duration of the nighttime awakening. The longer the period of wakefulness the greater the likelihood that nighttime dream recall will also be remembered in the morning. In summary, the dream literature indicates that memory decay is very rapid during sleep and relatively long periods of wakefulness are necessary to produce permanent memories.

Similarly, the literature on the proximity of stimulus input to sleep onset supports the memory consolidation hypothesis. The probability of remembering a stimulus presented during wakefulness decreases as the proximity of the stimulus to sleep onset becomes greater. In a study where stimuli were presented continuously to subjects lying in bed it was found that information presented to the subject within 5 min of sleep onset was not recalled later, while information presented earlier was recalled (GUILLEMINAULT and DEMENT 1977). In another study subjects were awakened several times during the night and presented verbal stimuli for recall. The results showed that when subjects spontaneously remained awake for a long period of time after a word was presented they were more likely to remember that word the next morning than when they returned to sleep more quickly (GOODENOUGH et al. 1971). Again, these data indicate that a relatively prolonged period of wakefulness is necessary for memory consolidation to take place.

On the clinical side specific sleep disorders which are associated with pathological sleepiness and hence frequent microsleeps during "wakefulness" have memory problems associated with them (MERLOTTI et al. 1987). In fact, many patients with sleep apnea or narcolepsy include memory problems among their symptom picture. When these patients are kept fully alert during neuropsychological testing of memory, they turn out to have normal memory functions. Similarly, clinical reversal of the daytime sleepiness in these patients also reverses the memory problems.

Finally, there is evidence from the literature on benzodiazepine-induced amnesia which suggests that sleep and memory are related. The observation that benzodiazepines produce amnesia emerged from reports of their clinical use as presurgery medications. While the initial reports involved intravenous diazepam and were anecdotal in nature, subsequent studies have demonstrated that amnesia is a characteristic of all the benzodiazepines, with the magnitude of the effect being a function of route of administration, dose, and the pharmacokinetics of the particular drug (ROTH et al. 1984).

A typical study involves the administration of the drug, presentation of stimuli for recall, followed by an immediate and possibly delayed recall of that information. In the delayed recall studies sleep often occurs (if it is a nighttime study) during the period between presentation of the stimuli and their subsequent recall, while in daytime studies subjects are normally awake. Most of the nighttime studies make an objective determination of sleep, but the daytime studies include, at best, a subjective rating of sedation. Clinically, the amnesia has been described as the anterograde effect of benzodiazepines. However, one might argue that, in fact, it is the retrograde effect of sleep (or sleepiness) produced by the drug that is being observed (ROTH et al. 1984).

There is strong evidence that the amnesia observed in the delayed recall paradigm is associated with a hastened sleep onset. It is hypothesized that the hastened sleep onset disrupts consolidation of information for long-term memory. Some of the evidence is indirect. First, the amnesia is systematically related to dose, as are hypnotic effects. In a study using nighttime administration, recall in the morning for a memory task presented during an awakening 3 h after drug administration was significantly better after 0.25 mg triazolam than after 0.50 mg (ROEHRS et al. 1983), and both doses produced poorer recall than placebo. Secondly, as hypnotic effects increase after repeated administration of long-acting benzodiazepines, so do amnesic effects (ROEHRS et al. 1983). Morning recall after one night when flurazepam 30 mg had been administered, was similar to that after 0.25 mg triazolam and different from that after 0.50 mg triazolam; after six nights of flurazepam treatment, recall was reduced to the same level as after 0.50 mg triazolam and was different from that after 0.25 mg triazolam. Furthermore, other drugs, nonbenzodiazepines, which produce hypnotic effects such as secobarbital also produce amnesia. Morning recall after 200 mg secobarbital was similar to that after 0.25 mg triazolam or 4 mg lorazepam and in all cases was recall poorer than after placebo (ROEHRS et al. 1983).

More directly, there is a correlation between the latency of falling asleep after the nighttime awakening and recall in the morning. This suggests that hastened sleep onset disrupts memory consolidation and hence morning recall. In one study the correlation among subjects between morning recall and latency to going back to sleep was 0.74 (ROTH et al. 1980). In another study the correlation among drug conditions (different drugs and doses) between mean sleep latency and mean morning recall was 0.86 (ROEHRS et al. 1984). In these studies amnesia was usually associated with latencies of 7 min or less and normal recall with latencies of more than 10 min. Consequently, in one study subjects who had received hypnotics were made to remain awake for 15 min after the material to be remembered was presented (ROEHRS et al. 1983), and the amnesic effect was removed. Thus, in part, the amnesia associated with benzodiazepines is a retrograde effect of sleep which disrupts long-term memory consolidation.

Whether the *immediate* amnesia associated with these drugs is due to the sleep (or sleepiness) they produce, remains controversial. It is argued that memory impairments are observed both in tests of immediate recall when there has obviously been no sleep and also when subjects have remained awake throughout the test session before the delayed recall (Lister 1985). However, although the subjects remain awake, there is no question that they are sleepy, both at the time of initial acquisition and over the retention interval before the delayed recall. After all, these drugs produce sleepiness. Sleep latency has become a well documented and validated objective measure of sleepiness (Roth et al. 1982). The studies discussed above all show decreased sleep latencies associated with the various hypnotics, implying an increased sleepiness. Thus, the question arises as to whether the immediate memory impairments observed are correlated with sleep latency. None of the earlier studies included that assessment.

We have analyzed data collected in several different experiments to assess the relation of immediate recall to subsequent sleep latency. Each study included an assessment of immediate recall on the Henry Ford Hospital memory task, described in earlier publications, and the measurement of sleep latency within 10 min of completing the task (Roth et al. 1980). One study was conducted at night (Roehrs et al. 1984), and the other during the day (Zwyghuizen-Doorenbos et al. 1987), and together they included various drug treatments producing different sleep latencies (flurazepam 30 mg, temazepam 30 mg, lormetazepam 1.5 mg, triazolam 0.50 mg, triazolam 0.50 mg plus 4 mg caffeine/kg, triazolam 0.50 mg plus 8 mg caffeine/kg, and two placebo conditions). Mean immediate recall for each drug condition was correlated to mean sleep latency for that condition ($r=0.86$, $p<0.001$). Thus, we would argue that the immediate amnesia associated with benzodiazepines is the result of the sleepiness they produce and if recipients of the drugs are allowed to sleep, subsequent recall (delayed) will be further impaired.

In summary, the various areas of research all seem to indicate that sleep or sleepiness inhibit the consolidation of memory. Therefore, investigations of the effects of any factor on memory have also to delineate the effect of that factor on sleep and sleepiness. Only in this way can the direct effects of the factor under investigation be separated from its indirect (i.e., secondary to sleep or sleepiness) effect on memory.

References

Baekeland F, Lasky R (1968) The morning recall of rapid eye movement period reports given earlier in the night. J Nerv Ment Dis 147:570–579

Carskadon MA, Dement WC, Mitler M, Guilleminault C, Zarcone VP, Spiegel R (1976) Self-reports versus sleep laboratory findings in 122 drug-free subjects with complaints of chronic insomnia. Am J Psychiatry 133:12

Dement WC, Kleitman N (1957) The relation of eye movements during sleep to dream activity: an objective method for the study of dreaming. J Exp Psychol 53:339–346

Foulkes D, Vogel G (1974) The current status of laboratory dream research. Psychiatr Ann 7:7–23

Goodenough DR (1978) Dream recall: history and current status of the field. In: Arkin A, Antrobus J, Ellman S (eds) The mind in sleep: psychology and psychophysiology. Wiley, Chichester

Goodenough DR, Sapan J, Cohen H, Portnoff G, Shapiro A (1971) Some experiments concerning the effects of sleep on memory. Psychophysiology 8:749–762

Guilleminault C, Dement WC (1977) Amnesia and disorders of excessive daytime sleepiness. In: Drucker-Colin RR, McGaugh JL (eds) Neurobiology of sleep and memory. Academic, London

Jenkins JG, Dallenbach KM (1924) Obliviscence during sleeping and waking. Am J Psychol 35:605–612

Lister RG (1985) The amnesic action of benzodiazepines in man. Neurosci Biobehav Rev 9:1–8

Merlotti L, Roehrs T, Young D, Zorick F, Fortier J, Roth T (1987) Symptom patterns of narcolepsy patients: a questionnaire study. Sleep Res 16:391

Roehrs T, Zorick F, Sicklesteel J, Wittig R, Hartse K, Roth T (1983) Effects of hypnotics on memory. J Clin Psychopharmacol 3:310–313

Roehrs T, McLenaghan A, Koshorek G, Zorick F, Roth T (1984) Amnesic effects of lormetazepam. In: Hindmarch I, Ott H, Roth T (eds) Sleep, benzodiazepines and performance. Springer, Berlin Heidelberg New York, pp 165–172

Roth T, Hartse KM, Saab PG, Piccione PM, Kramer M (1980) The effects of flurazepam, lorazepam, and triazolam on sleep and memory. Psychopharmacology 70:231–237

Roth T, Roehrs T, Wittig R, Zorick F (1984) Benzodiazepines and memory. Br J Clin Pharmacol 18:45S-49S

Roth T, Roehrs T, Zorick F (1982) Sleepiness: its measurement and determinants. Sleep 5 (Suppl 2):S128–S134

Simon CW, Emmons WH (1956) Responses to material presented during various levels of sleep. J Exp Psychol 51:89–97

Webb WB, Kersey J (1967) Recall of dreams and the probability of stage 1-REM sleep. Percept Mot Skills 24:627–630

Williams HL, Morlock HC Jr, Morlock JV (1966) Instrumental bchavior during sleep. Psychophysiology 2:208–216

Wilson WP, Zung WWK (1966) Attention, discrimination, and arousal during sleep. Arch Gen Psychiatry 15:523–528

Yaroush R, Sullivan JJ, Ekstrand BR (1971) Effect of sleep on memory. II. Differential effect of the first and second half of the night. J Exp Psychol 88:361–366

Zwyghuizen-Doorenbos A, Roehrs T, Smith D, Tietz E, Roth T (1987) Caffeine's reversal of triazolam-induced impairment of waking function. Sleep Res 16:161

Benzodiazepine-Induced Amnesia and Anaesthetic Practice: A Review

C. A. O'BOYLE[1]

Abstract

Anaesthetic practice is the only clinical context in which amnesia is a valued property of benzodiazepine drugs, since decreased recall considerably enhances patient tolerance and acceptance or surgical and diagnostic procedures. Research on the amnesic effects of diazepam, midazolam, lorazepam and flunitrazepam, administered via oral, i.v. or i.m. routes to patients undergoing surgical or diagnostic procedures is reviewed. The degree of anterograde amnesia is a function of the drug, the route of administration and the population of patients being assessed. Retrograde amnesia has not been conclusively demonstrated. Amnesia is more profound for cutaenous-tactile and auditory than for visual stimuli, but actual surgical events, or emotionally laden material, are more likely to be recalled than artificial stimuli. Evidence that the benzodiazepines prevent affective and cognitive processing under general anaesthesia and decrease traumatic postoperative recall of intra-operative events is reviewed. The explanatory value of modern theories of memory for research on benzodiazepine-induced amnesia, and the research potential of the surgical setting are outlined. The development of non-sedative anxiolytics and specific benzodiazepine antagonists provides the tools for assessing the contribution of sedative and anxiolytic properties of drugs to their amnesic effects.

1 Introduction

Surgical and diagnostic procedures can be extremely stressful both physically and psychologically (WILSON-BARNETT 1979; JOHNSTON 1980; VOLICER et al. 1977). The administration of benzodiazepine drugs has become the preferred technique for the management of patients in such situations (KANTO 1981; KANTO and KLOTZ 1982; PHILIP 1985). Central to the popularity of the benzodiazepines, in this context, is their ability to cause anterograde amnesia (KANTO 1981). Although amnesia is considered an unwanted and potentially serious side effect when these agents are used as anxiolytics or hypnotics (SHADER and GREENBLATT 1983; ESSMAN 1983), it is a very useful, indeed necessary, property in anaesthetic practice (DUNDEE 1979; KANTO 1981). This is because amnesia may considerably enhance patient tolerance and acceptance or surgical, dental and diagnostic procedures, particularly among those who are severely apprehensive or who must undergo multiple procedures.

The purpose of this review is to discuss the amnesic properties of benzodiazepine drugs specifically in relation to their use in anaesthesia. The review is limited

[1] Department of Psychology, Royal College of Surgeons in Ireland, St. Stephen's Green, Dublin 2, Ireland.

mainly to those studies in which the amnesic properties of diazepam, lorazepam, flunitrazepam or midazolam, administered alone or in combination with local anaesthetics or small doses of atropine, have been assessed in patients undergoing surgical, dental or endoscopic procedures. Laboratory studies of the effects of benzodiazepines on psychomotor performance and memory in healthy volunteers are not included, unless directly relevant to the anaesthetic situation, since these have been widely reviewed elsewhere (JANKE and DEBUS 1968; BERGER and POTTERFIELD 1969; MCNAIR 1973; KLEINKNECHT and DONALDSON 1975; WITTENBORN 1979; VOGEL 1979; HINDMARCH 1981; MURRAY 1984; LISTER 1985).

2 General Indications for Benzodiazepine Use in Anaesthesia

Benzodiazepines have a wide range of applications in anaesthetic practice (DUNDEE and HASLETT 1970; GREENBLATT and SHADER 1974; DUNDEE 1979; KANTO 1981; KANTO and KLOTZ 1982; PHILIP 1985). First, since anxiety about surgery can begin days or weeks before the scheduled procedure, benzodiazepines may be administered for their anxiolytic effects. Secondly, while some procedures such as endoscopy and minor surgery can be performed on an outpatient basis, patients must be admitted to hospital prior to the day of the operation when more major surgery is contemplated. Hypnotic medication is usually required to produce sleep in such patients, particularly on the night prior to surgery. The third major indication for benzodiazepines is as premedicants, normally given on the day of surgery, before transport to the operating theatre. The aims of premedication are to establish mental and emotional relaxation, reduce sensory input and metabolic rate, prevent autonomic activation and to ease the induction of anaesthesia (COLINS 1976). Despite heavy sedation on arrival at theatre, few patients will willingly accept application of the face mask for administration of a volatile general anaesthetic and, if endotracheal intubation is contemplated, adequate sedation is a prerequisite. Short-acting sedatives, including benzodiazepines, are therefore usually given to render the patient unconscious during induction of anaesthesia. Finally, benzodiazepines play a role in the postoperative period in the management of postanaesthetic agitation and also as hypnotics. In addition to their role as adjuncts to general anaesthesia, these drugs are increasingly widely used as supplements to local anaesthesia for outpatient ambulatory dental and endoscopic procedures (PHILIP 1985).

3 Benzodiazepine-Induced Amnesia in Anaesthetic Practice

The amnesic properties of benzodiazepine drugs were discovered soon after their introduction into anaesthetic practice (BRANDT and OAKES 1965; HASLETT and DUNDEE 1968; BROWN et al. 1968; ROGERS et al. 1965). In a placebo-controlled laboratory study, CLARKE and colleagues (1970) subsequently demonstrated that the that i.v. diazepam caused dense anterograde amnesia for approximately

10 min after administration, followed by a period of better but still impaired memory function. Amnesia was not accompanied by a marked decrease in the level of consciousness. Their conclusion that the impairment in memory resulted from a specific effect of the drug on the consolidation process, rather than a direct effect on registration, storage or retrieval of information, has found widespread support (BROWN and LEWIS 1982; CLARK et al. 1979; GHONEIM and MEWALDT 1975, 1977; O'BOYLE et al. 1983; GHONEIM et al. 1984; LISTER 1985). It is generally agreed that the benzodiazepines, administered alone, do not cause retrograde amnesia but may facilitate retrieval of information learned prior to drug administration (HINRICHS et al. 1984).

Research on the amnesic effects of benzodiazepines in anaesthesia has proceeded relatively independently of laboratory-based research in normal healthy volunteers. There are a number of reasons for this. First, amnesia is a desired main effect in the anaesthetic context, as opposed to an undesired side effect in the anxiolytic, hypnotic and anticonvulsant contexts. Secondly, whereas most laboratory studies have been conducted on young healthy volunteers, benzodiazepines are administered to patients of all ages suffering from a wide variety of illnesses. Thirdly, there has been considerably greater emphasis on the effect of different routes of administration on drug response. Finally, whereas memory stimuli in laboratory studies often have had little meaning or relevance for the subject, amnesia for painful and traumatic stimuli has been studied in patients undergoing surgical and diagnostic procedures. However, because of the limitations imposed by the setting, clinical studies, although ecologically more valid, have utilized less sophisticated methodology.

4 Amnesic Properties of Individual Benzodiazepines

4.1 Diazepam

The anterograde amnesic properties of diazepam were discovered soon after its introduction into anaesthetic practice (BRANDT and OAKES 1965; KAHLER et al. 1967; HASLETT and DUNDEE 1968) and have been demonstrated by numerous authors (Table 1). In controlled volunteer studies, CLARKE et al. (1970) demonstrated anterograde amnesia following i.v. diazepam 0.24 mg/kg, and GROVE-WHITE and KELMAN (1971) showed similar effects following an i.v. dose as low as 0.05 mg/kg. The percentage of volunteers experiencing amnesia for a given i.v. dose of diazepam was increased following a rapid rate of infusion (KORTTILA et al. 1976). Following i.v. administration of 5 or 10 mg as premedication, anterograde amnesia occurs in 50%–90% of patients respectively. The effect is present at 1 min, peaks at about 2–3 min and declines over the ensuing 30 min (DUNDEE and PANDIT 1972a, b; PANDIT et al. 1971). Following very high doses (20 mg loading dose and 15 mg maintenance dose) SEOW et al. (1985) reported a relapse of amnesia at 1.5–6 h into the postoperative period in some patients.

Diazepam-induced amnesia is highly dependent on the route of administration. PANDIT and DUNDEE (1970) found that only 4% of patients had anterograde amnesia following i.m. diazepam. Although some early reports failed to show am-

Table 1. Selected studies in which partial or complete anterograde amnesia has been found following diazepam in patients undergoing surgical and diagnostic procedures

Dose	Route	*n*	Frequency of amnesia	Study
Sedation of surgery or dental conservation under local anaesthesia				
0.15 mg/kg	i.v.	50	48% for local anaesthetic, 46% for procedure	DIXON et al. (1984)
10 mg	i.v.	20	22% for local anaesthetic	O'BOYLE et al. (1987a)
13 mg[a]	i.v.	14	64% for local anaesthetic, 81% for controlled stimuli	GREGG et al. (1974)
13.9 mg[a]	i.v.	35	91% for cardioversion	KAHLER et al. (1967)
0.2 mg/kg	i.v.	34	100% for local anaesthetic	HEALY et al. (1970)
0.2 mg/kg	i.v.	34	48% for local anaesthetic, 59% for procedure	AUN et al. (1984)
0.23 mg/kg	i.v.	18	50% for local anaesthetic, 61% and 34% for visual and auditory stimuli	BARCLAY et al. (1980)
18.8 mg[a]	i.v.	77	71% for local anaesthetic	DRISCOLL et al. (1972)
0.27 mg/kg[a]	i.v.	20	85% for cards	LUNDGREN et al. (1983)
0.29 mg/kg	i.v.	43	91% for fillings	DIXON et al. (1980)
0.29 mg/kg	i.v.	50	72% for local anaesthetic	BARKER et al. (1986)
0.33 mg/kg	i.v.	55	75% for local anaesthetic, 93% for extraction	O'NEILL et al. (1970)
Sedation for endoscopy				
12.1 mg[a]	i.v.	40	55% for procedure	DOUGLAS et al. (1980)
0.125 mg/kg		24	42% for procedure	KORTTILA et al. (1978a)
0.14 mg/kg	i.v.	20	20% for scope insertion, 30% for scope removal	HANNO and WEIN (1983)
0.15 mg/kg	i.v.	50	44% for scope insertion, 57% for scope removal	AL-KHUDHAIRI et al. (1982)
0.15 mg/kg	i.v.	31	7% for procedure	BERGGREN et al. (1983)
0.15 mg/kg	i.v.	50	30% for scope insertion, 44% for scope removal	WHITMAN et al. (1983)
12.5 mg[a]	i.v.	75	63% for scope insertion, 73% for scope removal	BARDHAN et al. (1984)
0.20 mg/kg	i.v.	27	67% for scope insertion, 70% for scope removal	KORTTILA and TARKKANEN (1985)
0.25 mg/kg	i.v.	23	70% for procedure	KORTTILA et al. (1978a)
Premedication for surgery under general anaesthesia				
5 mg	i.v.	25	50% maximum for cards	DUNDEE and PANDIT (1972a)
10 mg	i.v.	25	90% maximum for cards	DUNDEE and PANDIT (1972a)
5 mg	p.o.	20	5% maximum for cards	KOTHARY et al. (1981)
5 mg	p.o.	10	0–20% for cards	McKAY and DUNDEE (1980)
10 mg	p.o.	20	10%–85% for cards	McKAY and DUNDEE (1980)
10 mg[b]	p.o.	25	4% for 75 min card	STUDD and ELTRINGHAM (1980)
10 mg	p.o.	20	20% maximum for cards	KOTHARY et al. (1981)
20 mg	p.o.	20	10–80% for cards	McKAY and DUNDEE (1980)
20 mg	p.o.	20	30% maximum for cards	KOTHARY et al. (1981)
10 mg	i.m.	100	Insignificant amnesia	DUNDEE et al. (1979)
0.25 mg/kg	i.m.	20	10% for cards	LINDGREN et al. (1979)

[a] Mean dose. [b] Benzodiazepine night sedation also given.

nesia following oral administration of diazepam (HARRY and RICHARDS 1971; WILSON and ELLIS 1973), the effect is now well documented in both volunteers and patients, although not as profound as that seen following i.v. administration of the drug (BRANDT and OAKES 1965; HAQ and DUNDEE 1968; BAIRD and HAILEY 1972; RICHARDSON and MANFORD 1979; MCKAY et al. 1978; MALE et al. 1980; MCKAY and DUNDEE 1980; KOTHARY et al. 1981; O'BOYLE et al. 1983; MALE and JOHNSON 1984). Diazepam 5 mg, orally, has little effect on memory, but amnesia is found after 10 and 20 mg doses, beginning 20–30 min after administration and persisting for at least 120 min (MCKAY and DUNDEE 1980). Increasing the oral dose appears to increase the duration rather than the depth of amnesia (BRANDT and OAKES 1965; BAIRD and HAILEY 1972; MCKAY and DUNDEE 1980).

Rectal administration of diazepam has been recommended as a premedicant (SONANDER et al. 1985) but amnesia following this mode of administration appears to be considerably less than that following i.v. administration (LUNDGREN and ROSENQUIST 1986).

4.2 Midazolam

There has been increasing interest in the use of midazolam in anaesthetic practice as a premedicant, sedative and induction agent because of its potency and its water solubility, which results in minimal venous irritation (GAMBLE et al. 1981; KANTO and KLOTZ 1982; PHILIP 1985; REVES et al. 1985). Part of the popularity of midazolam is undoubtedly due to its marked effect on memory (Table 2). Amnesia following midazolam has been demonstrated in patients, following i.v. (CONNER et al. 1978a; FRAGEN et al. 1978; DUNDEE and WILSON 1980; BERGGREN and ERIKSSON 1981; AL-KHUDHAIRI et al. 1982; WHITE 1982; WHITMAN et al. 1983; MCCLURE et al. 1983; BERGGREN et al. 1983; COLE et al. 1983; MCGIMPSEY et al. 1983; AUN et al. 1984; BARKER et al. 1986), i.m. (MCATEER et al. 1984; REINHART et al. 1985; VAN WIJHE et al. 1985) and oral administration (O'BOYLE et al. 1987a).

Using standardized memory cards, anterograde amnesia has been demonstrated in 90%–96% of patients after a 5 mg i.v. dose of midazolam; the effect was present 2 min after injection, most profound between 2 and 5 min and declined over the ensuing 20–40 min (CONNER et al. 1978a; DUNDEE and WILSON 1980). Following i.m. injection, 0.1 mg/kg produced complete anterograde amnesia for urological procedures in 81% of patients, starting at 5–10 min after injection and lasting for 45 min (REINHART et al. 1985). Lower levels of amnesia have also been reported, however, following i.m. injection (VAN WIJHE et al. 1985; MCATEER et al. 1984). O'BOYLE et al. (1987a), using controlled stimuli and questionnaires, have recently demonstrated anterograde amnesia in 53% of patients at 35–40 min after 15 mg midazolam given orally. Although dose-response studies have not been conducted, it appears that the degree of anterograde amnesia following midazolam is dose related. For example, in patients undergoing upper gastrointestinal endoscopy the percentage amnesia for insertion of the endoscope was 64% following i.v. midazolam 0.07 mg/kg (WHITMAN et al. 1983), 88% following 0.1 mg/kg (AL-KHUDHAIRI et al. 1982) and 96% following a mean dose of 10.3 mg (range 5–15 mg) (BARDHAN et al. 1984).

Table 2. Studies in which partial or complete anterograde amnesia has been found following midazolam in patients undergoing surgical or diagnostic procedures

Dose	Route	*n*	Frequency of amnesia	Study
Sedation for surgery or dental conservation under local anaesthesia				
0.08 mg/kg	i.v.	48	92% for local anaesthetic, 82% for procedure	DIXON et al. (1984)
0.1 mg/kg	i.v.	22	82% for procedure	McGIMPSEY et al. (1983)
0.17 mg/kg[a]	i.v.	34	79% for local anaesthetic, 91% for dental procedure	AUN et al. (1984)
0.14 mg/kg	i.v.	50	98% for local anaesthetic	BARKER et al. (1986)
12 mg[b]	i.v.	10	90% for operation, 100% for post-op. card	McCLURE et al. (1983)
0.18 mg/kg[a]	i.v.	20	100% for procedure	HANNO and WEIN (1983)
0.1 mg/kg	i.m.	27	81% for procedure	REINHART et al. (1985)
15 mg	p.o.	20	42% for local anaesthetic	O'BOYLE et al. (1987a)
Sedation for endoscopy				
0.05 mg/kg	i.v.	29	60% for procedure, 25%–35% for cards	BERGGREN et al. (1983)
0.07 mg/kg	i.v.	50	64% for scope insertion, 74% for removal	WHITMAN et al. (1983)
0.1 mg/kg	i.v.	50	88% for scope insertion, 92% for removal	AL-KHUDHAIRI et al. (1982)
10.3 mg[a]	i.v.	75	96% for scope insertion and removal	BARDHAN et al. (1984)
Premedication for surgery under general anaesthesia				
5 mg	i.v.	24	91% for 1 min card, 61% for 32 min card, 57% for theatre	CONNER et al. (1978a)
5 mg	i.v.	40	90% for 1 min card, 30% for 60 min card	DUNDEE and WILSON (1980)
5–7.5 mg	i.m.	50	40% for induction	McATEER et al. (1984)
0.07 mg/kg	i.m.	67	47% for induction	VAN WIJHE et al. (1985)
Induction of general anaesthesia				
0.15 mg/kg	i.v.	25	100% for operation	FRAGEN et al. (1978)
0.3 mg/kg	i.v.	20	Marked amnesia for cards	WHITE (1982)
0.36 mg/kg[b]	i.v.	31	100% lasting 1 h	BERGGREN and ERIKSSON (1981)

[a] Mean dose. [b] Benzodiazepine premedication also given.

4.3 Lorazepam

Lorazepam is used mainly for night-before sedation and as a premedicant for long procedures (DUNDEE et al. 1977a). Because of its long duration of action, it is not recommended for induction of anaesthesia (KANTO and KLOTZ 1982; DUNDEE et al. 1977a; DUNDEE 1979) or as supplemental sedation for ambulatory outpatient procedures (PHILIP 1985). The profound amnesia produced by lorazepam has a number of distinct features (Table 3). First, after intravenous admin-

Table 3. Studies in which partial or complete anterograde amnesia has been found following lorazepam in patients undergoing surgical or diagnostic procedures

Dose	Route	*n*	Frequency of amnesia	Study
Sedation for surgery or dental conservation under local anaesthesia				
1 mg	i.v.	10	20% for 30 min card	
2 mg	i.v.	10	30% for 30 min card	
3 mg	i.v.	10	60% for 30 min card	
5 mg	i.v.	5	100% for 30 min card	HEISTERKAMP and COHEN (1975)
0.05 mg/kg	i.v.	18	34% for local anaesthetic, 28% for 15 min card, 22% for 20 min sound	BARCLAY et al. (1980)
0.03 mg/kg	i.m.	31	0–19% for procedure	KORTTILA et al. (1980)
0.06 mg/kg	i.m.	36	20%–43% for procedure	KORTTILA et al. (1980)
2 mg	p.o.	20	7.5% for childbirth, 25% for labour	McAULEY et al. (1982)
Premedication for surgery under general anaesthesia				
1 mg	p.o.	20	10% for induction	DUNDEE et al. (1977a)
2 mg	p.o.	20	20% for induction	DUNDEE et al. (1977a)
2 mg	p.o.	52	35% for journey to theatre	MAGBAGBEOLA (1974)
2 mg	p.o.	30	100% partial for cards	MALE and JOHNSON (1984)
2 mg	p.o.	20	30% for 2 h card	KOTHARY et al. (1981)
2.5 mg	p.o.	50	44% for induction	DUNDEE et al. (1977a)
2.5 mg	p.o.	50	58% for induction	RUBIN et al. (1980)
2.5 mg	p.o.	67	68% for procedure	DODSON and EASTLEY (1978)
2.5 mg[a, b]	p.o.	25	44% and 48% for 75 min card, 27% for epidural	STUDD and ELTRINGHAM (1980)
3 mg	p.o.	25	52.5% for pre-op. number	WILSON (1973)
3 mg	p.o.	20	45% for 2 h card	KOTHARY et al. (1981)
0.05 mg/kg	p.o.	25	28% for 90 min card	BURTLES and ASTLEY (1983)
0.05 mg/kg	p.o.	15	93% for 2 h card	BRADSHAW et al. (1981)
0.05 mg/kg	p.o.	112	20% for induction at 2 h	PETERS and BRUNTON (1982)
4 mg	p.o.	200	57% for induction	DUNDEE et al. (1977a)
4 mg	p.o.	20	70% for 2 h card	KOTHARY et al. (1981)
5 mg	p.o.	50	76% for induction	DUNDEE et al. (1977a)
8 mg	p.o.	15	77% for induction	DUNDEE et al. (1977a)
2 mg	i.v.	32	59% for card, 66% for induction	CONNER et al. (1975)
2 mg	i.v.	25	10%–50% for cards	PANDIT et al. (1976)
2 mg	i.v.	30	45% for 32 min card, 40% for operating theatre	L'ARMAND et al. (1980)
4 mg	i.v.	31	80% for card, 72% for induction	CONNER et al. (1975)
4 mg	i.v.	25	10%–80% for cards	PANDIT et al. (1976)
4 mg	i.v.	30	64% for 32 minute card, 74% for operating theatre	L'ARMAND et al. (1980)
4 mg	i.v.	20	95% for 2 h card	GEORGE and DUNDEE (1977)
2–4 mg	i.m.	75	45% for induction, 70% for phrase, 80% for cards	GALE et al. (1983)

Table 3 (continued)

Dose	Route	*n*	Frequency of amnesia	Study
0.04 mg/kg[a,b]	i. m.	22	56% for induction	Hudson et al. (1979)
0.055 mg/kg[a]	i. m.	100	64% for procedure	Hewitt and Barr (1978)
4 mg	i. m.	100	78% for induction	Dundee et al. (1977a)
4 mg	i. m.	9	66% for 90 minute card	Paymaster (1976)
4 mg	i. m.	20	35%–80% for specific events, 71% for 2 h card	Aleniewski et al. (1977)
8 mg	i. m.	13	100% for induction	Dundee et al. (1977a)

[a] Mean dose. [b] Benzodiazepine night sedation also given.

istration there is a delay of 15–20 min before the onset of the amnesic action, which then lasts for up to 6–8 h and is dose dependent (Pandit et al. 1976; Heisterkamp and Cohen 1975; L'Armand et al. 1980; Barclay et al. 1980; George and Dundee 1977; Long and Eltringham 1977; Conner et al. 1975; Taub and Eisenberg 1976; Dundee et al. 1977a, b, 1979). The onset latency appears to be dose related, being 30 min after 2 mg but only 15 min after 4 mg (Pandit et al. 1976). This time-effect relationship differs from that of diazepam or midazolam where the latency is of the order of 1–2 min. The latency of onset and long duration of action of lorazepam-induced amnesia has been attributed to slow and extensive tissue distribution of the drug (Greenblatt and Shader 1978).

A further distinct feature of lorazepam is that anterograde amnesia can also be reliably induced following oral and i.m. administration. A 1 mg oral dose has only a minimal effect on memory while increasing doses from 4 to 5 mg does not increase the frequency of anterograde amnesia (Dundee et al. 1977a). Varying degrees of anterograde amnesia have been demonstrated following oral doses of lorazepam ranging from 2 to 8 mg in a variety of patient populations (Magbagbeola 1974; Dodson and Eastley 1978; Wilson 1973; Kothary et al. 1981; Peters and Brunton 1982; Rubin et al. 1980; Dundee et al. 1977a, b; Burtles and Astley 1983; Studd and Eltringham 1980; McAuley et al. 1982; Male and Johnson 1984; Bradshaw et al. 1981). Depending on the duration of the procedure, 4 mg lorazepam appears to provide good amnesia in 50%–70% of patients (Dundee et al. 1977a; Kothary et al. 1981).

Lorazepam also produces reliable amnesia when administered i.m. (Paymaster 1976; Galloon et al. 1977; Aleniewski et al. 1977; Hewitt and Barr 1978; Hudson et al. 1979; Dundee et al. 1977a; Gale et al. 1983). For example, Dundee et al. (1977a) found 78% and 100% complete or partial amnesia for journey to the operating theatre and for induction of anaesthesia, following i.m. doses of 4 and 8 mg respectively.

The degree of amnesia following i.m. lorazepam appears to depend on the population studied. Korttila et al. (1980), despite using a relatively high dose of 0.06 mg/kg in patients undergoing lower abdominal and lower extremity surgery under local blockade, found that 63% of patients recalled the operation and 53% remembered their stay in the recovery room. In a further study, unpredictable amnesic effects were found following doses of 0.03 and 0.06 mg/kg, administered i.m. to patients undergoing neurosurgical procedures (KORTTILA et al. 1982). It was postulated that the unpredictable and variable effects of the drug were due to differences in penetration through the blood-brain barrier.

Lormetazepam, an analogue of lorazepam, has also been shown to produce anterograde amnesia following i.v. (OTT et al. 1980) and oral (ROEHRS et al. 1984) administration. Following sublingual administration of 2.5 mg lormetazepam in a novel wafer formulation (TAUBER et al. 1985), 30% and 48% of oral surgery patients were amnesic for the local anaesthetic and drilling respectively (O'BOYLE et al., in preparation).

4.4 Flunitrazepam

Flunitrazepam was first employed in anaesthesia in 1971 (VEGA 1971) and has been widely used in Europe and South America as a premedicant (FREUCHEN et al. 1981; MCGOWAN et al. 1980) and for induction of anaesthesia (DUNDEE et al.

Table 4. Studies in which partial or complete anterograde amnesia has been found following flunitrazepam in patients undergoing surgical and diagnostic procedures

Dose	Route	*n*	Frequency of amnesia	Study
Sedation for surgery or dental procedures under local anaesthesia				
0.014 mg/kg	i.v.	39	85% for local anaesthetic	DIXON et al. (1980)
Sedation for endoscopy				
0.01 mg/kg	i.v.	24	67% for insertion and removal of scope	KORTTILA et al. (1978a)
0.02 mg/kg	i.v.	21	75% for scope insertion, 81% for scope removal	KORTTILA et al. (1978a)
0.01 mg/kg	i.v.	79	40%–65% for insertion, 45%–75% for removal (effect age dependent)	KORTTILA et al. (1978b)
1.2 mg[a]	i.v.	25	96% partial or complete for procedure	NIMMO et al. (1978)
Premedication for surgery under general anaesthesia				
1 mg	i.v.	20	40%–90% for cards	GEORGE and DUNDEE (1977)
2 mg	i.v.	30	45%–100% for cards	GEORGE and DUNDEE (1977)
0.5 mg	p.o.	20	0%–30% for cards	MCKAY and DUNDEE (1980)
1 mg	p.o.	20	5%–65% for cards	MCKAY and DUNDEE (1980)
0.02 mg/kg	p.o.	49	10%–30% for cards	LINDGREN et al. (1980)
1–2 mg	p.o.	72	74% for induction, 81% for post-op. events	RICHARDSON and MANFORD (1979)
0.02 mg/kg	i.m.	50	33% for cards at 45 min, 84% for post-op. events	LINDGREN et al. (1979)

[a] Mean dose.

1976; KANTO and KLOTZ 1982; CLARKE and LYONS 1977; STOVNER 1973). Most reports show flunitrazepam to be similar to diazepam except that at comparable sedative doses it produces deeper and more prolonged amnesia. (GEORGE and DUNDEE 1977; DUNDEE et al. 1976; KORTTILA et al. 1978 a). Anterograde amnesia has been demonstrated (Table 4) in patients following flunitrazepam premedication via the oral (LINDGREN et al. 1980; MCKAY and DUNDEE 1980; RICHARDSON and MANFORD 1979), i.v. (KORTTILA et al. 1978 a; GEORGE and DUNDEE 1977; DIXON et al. 1980; NIMMO et al. 1978) and i.m. routes (LINDGREN et al. 1979). Interestingly, KORTTILA et al. (1978 b) have shown a linear increase in the speed of onset and duration of amnesic action with increasing age in patients undergoing bronchoscopy following i.v. flunitrazepam 0.01 mg/kg.

5 Benzodiazepine-Induced Amnesia in Different Sensory Modalities

The most widely used techniques for assessing amnesia in anaesthetic practice involve either showing the patient a series of pictures of objects and testing subsequent recall or simply questionning the patient postoperatively. Some authors have attempted to assess benzodiazepine-induced amnesia in different sensory modalities. GREGG et al. (1974) found equivalent amnesia following i.v. diazepam for visual, auditory and pain stimuli. However, SEOW et al. (1985) and HEALY et al. (1970) have demonstrated longer-lasting amnesia for verbal material than for visual stimuli following i.v. diazepam. A similar finding has been reported following rectal administration of diazepam (LUNDGREN and ROSENQUIST 1986). STUDD and ELTRINGHAM (1980) found amnesia more profound for taped music than for administration of an extradural anaesthetic following both lorazepam and diazepam. BARCLAY et al. (1980) found that whereas only 36% of patients recalled a memory card after i.v. diazepam, 50% and 60% respectively recalled administration of a local anaesthetic and taped auditory stimuli. In a controlled volunteer study BARCLAY (1982) showed that visual stimuli were recalled with significantly greater frequency than cuteanous-tactile or auditory stimuli presented at the same time following i.v. diazepam. Similar findings have been reported following i.m. lorazepam (ALENIEWSKI et al. 1977). GELFMAN et al. (1978) presented controlled visual, painful and cutaenous-tactile stimuli at predetermined intervals to patients after a combination of i.v. diazepam and methohexitone. Amnesia was significantly more profound for cutaenous-tactile than for visual stimuli. CHERKIN and HARROUN (1971), in a review of memory for events occurring under general anaesthesia, concluded that auditory and painful stimuli were most likely to be recalled, while JONES and KONIECZKO (1986) have pointed out that the auditory neural pathway has a high metabolic rate, and continues to function during certain stages of general anaesthesia.

Although disagreement exists, benzodiazepine-induced amnesia appears to be more profound for auditory and cutaenous-tactile stimuli than for visual stimuli. However, emotionally laden events such as painful or stressful procedures tend to be recalled with greater frequency than artificial stimuli, probably because of greater cognitive elaboration of such material (O'BOYLE et al. 1986).

6 Benzodiazepines and State-Dependent Retrieval

State-dependent retrieval occurs when recall of information learned in one psychological state is impaired when the subject is in a state different from that in which the information was originally learned (OVERTON 1978). In the case of drug-induced state dependency, amnesia occurs if the subject learns the information in, say, a drugged state but is asked to retrieve it in a drug state different from that in which the material was learned (e.g. a drug-free state). Research findings on state-dependent retrieval with the benzodiazepines are equivocal and appear to depend on the test of recall used (PETERSEN and GHONEIM 1980; JENSEN and POULSON 1982; LUNDGREN and ROSENQUIST 1986; LISTER 1985). The importance of state-dependent learning for anaesthetic practice is that a previously stressful procedure, originally experienced under benzodiazepine sedation but subsequently forgotten, might be recalled when the patient is resedated during a follow-up procedure. This would be particularly relevant for stressful procedures such as endoscopies and minor oral surgery which are often frequently repeated. Considerably greater research effort into state-dependency phenomena in patients undergoing repeat procedures under benzodiazepine sedation appears necessary. In particular, research is needed to determine whether painful or traumatic events are more likely to be recalled on resedation. The possibility that mood state-dependent learning (BOWER 1981) may occur is also of considerable relevance in this context since both the surgical setting and benzodiazepine medication can markedly alter mood.

7 Benzodiazepines and Recall of Intraoperative Events Under General Anaesthesia

In classical "balanced" anaesthesia the effects of premedication agents, induction agents, analgesic drugs and neuromuscular blocking agents are combined with those of inhalational anaesthetics to produce clinical anaesthesia. However, there are numerous reports of patients remembering specific events associated with surgery which were registered during the time they were considered to be unconscious (for reviews, see CHERKIN and HARROUN 1971; MAINZER 1979; BREKENRIDGE and AITKENHEAD 1983; JONES and KONIECZKO 1986). Besides being a terrifying experience for the patient, especially if he or she is paralysed and unable to communicate, this may lead to subsequent nightmares and neurosis (BERGSTROM and BERNSTEIN 1968; BLACHER 1975; WILSON et al. 1975). There is increasing medicolegal interest in the phenomenon and a number of legal actions are pending, especially with women who claim to have been conscious and able to recall events during anaesthesia for Caesarian section (JONES and KONIECZKO 1986).

Because the majority of patients appear profoundly amnesic when tested for postoperative recall or recognition of intraoperative events, such events are obviously not generally amenable to consciousness or accessable through intentional forms of remembering. It is possible, however, that intraoperative auditory or painful stimuli may be unconsciously retained in long-term memory and, though inaccessable to conscious postoperative recall, may influence subsequent behaviour or convalescence. BENNETT et al. (1985) instructed 11 anaesthetized patients under general anaesthesia to touch their ears during the postoperative interview. Despite being completely amnesic for the intraoperative suggestion, significantly more of the patients subsequently touched their ears than of a control group exposed to taped operating theatre sounds only. Recently, BONKE et al. (1986) showed that exposure to positive suggestions during general anaesthesia as compared with noise or operating theatre sounds protected patients over 55 years of age from prolonged postoperative stay in hospital. Thus, it appears that high-level stimuli may be processed during anaesthesia and stored in subconscious memory, and may not be retrieved into working memory but may subsequently influence behaviour.

When memory retention after an operation does occur, pain and sounds are recalled most often (CHERKIN and HARROUN 1971). It seems likely that the coincidence of lightening of anaesthesia with acute pain or with auditory information of high emotional content increases arousal and heightens the possibility of recall. It is known that affective reactions can occur without extensive perceptual or cognitive encoding and that reliable affective discriminations can be made in the total absence of recognition memory. ZAJONC (1980) has concluded that affect and cognition are under the control of separate and partially independent systems that can influence each other in a variety of ways, and that both constitute independent sources of effects in information processing. Preventing intraoperative affective reactions as well as cognitive processing may decrease awareness and subsequent recall. MATTILA et al. (1979) administered either 10 mg diazepam or 1 mg flunitrazepam i.v. during general anaesthesia, to 90 gynaecological or abdominal surgery patients, immediately before the skin incision was made. The hypnotics significantly decreased movements and arterial blood pressure responses to the skin incision. One patient in the flunitrazepam group experienced unpleasant recall of intraoperative events. CORMACK (1979) showed that the time taken for a patient to respond to command (TCR) when nitrous oxide was terminated at the end of surgery was significantly increased by lorazepam (4 mg/70 kg) premedication, compared to 10 mg morphine. The author suggested lorazepam as the ideal supplement for decreasing the possibility of awareness and subsequent recall of intraoperative events. BARR et al. (1977) gave 20 mg lorazepam or 10 mg diazepam i.v. immediately following delivery under general anaesthetic to 222 elective or emergency Caesarian patients. Postoperative recall of intraoperative events was 2.2% (two patients) in the lorazepam group and 1.1% (one patient) in the diazepam group. The long duration of anterograde amnesia following lorazepam probably makes this the ideal agent for such suplementation.

8 Theoretical and Methodological Considerations

The anterograde amnesic effects of benzodiazepines have been well established in anaesthetic studies. There are, however, a number of theoretical and methodological considerations pertaining to further research in this area. The majority of authors interpret the amnesic properties of benzodiazepines in terms of the two-component model of memory (Broadbent 1958; Atkinson and Shiffrin 1968). While the explanatory utility of this model is undoubted, there is increasing evidence that this view represents an oversimplification (see Craik and Lockhart 1972; Broadbent 1984; Baddeley 1978, 1984; Cohen et al. 1986; Gruneberg and Morris 1978). Although more modern conceptualizations of memory function have been developed (Baddeley 1984) they have had little impact on anaesthetic research. One such formulation of considerable relevance is the "levels of processing" approach developed by Craik and Lockhart (1972) which highlights the association between the encoding of material and its subsequent memorability. More specifically it is suggested that the more deeply the material is encoded, the more durable will be the memory trace. Many studies of memory function in anaesthesia have shown that painful and emotionally laden materials are more likely to be recalled than less threatening material, despite sedation or general anaesthesia (e.g. Cherkin and Harroun 1971). Likewise, actual surgical events are more likely to be recalled at a given level of benzodiazepine sedation than are controlled stimuli such as memory cards (e.g. Gregg et al. 1974; Conner et al. 1975; Gale et al. 1983; O'Boyle et al. 1987b). The most likely explanation of such phenomena is that emotionally laden or traumatic stimuli are processed to a greater degree and consequently are more likely to be recalled. O'Boyle et al. (1987b), using the model of memory proposed by Tulving (1972, 1983), have shown that the recall of a stressful event occurring during benzodiazepine sedation was positively correlated with the degree of cardiovascular arousal provoked by the event. It was hypothesised that traumatic events such as drilling were subjected to greater cognitive elaboration (see Warrington 1986) and therefore more likely to be recalled for a given degree of sedation.

The clinical situation places considerable limitations on research possibilities in anaesthetic practice. However, the availability of a large heterogeneous population experiencing personally meaningful and potentially traumatic events in a relatively controlled "real-life" setting makes this an important research domain (O'Boyle et al. 1984, 1985). However, an increase in methodological sophistication appears warranted. Given the importance and possible traumatising effects of recall of disturbing auditory and painful events, the continuing reliance on testing memory in one sensory modality with relatively meaningless stimuli (memory cards) seems inadequate. When testing amnesia in different sensory modalities the practice of asking the patient to identify, verbally, stimuli administered in other modalities enforces semantic representation of such stimuli and may not provide a true measure of memory in the other modalities. There is a tendency to ignore the possible interactions between the many drugs which are often given concurrently to a patient. For example, it may be difficult to distinguish, in some situations, the anterograde amnesic effects of a premedicant from the possible retrograde effects of the anaesthetic. Furthermore, in the absence of a placebo or control group, care should be exercised in attributing poor recall of predrug stimuli to retrograde amnesia. LUNDGREN and ROSENQUIST (1983), for example, found poor recall for cards shown before the start of an operation in an unsedated group. Such effects may be due simply to individual differences or to the deleterious consequences of preoperative stress and anxiety on the registration of information (EYSENCK 1977). Predrug anxiety levels may further influence the extent and type of response to benzodiazepines (DESAI et al. 1983; O'BOYLE et al. 1985).

9 Concluding Remarks

The recent development of specific benzodiazepine antagonists such as RO 15-1788 (HUNKELER et al. 1981) and ZK 93426 (DUKA et al. 1986) provides extremely useful tools for detailed study of the receptor substrates for the effects of the benzodiazepines. RO 15-1788 has already been shown to antagonise the performance and amnesic effects of the benzodiazepines (DARRAGH et al. 1981, 1982; O'BOYLE et al. 1983), indicating involvement of the benzodiazepine receptor in these effects. The development of anxiolytic compounds such as PK 8165 (POGGIOLI et al. 1985; WILLER et al. 1986) and buspirone (SKOLNICK et al. 1984; GOLDBERG 1984) which are non-sedative, provides the means for assessing the relative contributions of the sedative and anxiolytic effects of drugs to their effects on memory.

References

Aleniewski MI, Bulas BJ, Maderazo L, Mendoza C (1977) Intramuscular lorazepam versus pentobarbital premedication. A comparison of patient sedation, anxiolysis and recall. Anesth Analg 56:489–492

Al-Khudhairi D, Whitman JG, McCloy RF (1982) Midazolam and diazepam for gastroscopy. Anaesthesia 37:1002–1006

Atkinson RC, Shiffrin RM (1968) Human memory: a proposed system and its control processes. In: Spence KW, Spence JT (eds) The psychology of learning and motivation: Advances in research and theory, vol 2. Academic Press, New York, pp 89–195

Aun C, Flynn PJ, Richards J, Major E (1984) A comparison of midazolam and diazepam for intravenous sedation in dentistry. Anaesthesia 39:589–593

Baddeley AD (1978) The psychology of memory. Harper and Row, New York

Baddeley AD (1984) The fractionation of human memory. Psychol Med 14:259–264

Baird ES, Hailey DM (1972) Delayed recovery from a sedative: correlation of the plasma levels of diazepam with clinical effects after oral and intravenous administration. Br J Anaesth 44:803–805

Barclay JK (1982) Variations in amnesia with intravenous diazepam. Oral Surg 53:329–334

Barclay JK, Hunter K MacD, Jones H (1980) Diazepam and lorazepam compared as sedatives for outpatient third molar surgery. Br J Oral Surg 18:141–149

Bardhan KD, Morris P, Taylor PC, Hinchliffe RFC, Harris PA (1984) Intravenous sedation for upper gastrointestinal endoscopy: diazepam versus midazolam. Br Med J 288:1046

Barker I, Butchart DGM, Gibson J, Lawson JIM, MacKenzie N (1986) I.V. sedation for conservative dentistry. Br J Anaesth 58:371–377
Barr AM, Moxon A, Wollam CHM, Freyer ME (1977) The effect of diazepam and lorazepam on awareness during anaesthesia for Caesarian section. Anaesthesia 32:873–878
Bennett HL, Davis HS, Giannini JA (1985) Non-verbal response to intraoperative conversation. Br J Anaesth 57:174–179
Berger FM, Potterfield J (1969) The effect of antianxiety tranquilizers on the behaviour of normal persons. In: Evans WO, Kline NS (eds) The psychopharmacology of the normal human. Springfield, New York, pp 83–113
Berggren L, Eriksson I (1981) Midazolam for induction of anaesthesia in outpatients: a comparison with thiopentone. Acta Anaesthesiol Scand 25:492–496
Berggren L, Eriksson P, Mollenholt P, Wickbom G (1983) Sedation for fibreoptic gastroscopy: a comparative study of midazolam and diazepam. Br J Anaesth 55:289–296
Bergstrom H, Bernstein K (1968) Psychic reactions after analgesia with nitrous oxide for Caserian section. Lancet ii:541–542
Blacher RS (1975) On awakening paralyzed during surgery. A syndrome of traunatic neurosis. JAMA 234:67–68
Bonke B, Schmitz PIM, Verhage F, Zwaveling A (1986) Clinical study of so-called unconscious perception during general anaesthesia. Br J Anaesth 58:957–964
Bower GH (1981) Mood and memory. Am Psychol 36:129–148
Bradshaw EG, Ali AA, Mulley BA, Rye RM (1981) Plasma concentrations and clinical effects of lorazepam after oral administration. Br J Anaesth 53:517–521
Brandt AL, Oakes FF (1965) Preanaesthesia medication: double-blind study of a new drug, diazepam. Anesth Analg 44:125–129
Brekenridge JL, Aitkenhead AR (1983) Awareness during anaesthesia – A review. Ann R Coll Surg Engl 65:93–96
Broadbent DE (1958) Perception and communication. Pergammon, Oxford
Broadbent DE (1984) The Maltese cross: a new simplistic model for memory. Behav Brain Sci 7:55–94
Brown JD, Lewis V (1982) A comparison between transient amnesias induced by two drugs (diazepam or lorazepam) and amnesia of organic origin. Neuropsychologia 20:55–69
Brown PRH, Main DMG, Lawson JIM (1968) Diazepam in dentistry. Br Dent J 125:498–501
Burtles R, Astley B (1983) Lorazepam in children. Br J Anaesth 55:275–279
Cherkin A, Harroun P (1971) Anaesthesia and memory process. Anaesthesiology 34:469–473
Clark EO, Glanzer M, Turndorf H (1979) The pattern of memory loss resulting from intravenously administered diazepam. Arch Neurol 36:296–300
Clarke PRF, Eccersley PS, Frisby JP, Thornton JA (1970) The amnesic effect of diazepam. Br J Anaesth 42:690–697
Clarke RSJ, Lyons SM (1977) Diazepam and flunitrazepam as induction agents for cardiac surgical operations. Acta Anaesthesiol Scand 21:282–292
Cohen G, Eysenck MW, LeVoi ME (1986) Memory: a cognitive approach. Open University Press, Milton Keynes
Cole SG, Brozinsky S, Isenberg JI (1983) Midazolam, a new more potent benzodiazepine, compared with diazepam: a randomized, double-blind study of preendoscopic sedatives. Gastrointest Endosc 29:219–222
Colins VJ (1976) Principles of anaesthesiology, 2nd edn. Lea and Febiger, Philadelphia
Conner JT, Parson N, Katz RL, Wapner S, Bellville JW (1975) Evaluation of lorazepam and pentobarbital as surgical premedicants. Clin Pharmac Ther 19:24–29
Conner JT, Katz RL, Pagano RR, Graham CW (1978a) RO21–3981 for intravenous premedication and induction of anaesthesia. Anesth Analg 57:1–5
Conner JT, Katz RL, Bellville JW, Graham C, Pagano R, Dorey F (1978b) Diazepam and lorazepam for intravenous surgical premedication. J Clin Pharmacol 18:285–292
Cormack RS (1979) Awareness during surgery – a new approach. Br J Anaesth 51:1051–1054
Craik FIM, Lockhart RS (1972) Levels of processing: a framework for memory research. J Verb Learn Verb Behav 11:671–684
Darragh A, Lambe R, Scully M, O'Boyle CA, Brick I, Downie WW (1981) Investigation in man of the efficacy of a benzodiazepine antagonist. Lancet ii:8–10

Darragh A, Lambe R, O'Boyle CA, Kenny M, Brick I, Taffe W (1982) RO 15–1788 antagonises the central effects of diazepam in man without altering bioavailability. Br J Clin Pharmacol 14:677–682
Desai N, Taylor-Davies, Barnett DB (1983) The effects of diazepam and oxprenolol on short-term memory in individuals of high and low state anxiety. Br J Clin Pharmacol 15:197–202
Dixon RA, Bennett NR, Harrison MJ, Kenyon C, Thornton JA (1980) I.V. flunitrazepam and i.v. diazepam in conservative dentistry: a cross-over trial. Br J Anaesth 52:517–526
Dixon J, Power SJ, Grundy EM, Lumley J, Morgan M (1984) Sedation for local anaesthesia. Comparison of intravenous midazolam and diazepam. Anaesthesia 39:372–376
Dodson ME, Eastley RJ (1978) Comparative study of two long-acting tranquilizers for oral premedication. Br J Anaesth 50:1059–1063
Douglas JG, Nimmo WS, Wanless R, Jarvie DR, Heading RC, Finlayson NDC (1980) Sedation for upper gastrointestinal endoscopy. A comparison of oral temazepam and i.v. diazepam. Br J Anaesth 52:811–815
Driscoll P, Smilack ZH, Lightbody PM, Fiorucci RD (1972) Sedation with intravenous diazepam. J Oral Surg 30:332–343
Duka T, Holler L, Obeng-Gyan R, Dorrow R (1986) Initial human pharmacology of the neutral benzodiazepine receptor antagonist β-carboline ZK 93 426. Br J Clin Pharmacol 22:228P
Dundee JW (1979) Benzodiazepine sedation-amnesia. In: Dundee JW (ed) Intravenous anaesthetic agents. Edward Arnold, London, pp 670–681 (Current topics in anasthesia, vol. 1)
Dundee JW, Haslett WHK (1970) The benzodiazepines: a review of their actions and uses relevant to anaesthetic practice. Br J Anaesth 42:217–227
Dundee JW, Pandit SK (1972a) Anterograde amnesic effects of pethidine, hyoscine and diazepam in adults. Br J Pharmacol 44:140–144
Dundee JW, Pandit SK (1972b) Studies on drug-induced amnesia with intravenous anaesthetic agents in man. Br J Clin Pract 26(4):164–166
Dundee JW, Wilson DB (1980) Amnesic action of midazolam. Anaesthesia 35:459–461
Dundee JW, Varadarajan CA, Gaston JH, Clarke RSJ (1976) Clinical studies with induction agents. XLII Flunitrazepam. Br J Anaesth 48:551–555
Dundee JW, Lilburn JK, Nair SG, George KA (1977a) Studies of drugs given before anaesthesia. XXVI: Lorazepam. Br J Anaesth 49:1047–1056
Dundee JW, Johnston HML, Lilburn JK, Nair SG, Scott MG (1977b) A placebo-controlled comparison of the sedative properties of three benzodiazepines, lorazepam, flunitrazepam and fosazepam. Br J Clin Pharmacol 4:706–708
Dundee JW, McGowan WAW, Lilburn JK, McKay AC, Hegarty JE (1979) Comparison of the actions of diazepam and lorazepam. Br J Anaesth 51:439–445
Essman WB (1983) Clinical pharmacology of learning and memory. MTP Press, Lancaster
Eysenck MW (1977) Human memory: theory, research and individual differences. Pergammon, Oxford
Fragen RJ, Gahl F, Caldwell N (1978) A water-soluble benzodiazepine, RO 21–3981, for induction of anaesthesia. Anesthesiology 49:41–43
Freuchen I, Ostergaard J, Mikkelsen BO (1981) Anaesthesia with flunitrazepam and ketamine. Br J Anaesth 53:827–830
Gale GD, Galloon S, Porter WR (1983) Sublingual lorazepam: a better premedication? Br J Anaesth 55:761–765
Galloon S, Gale GD, Lancee WJ (1977) Comparison of lorazepam and diazepam as premedicants. Br J Anaesth 49:1265–1268
Gamble JAS, Kawar P, Dundee JW, Moore J, Briggs LP (1981) Evaluation of midazolam as an intravenous induction agent. Anaesthesia 36:868–873
Gelfman SS, Gracely RH, Driscoll EJ, Wirdzek PR, Sweet JB, Butler DP (1978) Conscious sedation with intravenous drugs: a study of amnesia. J Oral Surg 36:192–197
George KA, Dundee JW (1977) Relative amnesic actions of diazepam, flunitrazepam and lorazepam in man. Br J Clin Pharmacol 4:45–50
Ghoneim MM, Mewaldt SP (1975) Effects of diazepam and scopolamine on storage and retrieval processes in memory. Psychopharmacology 44:257–262
Ghoneim MM, Mewaldt SP (1977) Studies on human memory: the interactions of diazepam, scopolamine and physostigmine. Psychopharmacology 52:1–6

Ghoneim MM, Hinrichs JV, Mewaldt SP (1984) Dose-response analysis of the behavioral effects of diazepam: I Learning and memory. Psychopharmacology 82:291–295
Goldberg HL (1984) Buspirone hydrochloride: a unique new anxiolytic agent. Pharmacotherapy 4:315–324
Greenblatt DJ, Shader RI (1974) Anaesthesia and surgery. In: Greenblatt DJ, Shader RI (eds) Benzodiazepines in clinical practice. Raven, New York, pp 197–216
Greenblatt DJ, Shader RI (1978) Prazepam and lorazepam: two new benzodiazepines. N Engl J Med 229:1342–1344
Gregg JM, Ryan DE, Levin KH (1974) The amnesic actions of diazepam. J Oral Surg 32:651–664
Grove-White IG, Kelman GR (1971) Effect of methohexitone, diazepam and sodium 4-hydroxybutyrate on short-term memory. Br J Anaesth 43:113–116
Gruneberg MM, Morris P (1978) Aspects of memory. Methuen, London
Hanno PM, Wein AJ (1983) Anesthetic techniques for cytoscopy in man. J Urol 130:1070–1072
Harry FVA, Richards DJ (1971) Lorazepam: a study in psychomotor depression. Br J Clin Pract 26:371–372
Haslett WHK, Dundee JW (1968) Studies of drugs given before anaesthesia. XIV: Two benzodiazepine derivatives, chlordiazepoxide and diazepam. Br J Anaesth 40:250–258
Haq IU, Dundee JW (1968) Studies of drugs given before anaesthesia. XVI: Oral diazepine and trimeprazine for adenotonsillectomy. Br J Anaesth 40:972–979
Healy TEJ, Lautch H, Hall N, Tomlin PJ, Vickers MD (1970) Interdisciplinary study of diazepam sedation for outpatient dentistry. Br Med J 3:13–17
Heisterkamp DV, Cohen PJ (1975) The effect of intravenous premedication with lorazepam (Ativan), pentobarbitone or diazepam on recall. Br J Anaesth 47:79–81
Hewitt JM, Barr AM (1978) Premedication with lorazepam for bronchoscopy under general anaesthesia. Br J Anaesth 50:1149–1153
Hindmarch I (1981) Psychomotor function and psychoactive drugs. In: Lader M, Richens A (eds) Methods in clinical pharmacology – Central nervous system. Macmillan, London, pp 29–50
Hinrichs JV, Ghoneim MM, Mewaldt SP (1984) Diazepam and memory: retrograde facilitation produced by interference reduction. Psychopharmacology 84:158–162
Hudson IN, Davidson IA, Hider CF, Wright ADG (1979) The use of lormetazepam in cardiac bypass surgery: A comparative study. Curr Med Res Opin 6:303–308
Hunkeler W, Mohler H, Pieri L, Polc P, Bonetti EP, Cumin R, Schaffner R, Haefely W (1981) Selective antagonists of benzodiazepines. Nature 290:514–516
Janke W, Debus G (1968) Experimental studies of antianxiety agents with normal subjects: methodological considerations and a review of main effects. In: Efron DH (ed) Psychopharmacology – A review of progress. US Government Printing Office, Washington DC, pp 205–230
Jensen HH, Poulsen JC (1982) Amnesic effects of diazepam: "Drug-dependence" explained by state dependent learning. Scand J Psychol 23:107–111
Johnston M (1980) Anxiety in surgical patients. Psychol Med 10:145–152
Johnstone M (1976) The effect of lorazepam on the vasoconstriction of fear. Anaesthesia 31:868–872
Jones JG, Konieczko K (1986) Hearing and memory in anaesthetised patients. Br Med J 1291–1292
Kahler RL, Burrow GN, Felig P (1967) Diazepam induced amnesia for cardioversion. JAMA 200:997–998
Kanto J (1981) Benzodiazepines as oral premedicants. Br J Anaesth 53:1179–1187
Kanto J, Klotz U (1982) Intravenous benzodiazepines as anaesthetic agents: pharmacokinetics and clinical consequences. Acta Anaesthesiol Scand 26:554–569
Kleinknecht RA, Donaldson D (1975) A review of the effects of diazepam on cognitive and psychomotor performance. J Nerv Ment Dis 161:399–411
Korttila K, Linnoila M (1976) Amnesic action and skills related to driving after intravenous flunitrazepam. Acta Anaesthesiol Scand 20:160–168
Korttila K, Tarkkanen J (1985) Comparison of diazepam and midazolam for sedation during local anaesthesia for bronchoscopy. Br J Anaesth 57:581–586

Korttila K, Mattila MJ, Linnoila M (1976) Prolonged recovery after diazepam sedation: the influence of food, charcoal ingestion and injection rate on the effects of intravenous diazepam. Br J Anaesth 48:333–340

Korttila K, Saarnivarra L, Tarkkanen J, Himberg JJ, Hytonen M (1978 a) Comparison of diazepam and flunitrazepam for sedation during local anaesthesia for bronchoscopy. Br J Anaesth 50:281–287

Korttila K, Saarnivarra L, Tarkkanen J, Himberg JJ, Hytonen M (1978 b) Effect of age on amnesia and sedation induced by flunitrazepam during local anaesthesia for bronchoscopy. Br J Anaesth 50:1211–1217

Korttila K, Levanen J, Auvinen J (1980) Failure of intramuscularly administered lorazepam and scopolamine-morphine premedication to produce amnesic effects to supplement conduction anaesthesia. Acta Anaesthesiol Scand 24:325–330

Korttila K, Tarkkanen L, Kuurne T, Hamberg JJ, Abbondati G (1982) Unpredictable central nervous system effects after lorazepam premedication for neurosurgery. Acta Anaesthesiol Scand 26:213–216

Kothary SP, Brown ACD, Pandit UA, Samra SK, Pandit SK (1981) Time course of antirecall effect of diazepam and lorazepam following oral administration. Anesthesiology 55:641–644

L'Armand J, Vredevoe LA, Conner JT, Herr GP, Schehl D (1980) Lorazepam and morphine for i.v. surgical premedication. Br J Anaesth 52:1259–1263

Lindgren L, Saarnivaara L, Himberg JJ (1979) Comparison of i.m. pethidine, diazepam and flunitrazepam in children undergoing otolaryngological surgery. Br J Anaesth 51:321–327

Lindgren L, Saarnivaara L, Himberg JJ (1980) Comparison of oral triclofos, diazepam and flunitrazepam as premedicants in children undergoing otolaryngological surgery. Br J Anaesth 52:283–289

Lister RG (1985) The amnesic action of benzodiazepines in man. Neurosci Behav Rev 9:87–94

Long DH, Eltringham RJ (1977) Lorazepam as night sedation and premedication: a comparison with dichloralphenazone and papaveretum. Anaesthesia 32:649–653

Ludlam R, Bennett JR (1971) Comparison diazepam and morphine as premedication for gastrointestinal endoscopy. Lancet:1397–1399

Lundgren S, Rosenquist JB (1983) Amnesia, pain experience and patient satisfaction after sedation with intravenous diazepam. J Oral Maxillofac Surg 41:99–103

Lundgren S, Rosenquist JB (1986) Amnesia, sedation effect and patient comfort. An intraindividual comparison between intravenous diazepam and rectal diazepam solution. J Oral Maxillofac Surg (in press)

Mainzer J (1979) Awareness, muscle relaxants and balanced anaesthesia. Can Anaesth Soc J 26:386–393

Magbagbeola JAO (1974) A comparison of lorazepam and diazepam as oral premedicants for surgery under regional anaesthesia. Br J Anaesth 46:449–451

Male CG, Johnson HD (1984) Oral benzodiazepine premedication in minor gynaecological surgery. Br J Anaesth 56:499–507

Male CG, Lim YT, Male M, Stewart JM, Gibbs JM (1980) Comparison of three benzodiazepines for oral premedication in minor gynaecological surgery. Br J Anaesth 52:429–436

Mattila MAK, Saila K, Kokko T, Karkkainen T (1979) Comparison of diazepam and flunitrazepam as adjuncts to general anaesthesia in preventing arousal following surgical stimuli. Br J Anaesth 59:329–337

McAteer EJ, Dixon J, Whitman JG (1984) Intramuscular midazolam: A comparison of midazolam with papaveretum and hyoscine for intramuscular premedication. Anaesthesia 39:1177 1182

McAuley DM, O'Neill MP, Moore J, Dundee JW (1982) Lorazepam premedication for labour. Br J Obstet Gynaecol 89:149–154

McClure JH, Brown DT, Wildsmith JAW (1983) Comparison of the i.v. administration of midazolam and diazepam as sedation during spinal anaesthesia. Br J Anaesth 55:1089–1093

McGimpsey JG, Kawar P, Gamble JAS, Browne ES, Dundee JW (1983) Midazolam in dentistry. Br Dent J (July):47–50

McGowan WAW, Dundee JW, Clarke RSJ, Howard PJ (1980) Comparison of the subjective effects and plasma concentrations following oral and i.m. administration of flunitrazepam in patients. Br J Anaesth 52:447–451

McKay AC, Dundee JW (1980) Effect of oral benzodiazepines on memory. Br J Anaesth 52:1247–1257

McKay AC, Dundee JW, George KA (1978) The amnesic effect of orally administered benzodiazepines. Br J Anaesth 50:1080P–1081P

McNair DM (1973) Antianxiety drugs and human performance. Arch Gen Psychiatry 29:611–617

Murray JB (1984) Effects of valium and librium on human psychomotor and cognitive functions. Genet Psychol Monogr 109:167–169

Nimmo WS, Forrest JA, Heading RC, Finlayson NDC, Prescott LF (1978) Premedication for upper gastrointestinal endoscopy: a comparative study of flunitrazepam, diazepam and neuroleptanalgesia. Endoscopy 10:183–186

O'Boyle CA, Lambe R, Darragh A, Taffe I, Kenny M (1983) RO 15–1788 antagonizes the effects of diazepam in man without affecting its bioavailability. Br J Anaesth 55:349–355

O'Boyle CA, Harris D, Barry H, Cullen JH (1984) Benzodiazepine anxiolysis in "real-life" stress: oral temazepam and intravenous diazepam in oral surgery. Br J Clin Pharmacol 19:582P

O'Boyle CA, Harris D, Barry H, Cullen JH (1985) Differential effect of benzodiazepine sedation in high and low anxious patients in a real-life stress setting. Psychopharmacology 88:226–229

O'Boyle CA, Harris D, Barry H (1986) Sedation in outpatient oral surgery. Comparison of temazepam by mouth and diazepam i.v. Br J Anaesth 58:378–384

O'Boyle CA, Harris D, Barry H, McCreary C, Bewley A, Fox E (1987 a) Comparison of midazolam by mouth and diazepam i.v. in outpatient oral surgery. Br J Anaesth 59:746–754

O'Boyle CA, Barry H, Fox E, Harris D, McCreary C (1987 b) Benzodiazepine induced event amnesia following a stressful surgical procedure. Psychopharmacology 91:244–247

O'Neill R, Verrill PJ, Aellig WH, Laurence DR (1970) Intravenous diazepam in minor oral surgery. Further studies. Br Dent J 128:15–18

Ott H, Hemmerling KG, Kugler J, Suttman H, Doenicke A, Tesch C, Strassner G (1980) Amnestische Begleitwirkungen nach i.v.-Gabe von Lormetazepam und Flunitrazepam. Anaesthesiol Intensivmed 133:13.1–13.10

Overton DA (1978) Major theories of state-dependent learning – 1978. In: Colpaert FC, Rosecrans (eds) Stimulus properties of drugs: ten years of progress. Elsevier/North Holland, Amsterdam

Pandit SK, Dundee JW (1970) Preoperative amnesia: the incidence following the intramuscular injection of commonly used premedicants. Anaesthesia 26:493–499

Pandit SK, Dundee JW, Keilty SR (1971) Amnesia studies with intravenous premedication anaesthesia. Anaesthesia 26:421–428

Pandit SK, Heisterkamp DV, Cohen PJ (1976) Further studies on the anti-recall effect of lorazepam: a dose-time-effect relationship. Anaesthesiology 45:495–500

Paymaster NJ (1976) Evaluation of anxiolytic and amnesic effects of intramuscular lorazepam as a pre-operative medication. Curr Med Res Opin 1:388–394

Peters CG, Brunton JT (1982) Comparative study of lorazepam and trimeprazine for oral premedication in paediatric anaesthesia. Br J Anaesth 54:623–627

Petersen RC, Ghoneim MM (1980) Diazepam and human memory: influence on acquisition, retrieval and state-dependent learning. Prog Neuropsychopharmacol 4:81–89

Philip BK (1985) Supplemental medication for ambulatory procedures under regional anaesthesia. Anesth Analg 64:1117–1125

Poggioli JA, Bonnet D, Von Frenckell R (1985) Activity of PK8165 in anxiety induced by dental surgery: a dose-ranging, double-blind study. Curr Ther Res 38:423–431

Reves JG, Fragen RJ, Vinik HR, Greenblatt DJ (1985) Midazolam: pharmacology and uses. Anesthesiol 62:310–324

Reinhart K, DallingerStiller G, Dennhardt R, Heinemeyer G, Eyrich K (1985) Comparison of midazolam, diazepam and placebo i.m. as premedication for regional anaesthesia. Br J Anaesth 57:294–299

Richardson FJ, Manford MLM (1979) Comparison of flunitrazepam and diazepam for oral premedication in older children. Br J Anaesth 51:313–319

Roehrs T, McLeneghan A, Koshorek G, Zovick F, Roth T (1984) Amnesic effects of lormetazepam. In: Hindmarch I, Ott H, Roth T (eds) Sleep, benzodiazepines and performance. Springer, Berlin Heidelberg New York, pp 165–172

Rogers WK, Waterman DH, Domm SE, Sunay A (1965) Efficacy of a new psychotropic drug in bronchoscopy. Dis Chest 47:280–283
Rubin J, Schweggmann I, Uys P (1980) Lorazepam as a premedicant in dental surgery. S Afr Med J 58:124–126
Seow LT, Mather LE, Cousins MJ (1985) Comparison of the efficacy of chlormethiazole and diazepam as i.v. sedatives for supplementation for extradural anaesthesia. Br J Anaesth 57:747–752
Shader RI, Greenblatt DJ (1983) Triazolam and anterograde amnesia: all is not well in the z-zone. J Clin Psychopharmacol 3:273
Skolnick P, Paul SM, Weissman BA (1984) Preclinical pharmacology of buspirone hydrochloride. Pharmacotherapy 4:308–314
Sonander H, Arnold E, Nilsson K (1985) Effects of rectal administration of diazepam. Diazepam concentrations in children undergoing regional anaesthesia. Br J Anaesth 57:578–580
Stovner J, Endresen R, Osterud A (1973) Intravenous anaesthesia with a new benzodiazepine Ro 5–4200. Acta Anaesthesiol Scand 17:163–169
Studd C, Eltringham RJ (1980) Lorazepam as night sedation and premedication: a comparison with diazepam. Anaesthesia 35:60–64
Taub HA, Eisenberg L (1976) An evaluation of memory under regional anesthesia with i.v. lorazepam as a premedicant. Anesth Analg 55:368–373
Tauber U, Tack JW, Dorrow R, Hilman J (1985) Plasma levels of lormetazepam after sublingual and oral administration of 1 mg to humans. Indust Pharmacol (in press)
Tulving E (1972) Episodic and semantic memory. In: Tulving E, Donaldson W (eds) Organization of memory. Academic, New York, pp 381–403
Tulving E (1983) Elements of episodic memory. Clarendon, Oxford
Van Wijhe M, Voogt-Frenkel E, Stijnen T (1985) Midazolam versus fentanyl/droperidol and placebo as intramuscular premedicants. Acta Anaesthesiol Scand 29:409–414
Vega DE (1971) Induction of anaesthetic sleep with by means of a new benzodiazepine derivative. Rev Urug Anaesth 5:41–44
Volicer BJ, Isenberg MA, Burns MW (1977) Medical-surgical differences in hospital stress factors. J Hum Stress (June):3–13
Vogel JR (1979) Objective measurement of human performance changes produced by anti-anxiety drugs. In: Fielding J, Lal S (eds) Anxiolytics. Futura, New York, pp 343–374
Warrington EK (1986) Memory for facts and memory for events. Br J Clin Psychol 25:1–12
White PF (1982) Comparative evaluation of intravenous agents for rapid sequence induction – thiopental, ketamine and midazolam. Anesthesiology 57:279–284
Whitman JG, Al-Khudhairi D, McCloy RF (1983) Comparison of midazolam and diazepam in doses of comparable potency during gastroscopy. Br J Anaesth 55:773–777
Willer JC, Von Frenkell R, Bonnet D, LeFur G (1986) The ability of PK 8165, a quinoline derivative, to reduce responses to a stressful situation in a double-blind study in man. Neuropharmacology 25:275–281
Wilson J (1973) Lorazepam as a premedicant for general anaesthesia. Curr Med Res Opin 1:308–315
Wilson J, Ellis FR (1973) Oral premedication with lorazepam (Ativan): a comparison with heptabarbitone (Medomin) and diazepam (Valium). Br J Anaesth 45:738–744
Wilson SL, Vaughan RW, Stephen CR (1975) Awareness, dreams and hallucinations associated with general anaesthesia. Anesth Analg 54:609–617
Wilson-Barnett J (1979) Stress in hospital. Churchill Livingstone, Edinburgh
Wittenborn JR (1979) Effects of benzodiazepines on psychomotor performance. Br J Clin Pharmacol 7:61S–67S
Zajonc RB (1980) Feeling and thinking: preferences need no inferences. Am Psychol 35:151–175

Benzodiazepines, Memory and Information Processing

Lormetazepam, Memory and Information Processing: A Review

J. Z. BHATTI[1], C. A. ALFORD and I. HINDMARCH

Abstract

A review was made of the literature exploring the psychopharmacology of lormetazepam. Results from studies show that there can be a general hangover the morning following nocturnal doses of 2 mg. Findings from more recent work have shown that the 1.5 mg dose can disrupt retrieval of information from short-term memory. Lormetazepam 1 mg has no residual sedative effects. A daytime psychopharmacodynamic study showed lormetazepam 1–2 mg to have no disruptive effects 5 h after drug intake.

1 Introduction

There are a considerable number of benzodiazepines currently available as sedatives or hypnotics which can be effectively used to treat insomnia. The extent to which they produce residual sedation, impairing early morning behaviour, differs from one benzodiazepine to the next. Benzodiazepine "hangover" can be detected using tests that measure subjective aspects of drug action, and objective tests that measure changes in performance. HINDMARCH (1984) has clearly shown that a wide range of benzodiazepines possess unwanted residual sedative side effects, which are manifested as perceived impairment of early morning behaviour. These subjective reports of residual sedation have been confirmed by findings from studies that have employed objective tests of psychomotor function (HINDMARCH 1980).

A number of benzodiazepines are known to cause anterograde amnesia, and are of use as surgical premedicants. Impairment of mnemonic processes in patients undergoing surgery have been reported for diazepam (CLARKE et al. 1970), lorazepam (HEISTERKAMP and COHEN 1974) and flunitrazepam (DUNDEE and GEORGE 1976). Benzodiazepine-induced amnesia is an unwanted side effect when that benzodiazepine is to be used as a sedative hypnotic. Disruption of memory has been reported the morning after night-time administration of lorazepam (ROTH et al. 1980), flunitrazepam (BIXLER et al. 1979; HARRISON et al. 1985), flurazepam (ROTH et al. 1980) and triazolam (SPINWEBER and JOHNSON 1983). The residual effects of benzodiazepines on memory performance is an important parameter that needs to be borne in mind when considering the safety of patients.

[1] Human Psychopharmacology Research Unit, Department of Psychology, University of Leeds, Leeds LS2 9JT, UK.

Researchers who explore aspects of memory are faced with the difficulty that there are no widely accepted models of memory. The concept that dominated most of the early work on memory was that of ATKINSON and SHIFFRIN (1971). According to their model of memory, information is placed into a limited-capacity short-term store, in which it must be maintained by rehearsal if it is not to be replaced by other items of incoming information. In addition to exploring the structure of memory it is important to examine the way in which information is processed within it. Information processing models provide useful frameworks within which it is possible to explore a number of aspects of cognitive functioning. The following illustrates some of the processes involved in this model:

Information input "Stimulus" → Encoding → Storage → Retrieval
→ Information output "Response"

There are two processing stages in this model, the encoding and the retrieval stages. Storage is considered to be relatively passive. The encoding stage involves the processing of incoming information to a form appropriate for storage and classification. The retrieval process involves a search for that memory trace followed by the extraction of useful information.

This accords well with STERNBERG's model of short-term memory (STERNBERG 1969) which requires incoming information to proceed through a number of stages such as coding the sensory input, searching memory for relevant information, selecting a response and then executing that response. These different stages have been isolated using an additive factor model, where a given manipulation that affects a single stage will have an additive effect on overall response time but one that effects more than one stage will have multiplicative effects. The logical basis for confining drug action to specific components of cognition as defined by the Sternberg model may be open to question. Memory tasks based on the Sternberg model may just consist of a series of components which are differentially sensitive to different drugs (BROADBENT 1984). In spite of these criticisms, tasks based on the Sternberg methodology have been useful in detecting drug effects (SUBHAN and HINDMARCH 1984a, b).

The effects of a drug on everyday activity is difficult to gauge, but it is possible to measure aspects of daily living under controlled laboratory conditions. It is not unreasonable to claim that impaired performance in the laboratory would be manifest in terms of impaired everyday activity. Clearly, normal functioning of information processing skills is required for everyday tasks that range from being able to store and recall a series of digits when dialling a telephone number, to being able to integrate complex information from a variety of sources when driving a car. It is therefore necessary when looking for possible residual effects of hypnotics to investigate performance on tasks that explore a number of aspects of information processing.

2 Earlier Studies

OSWALD et al. (1979) evaluated the effects of lormetazepam 1 mg and 2.5 mg, flurazepam 30 mg and placebo administered nightly for 3 weeks on daytime skilled performance. Four tasks were used to assess performance: a task of manual dexterity which involved placing awkwardly shaped pellets as quickly as possible into a small tube, a card-sorting task, a digit-symbol substitution task (DSST), and an auditory vigilance task of 1 h duration. Testing took place at 08.30, 12.30 and 16.30. Both the 1-mg and 2.5-mg doses of lormetazepam were found to have no residual effects on all tasks that involved memory. Performance on the manual dexterity task, which is primarily a measure of motor skills, was found to be impaired on the morning of the first week following the 2.5-mg dose when compared to placebo ($p<0.025$). Flurazepam 30 mg was found to impair performance on all tests at all times of the day throughout the study period.

NICHOLSON and STONE (1982) investigated the effects of lormetazepam 0.5 mg, 1 mg and 2 mg when given at night-time. A test of visuomotor coordination and a DSST were administered the following morning. The 2-mg dose was found to impair visuomotor coordination and performance on the DSST, whilst the 0.5-mg and 1-mg doses had no residual effects.

SUBHAN and HINDMARCH (1983) evaluated the effects of night-time administration of lormetazepam 0.5 mg, 1 mg and 2 mg on early morning performance. The tests used were the DSST and critical flicker fusion (CFF) test. Lormetazepam 0.5 mg was free from residual effects. The 1-mg dose was found to impair performance on the DSST. This was not accompanied by any change in CFF threshold and was not consistent with earlier findings (NICHOLSON and STONE 1982). The authors suggested that this finding may have been an erroneous result. Lormetazepam 2 mg was found to impair performance on the DSST and to significantly lower CFF threshold.

ROEHRS et al. (1984) investigated the effects of lormetazepam 1.5 mg, temazepam 30 mg and flurazepam 30 mg on results of tests of memory span, the digit-symbol copying test (DSCT) and the DSST the following morning. In addition, subjects were awakened 3 h after drug intake and were tested on a 16-item memory recall task. Recall was again tested, without further item presentation, the following morning. There were no residual effects on either memory span or DSCT following any of the drug conditions. Both temazepam and flurazepam were found to impair performance on DSST, whilst lormetazepam 1.5 mg had no residual effects. Recall 3 h after drug administration was impaired following temazepam and flurazepam, but not following lormetazepam. All the drug conditions impaired recall the following morning.

SUBHAN (1984) explored the effects of an acute night-time dose of lormetazepam 1 mg, triazolam 0.25 mg and flunitrazepam 1 mg on stages of information processing in a STERNBERG memory scanning task. Overall reaction times were found to be increased when tested 10 h after administration of flunitrazepam 1 mg. Lormetazepam 1 mg had no residual effects.

SUBHAN and HINDMARCH (1984a, b) investigated a range of benzodiazepines, lormetazepam 1 mg, triazolam 0.25 mg, nitrazepam 5 mg, temazepam 20 mg and flurazepam 15 mg. Two tests of working memory were given 12 h after drug administration. Memory span for digits was measured by requiring subjects to recall, in reverse order, a series of 10 random digits presented at the rate of one digit every second. The score was expressed as the maximum number of digits recalled in the correct order. Subjects ability to learn and recall information was tested using a list of 16 names and associated four-figure telephone numbers. All the hypnotics, with the exception of lormetazepam, impaired performance on both tasks.

3 Recent Studies

BHATTI and HINDMARCH (unpublished observations) studied lormetazepam 1 mg, 1.5 mg and 2 mg and triazolam 0.5 mg (used as a reference sedative hypnotic) when taken during the day (10.00). Drug effects were assessed 1, 3 and 5 h after administration using the CFF test, a STERNBERG memory scanning task and a word recognition task.

The results from the test of CFF threshold showed that all the active conditions, except lormetazepam 1 mg, reduced an individual's ability to process information. It is interesting to note that even the placebo group experienced a change in information processing rate over the course of the day. This would appear to support recent findings that show the CFF test to be capable of detecting changes across the circadian cycle (FREWER 1986). After 3 h lormetazepam 2 mg and triazolam 0.5 mg were still having a significant effect. The results showed that after 5 h all the active treatments had no residual effects. In terms of the information processing model, CFF is exploring the encoding process.

The STERNBERG memory scanning task, which involves encoding, storage and retrieval of information, also showed dose dependent effects of drugs which seemed to decay with the passage of time. At the test session 1 h after administration, lormetazepam 2 mg and triazolam 0.5 mg both caused a significant impairment in performance. After 3 h the only significant effect was caused by triazolam, and 5 h after drug administration there were no significant effects. The skills involved in performing the Sternberg memory scanning task would appear to be less liable to disruption by drug action than those involved in the CFF measure. Performance on the positive part of the word recognition task was unaffected by treatment condition. However, the negative part of the test, the responses to words that were "absent" clearly showed dose-dependent effects. There was a significant drop in performance 1 h after lormetazepam 1.5 mg or 2 mg and after triazolam 0.5 mg. After 3 h lormetazepam 2 mg and triazolam 0.5 mg were still having a significant but much reduced effect on performance. After 5 h there were no significant findings.

Analysing the two parts of this memory test, it is clear that they involve very different cognitive functions. Test words that are present are copy cues of information stored, and subjects are simply being asked for a recognition response. Test words that were not in the set presented are non-useful retrieval cues which are used to explore the memory set prior to a decision being made. Clearly, the former is easier than the latter (ANDERSON and BOWER 1972).

Table 1. Mean changes (SEM in parentheses) from predrug baseline scores 1, 3 and 5 h after medication, on critical flicker fusion test and Sternberg memory scanning task

Time (h)	Placebo	LOR 1.0 mg	LOR 1.5 mg	LOR 2.0 mg	TRI 0.5 mg
Critical flicker fusion (Hz; LSD = 1.1)					
Baseline	27.1	26.5	27.0	27.5	27.1
1	− 0.6	− 1.3	− 1.8*	− 2.3*	− 1.9*
	(0.8)	(0.8)	(1.3)	(1.6)	(2.1)
3	− 1.3	− 1.4	− 1.2	− 2.4*	− 2.4*
	(1.1)	(0.5)	(1.0)	(1.6)	(2.2)
5	− 0.9	− 1.0	− 1.6	− 1.9	− 1.2
	(0.9)	(0.6)	(0.9)	(0.9)	(1.5)
Sternberg memory scanning (ms; LSD = 90.8)					
Baseline	445.2	445.2	446.0	443.4	436.2
1	12.7	62.5	85.3	209.7*	158,6*
	(31.4)	(59.1)	(90.2)	(276.7)	(138.5)
3	8.2	13.6	28.1	78.0	150.8*
	(49.0)	(32.4)	(43.9)	(35.2)	(119.4)
5	− 11.9	− 12.3	3.3	18.6	32.4
	(32.8)	(34.8)	(37.5)	(46.9)	(51.2)

LOR, lormetazepam; TRI, triazolam; LSD, least significant difference.
* Significant difference from placebo at 5% on a Tukey least significant difference test.

Table 2. Mean changes (SEM in parentheses) from predrug baseline scores 1, 3 and 5 h after medication, on the components of the word recognition test

Time (h)	Placebo	LOR 1.0 mg	LOR 1.5 mg	LOR 2.0 mg	TRI 0.5 mg
Positive responses (%; LSD = 0.9)					
Baseline	29.1	28.8	29.1	29.0	29.2
1	0.0	0.0	0.3	0.5	0.6
	(0.9)	(0.6)	(0.8)	(1.4)	(2.0)
3	0.1	0.3	0.0	0.6	0.1
	(0.7)	(1.1)	(0.8)	(1.3)	(0.7)
5	− 0.3	− 0.1	− 0.3	0.3	0.1
	(1.0)	(0.8)	(0.9)	(1.5)	(1.5)
Negative responses (%; LSD = 1.7)					
Baseline	29.2	29.0	29.5	29.1	29.1
1	0.1	1.6	2.5*	4.1*	3.9*
	(0.7)	(1.1)	(1.4)	(2.6)	(3.5)
3	0.4	1.6	1.9	2.7*	2.7*
	(1.3)	(1.3)	(3.1)	(2.3)	(2.1)
5	0.4	1.3	1.6	1.8	1.8
	(0.7)	(1.4)	(2.7)	(1.9)	(1.7)

LOR, lormetazepam; TRI, triazolam; LSD, least significant difference.
* Significant difference from placebo at 5% on a Turkey least significant difference test.
Performance score (%) takes account of accuracy of response:

$$\text{Performance score} = 10 \log \left(\frac{\text{mean response time (ms)} \times \text{maximum correct responses}}{\text{actual correct responses}} \right)$$

A full account of these results is shown in Tables 1 and 2, and they are illustrated in Figs. 1–3.

DYE, ALFORD and HINDMARCH (unpublished observations) investigated the effects of lormetazepam 1 mg, 1.5 mg and 2 mg and triazolam 0.5 mg (used as a reference sedative hypnotic) when taken at bed time. Drug action was explored 1 h after medication by CFF test and immediate recall of word list, and after 10 h by CFF test, delayed recall of word list and a STERNBERG memory scanning task.

The results showed CFF threshold to be decreased 1 h after all the active treatments. None of the lormetazepam treatments were found to impair the ability to process information as measured by CFF the following morning at the 10-h test session, indicating absence of an effect of the drug on the encoding stage of information processing. Triazolam 0.5 mg was found to have a sedative effect as measured by CFF at 10 h.

The immediate recall of the word list (1-h test) showed that recall was significantly poorer for the lormetazepam 1.5 mg and the triazolam 0.5 mg conditions. It is of interest to note that at the 10-h test session all the active treatment conditions showed significantly less recall than placebo, this even being the case for treatments where there appeared to be good recall at the 1-h test.

There are two points that emerge from these results; first that it is possible to differentiate between treatments at the learning and immediate recall phase (1-h test), and secondly that all subjects on the active treatments have significantly poorer recall after 10 h. Interpreting these findings in terms of strength theory (KINTSCH 1968), it would appear that the strength of a memory trace is impaired by all the active treatments. Performance of the Sternberg memory scanning task

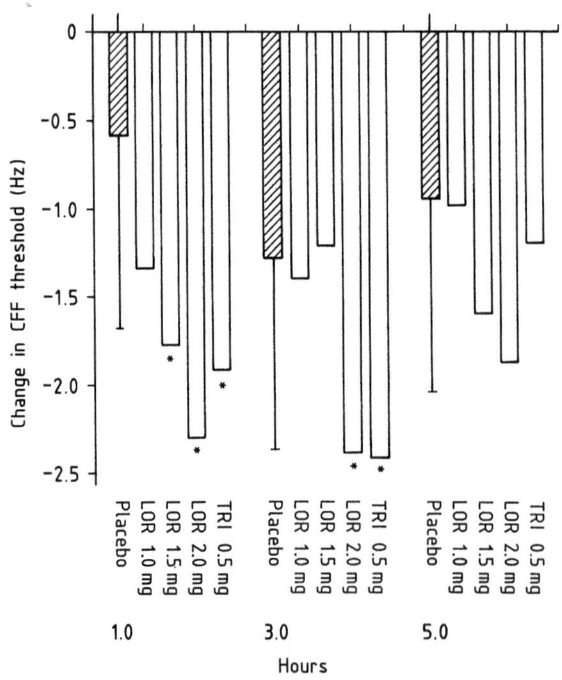

Fig. 1. Mean differences (+SEM for placebo) from baseline scores on the critical flicker fusion (CFF) test, 1, 3 and 5 h after drug administration (*LOR*, lormetazepam; *TRI*, triazolam)

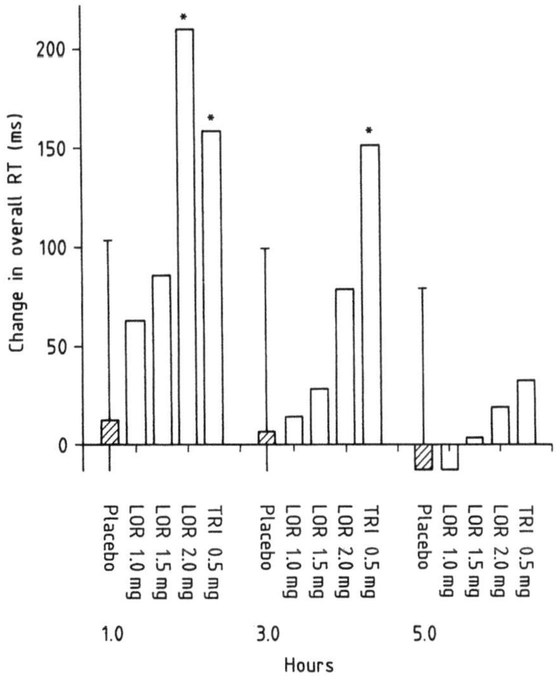

Fig. 2. Mean differences (+SEM for placebo) from baseline scores for the overall information retrieval reaction times (*RT*) on the Sternberg memory scanning task, 1, 3, and 5 h after drug administration

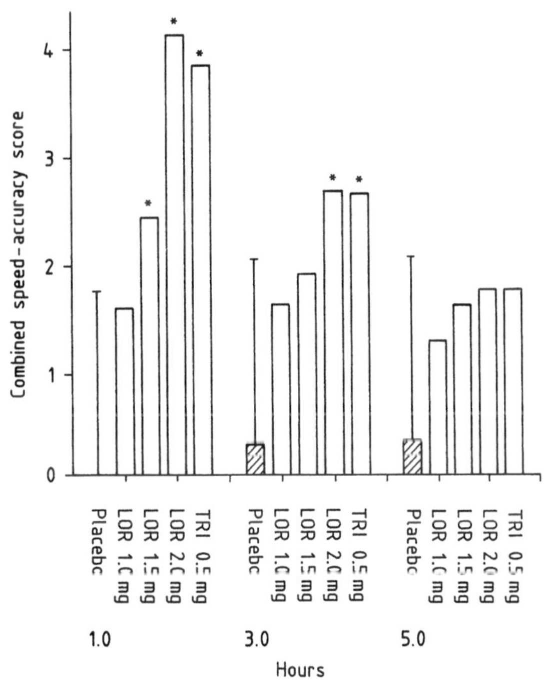

Fig. 3. Mean differences (+SEM for placebo) from baseline scores for the overall recognition score for words that were absent in the word recognition test, 1, 3 and 5 h after drug administration

was only tested at the 10-h test session when residual sedative effects were evident following lormetazepam 1.5 mg and 2 mg. Taken together with the results from CFF testing, this finding would suggest that any lormetazepam-induced hangover is localised in the storage and retrieval phases of information processing. Results are listed in full in Table 3 and illustrated in Figs. 4–6.

Table 3. Mean scores (SEM in parentheses) 1 h and 10 h after medication on critical flicker fusion test, word recall and Sternberg memory scanning task

Time (h)	Placebo	LOR 1.0 mg	LOR 1.5 mg	LOR 2.0 mg	TRI 0.5 mg
Critical flicker fusion (Hz; LSD=0.5)					
1	27.9	27.0*	27.2*	27.3*	26.0*
	(2.6)	(3)	(2.9)	(3.3)	(3.1)
10	28.4	28.2	28.6	28.5	27.9*
	(2.3)	(2.9)	(3.2)	(3)	(3.4)
Word recall (*n*; LSD=1.5)					
1	9.3	8.3	7.6*	8.5	5.9*
	(2.3)	(2.9)	(3.2)	(3)	(3.4)
10	6.3	3.6*	4.2*	3.3*	2.2*
	(2.5)	(2.3)	(3.1)	(3)	(3.2)
Sternberg memory scanning (ms; LSD=25.5)					
10	470.1	487.1	506.1*	501.1*	491.5
	(67)	(78)	(72)	(52)	(69)

LOR, lormetazepam; TRI, triazolam; LSD, least significant difference.
* Significant difference from placebo at 5% on a Tukey least significant difference test.

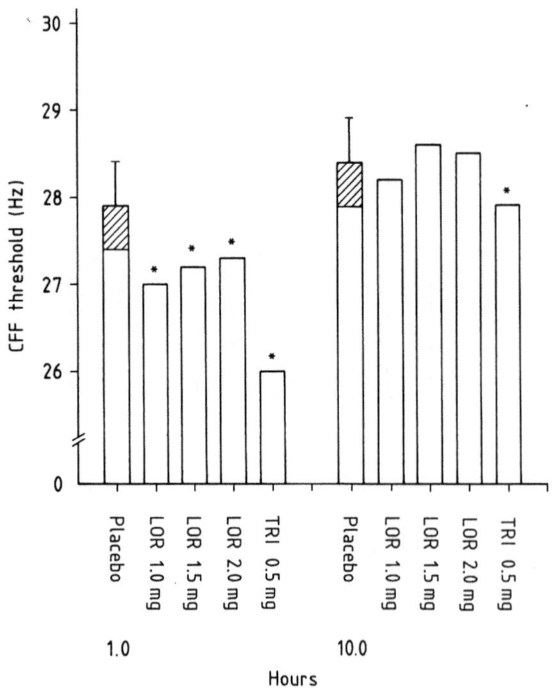

Fig. 4. Mean scores (+SEM for placebo) on the critical flicker fusion (CFF) test, 1 and 10 h after drug administration

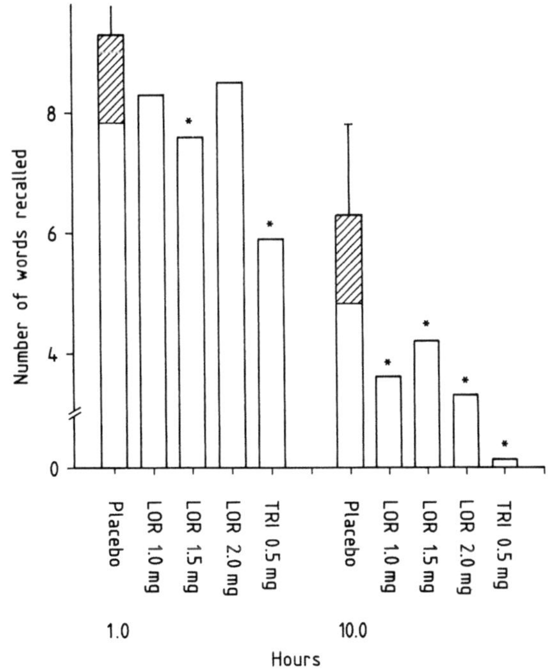

Fig. 5. Mean number (+SEM for placebo) of words recalled on the word recall test, 1 and 10 h after drug administration

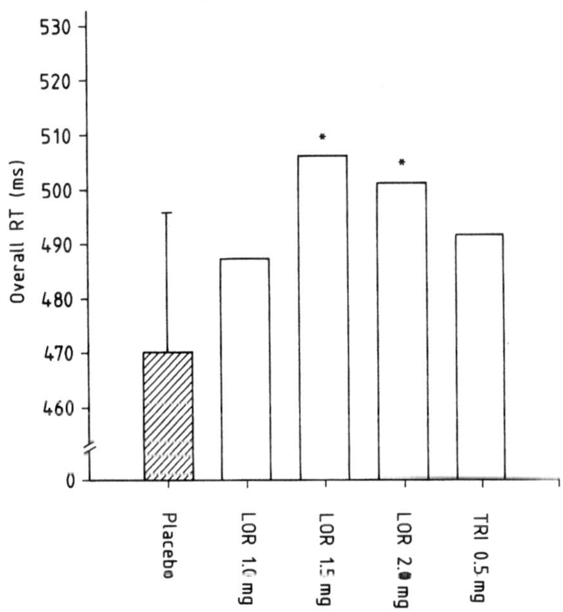

Fig. 6. Mean scores (+SEM for placebo) for the overall information retrieval reaction times (*RT*) on the Sternberg memory scanning task, 1 and 10 h after drug administration

4 Conclusions

Any reports of residual effects following the use of a sedative-hypnotic are cause for concern. Any measurable impairment in cognitive skills will make an individual less capable of dealing with the rigours of daily life. Lormetazepam is a sedative-hypnotic which, it is claimed, is free from "hangover" the following morning. On the whole the review of the literature supported this claim with the rider that the dose used should remain below 2 mg. The data from more recent studies suggest that a minimum dose of lormetazepam 1.5 mg is required to cause daytime sedation, and that a 2 mg dose is free from effect 5 h later. The data show that there is no gross hangover the day following lormetazepam 1 mg, 1.5 mg or 2 mg as measured by CFF. The results do, however, warn of disruption of cognitive tasks that place demands on mnemonic skills, with doses of lormetazepam of 1.5 mg or more.

References

Anderson JR, Bower GH (1972) Recognition and retrieval processes in free recall. Psychol Rev 79:97–123

Atkinson RC, Shiffrin RM (1971) The control of short-term memory. Sci Am 225:82–90

Bixler EO, Scharf MB, Soldatos CR, Mitsky DJ, Kales A (1979) Effects of hypnotic drugs on memory. Life Sci 25:1379–1388

Broadbent DE (1984) Performance and its measurements. Br J Clin Pharmacol (Suppl 1) 18:5–9

Clarke PRF, Eccersley PS, Frisby JP, Thornton JA (1970) The amnesic effect of diazepam (Valium). Br J Anaesth 42:690–697

Dundee JW, George DA (1976) The amnesic action of diazepam, flunitrazepam and lorazepam in man. Acta Anaesth Belg 27:3–11

Frewer LJ (1986) Some psychopharmacological variables affecting the critical flicker fusion threshold. PhD Thesis, University of Leeds

Harrison C, Subhan Z, Hindmarch I (1985) Residual effects of zopiclone and benzodiazepine hypnotics on psychomotor performance related to car driving. Drugs Exp Clin Res XI:823–829

Heisterkamp DV, Cohen PJ (1974) The effect of intravenous premedication with lorazepam (Ativan), pentobarbitone or diazepam on recall. Br J Anaesth 47:79–81

Hindmarch I (1980) Early morning performance after hypnotics. Sleep Top 1:4–8

Hindmarch I (1984) Subjective aspects of the effects of benzodiazepines on sleep and early morning behaviour. Irish J Med Sci 153(8):272–278

Kintsch W (1968) Recognition and free recall of organised lists. J Exp Psychol 78:481–487

Nicholson AN, Stone BM (1982) Hypnotic activity and effects on performance of lormetazepam and camazepam – analogues of temazepam. Br J Clin Pharmacol 13:433–439

Oswald I, Adam K, Borrow S, Idzikowski C (1979) The effects of two hypnotics on sleep, subjective feelings and skilled performance. In: Passouant P, Oswald I (eds) Pharmacology of the states of alertness. Pergamon, Oxford, pp 51–63

Roehrs T, McLenaghan A, Koshorek G, Zorick F, Roth T (1984) Amnesic effects of lormetazepam. In: Hindmarch I, Ott H, Roth T (eds) Sleep, benzodiazepines and performance. Springer, Berlin Heidelberg New York, pp 165–172

Roth T, Hartse KM, Saab PG, Piccione PM, Kramer M (1980) The effects of flurazepam, lorazepam and triazolam on sleep and memory. Psychopharmacology 70:231

Spinweber CL, Johnson LC (1983) Effects of triazolam (0.5 mg) on sleep, performance, memory and arousal threshold. Psychopharmacology 75:5–12

Sternberg S (1969) Memory scanning: mental processes revealed by reaction time experiments. Am Sci 57:421–457

Subhan Z (1984) The effects of benzodiazepines on short-term memory and information processing. In: Hindmarch I, Ott H, Roth T (eds) Sleep, benzodiazepines and performance. Springer, Berlin Heidelberg New York, pp 173–181

Subhan Z, Hindmarch I (1983) The effects of lormetazepam on aspects of sleep and early morning performance. Eur J Clin Pharmacol 25:47–51

Subhan Z, Hindmarch I (1984a) Effects of zopiclone and benzodiazepine hypnotics on search in short-term memory. Neuropsychobiology 12:244–248

Subhan Z, Hindmarch I (1984b) The psychopharmacological effects of vinpocetine in normal healthy volunteers. Eur J Clin Pharmacol 28:567–571

Anterograde and Retrograde Amnesia after Lormetazepam and Flunitrazepam

H. Ott[1], A. Rohloff, B. Aufdembrinke and K. Fichte

Abstract

In a pharmacopsychological study, memory impairments after single oral doses of benzodiazepines or placebo were investigated in 40 healthy men aged 20–40 years. The study was designed as a double-blind and placebo-controlled trial. Four independent groups of 10 subjects randomly received either 1 mg lormetazepam, 2 mg lormetazepam, 2 mg flunitrazepam, or placebo. The tests consisted of word lists, picture tests, and syllable pairs (consonant-vowel-consonant trigrams). Tests were performed before drug ingestion, and 1, 2, 3, and 5 h after application. Different test versions were used on each occasion. The target variables were immediate recall (after presentation and a 10-s distraction task) and delayed recall and recognition (after 30 min). Recognition was also tested after 24 h for all five versions.

A distinction must be made between anterograde amnesic effects and retrograde amnesic effects. The greatest anterograde memory impairments were observed after 2 mg flunitrazepam ($p < 0.05$). Lormetazepam 2 mg produced less marked impairments than flunitrazepam. Results after 1 mg lormetazepam did not differ from those after placebo. Performance in the memory tests was better under benzodiazepines than under placebo as regards material learned before drug ingestion, i.e. the benzodiazepines had not negative retrograde amnestic effects, but rather "promnesic" effects.

The results suggest that the extent of the benzodiazepines' amnesic effects – both negative (anterograde) and positive (retrograde) – depends on the dosage and type of substance.

1 Introduction

The basic conditions for normal memory functioning are adequate degrees of wakefulness and vigilance. In general, then, pharmacological memory research investigates changes of memory functions by exposing *awake* subjects to pharmacological influences, as is also the case for the study of other mental and psychomotor functions (Ott 19841984a, b).

Atkinson and Shiffrin (1971) have presented a simple concept of information processing and storage processes, yet they do not refer to individual variations of such factors as vigilance, perceptual selection, disposition, motivation, emotion, reaction state, or to environmental influences such as stimulus quality, stimulus intensity, or stimulus background (Fig. 1). The central feature of this two-component theory is based on the idea that after registration of sensory stimuli (acquisition phase), the information is processed in a so-called working memory (first component) by various control processes such as rehearsal, encoding, association

[1] Research Laboratories, Department of Pharmacopsychology, Schering AG, 1000 Berlin 65, FRG.

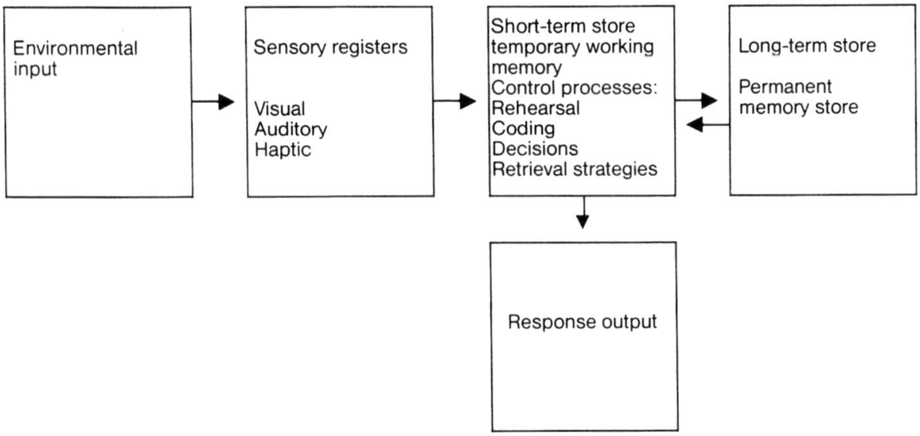

Fig. 1. The operation of information processing after ATKINSON and SHIFFRIN (1971), which forms the basis of the memory model

and decision making, and then transferred to long-term memory (second component). The recall of material stored in long-term memory to working memory is again an active process involving different retrieval strategies. If the strategy is successful, memory is used to control the response; otherwise the search continues or is called off.

Pharmacopsychological investigations of tranquilizers and hypnotics of the benzodiazepine type are frequently based on the above memory model (e.g., LILJEQUIST et al. 1978; OTT et al. 1980; SCHRATZER and BISCHOFF 1984; SUBHAN 1984). While the main interest in these works has focused on recall and recognition performance at various times AFTER drug intake, the following study concentrates on the ACQUISITION, i.e., the learning, of stimulus material.

2 Methods

The aim of this prospective empirical double-blind investigation was to study impairments of memory performance after single oral doses of benzodiazepines. The test substances administered were lormetazepam (Noctamid) 1 mg and 2 mg, flunitrazepam (Rohypnol) 2 mg, and placebo. Forty healthy male subjects were randomly allocated to four equal treatment groups. A number of different tests were performed to assess anterograde impairments of memory for up to 24 h after drug administration. In order to investigate phenomena of retrograde amnesia, information acquired before drug intake was tested 90 min, 270 min, and 24 h after ingestion.

Five versions of a word list, a picture test, and a syllable-pair test – each consisting of six items – were selected to assess memory performance. In the following, the results of the word lists are taken to illustrate the findings, as the findings of the pictures and syllable-pair tests can essentially be interpreted in the same

Table 1. Time schedule of the investigation of anterograde amnesia

Time		Test version	Test
PRE	−30 min	1 }	Immediate recall (learning phase)
	− 5 min	1 }	Delayed recall
			Recognition
	0 min	Medication	
POST	30 min p.m.	2	Immediate recall (learning phase)
	60 min p.m.	2	Delayed recall
			Recognition
	90 min p.m.	3	Immediate recall (learning phase)
	120 min p.m.	3	Delayed recall
			Recognition
	150 min p.m.	4	Immediate recall (learning phase)
	180 min p.m.	4	Delayed recall
			Recognition
	270 min p.m.	5	Immediate recall (learning phase)
	300 min p.m.	5	Delayed recall
			Recognition
	24 h p.m.	1–5	Recognition

p.m., postmedication
Each version of the test consisted of six words, six pictures and six syllable pairs, and each of the five different versions followed the same course. Thirty minutes after the learning phase (immediate recall), late recall and recognition were tested.

Table 2. Time schedule of the investigation of retrograde amnesia

Time		Version	Test
PRE	−30 min	1 }	Immediate recall (learning phase)
	− 5 min	1 }	Delayed recall
			Recognition
	0 min	Medication	
POST	90 min p.m.	1	Delayed recall
			Recognition
	270 min p.m.	1	Delayed recall
			Recognition
	24 h p.m.	1	Delayed recall
			Recognition

p.m., postmedication.
Each version of the test consisted of six words, six pictures and six syllable pairs. Version 1 was learnt before medication (immediate recall). Thirty minutes later, late recall and recognition were tested (prevalues). Postvalues for late recall and recognition of version 1 were recorded 90 min, 270 min, and 24 h postadministration.

Table 3. Test procedure during the learning phase with the distraction method (Randt et al. 1980) and the restrictive reminder principle (Buschke and Fuld 1974)

Items	Trial 1			Trial 2			Trials 3–5
	1st step	2nd step	3rd step	4th step	5th step	6th step	
	verbal presentation of all 6 items (●), 2 s for each item	10 s subtraction (distraction method of Randt et al. 1980)	(immediate) recall (+)	selective presentation of non-recalled items (restrictive reminder method of Buschke and Fuld 1974)	10 s subtraction	recall of *all* items (irrespective of sequence)	as trial 2 if necessary
1 Großstadt	●		+			+	
2 Pfirsich	●			●		+	
3 Mädchen	●		+			+	
4 Feier	●		+			+	
5 Aufstand	●			●		+	
6 Liebe	●		+			+	

●, presentation; +, recalled item; blank, missing.

Table 4. Response pattern that formed the basis for the parameter number of reproduced words (see Fig. 2)

Item	Trials				
	1	2	3	4	5
Großstadt (city)	x/0	0		0	
Pfirsich (peach)	x/0	0		0	0
Mädchen (girl)		x/0	0	0	0
Feier (party)	x/0		0	0	0
Aufstand (insurrection)	x/0		0		0
Liebe (love)	x/0		0		
Total	5	3	4	4	4

x, presentation of item; 0, recall of item.
In this example, the number of immediately recalled words is 20.

way. Tables 1 and 2 show the design of the trial and the times of stimulus presentation and testing for the investigation of both anterograde and retrograde amnesia. It is important that after drug intake four different versions of the stimulus material (versions 2–5) are used for testing at the different measurement points (90–270 min after drug administration for studying anterograde amnesia). This ensures that no test item is used twice. The time between the learning procedure and delayed recall was always 30 min when testing anterograde amnesia.

An adaptation of the method used by RANDT et al. (1980) was used; this allowed a detailed assessment of storage and retrieval processes in the learning phase. We started out from the theoretical assumption that items which are presented once and which can then be retrieved after a 10-s distraction phase have already been transferred from short-term to long-term memory. The acquisition phase progressed as shown in Table 3. At each measurement time, the subjects were presented a list of six words that were read out at 2-s intervals. Immediately afterwards the subjects were asked to count backwards in threes from a given number, e.g., 82. After 10 s the investigator stopped this procedure and asked the subject to recall the presented words. It was irrelevant in what order the words were recalled. If the subject paused for 20 s, the investigator repeated the words which could not be recalled (principle of restrictive reminders; BUSCHKE and FULD 1974). The subject was then asked to count backwards in threes from another number, say 77, for 10 s and had then to recall all six words. If he was still unable to remember all six words after the second trial, the investigator repeated the forgotten words from the first and second trials, and the above procedure was repeated. A maximum of five trial runs were permitted. The test was terminated when the subject recalled all six words in one trial run.

Table 4 shows an example of a response pattern that forms the basis for quantifying the characteristic parameters of immediate recall.

3 Results and Interpretation

3.1 Anterograde Amnesia

Figures 2 and 3 show the results recorded under the different conditions and at the different times. Figure 2 shows the total number of words reproduced in all trial runs. The four groups evidently had similar baseline scores, indicating the homogeneity of memory performance of the whole sample ($n = 40$). A drop in performance can be seen in all groups at the later measurement times, and this is most marked in the flunitrazepam group ($p < 0.05$).

Figure 3 shows the cumulative number of newly stored words and thus gives an indication of the transfer from short-term to long-term memory (for all five trials). The mean differences are not as great as for the total number of words reproduced, yet the paired comparisons with flunitrazepam show significant differences (least significant difference test; $p < 0.05$). Two additional measures (results not shown), the number of inconsistently reproduced items and the number of repeated presentations confirm the drop in performance under flunitrazepam in the acquisition phase ($p < 0.05$).

On the basis of the theory outlined above, the results can be interpreted as follows: unlike lormetazepam, flunitrazepam interferes with the learning process. On the one hand, it delays or blocks the transfer from working memory to long-term memory – expressed by the rise in the number of repeated presentations and

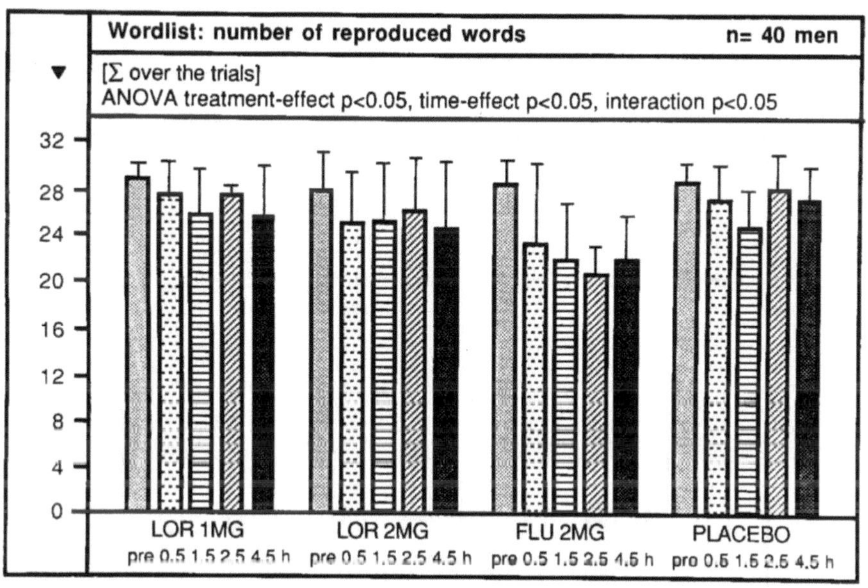

Fig. 2. Means and standard deviations of the number of reproduced words for each measurement in the five trial runs. For all posttreatment measurements, performance after 2 mg flunitrazepam was impaired compared with placebo ($p < 0.05$; LSD tests). Performance was not impaired under 1 mg lormetazepam and only at 4.5 h after administration of 2 mg lormetazepam

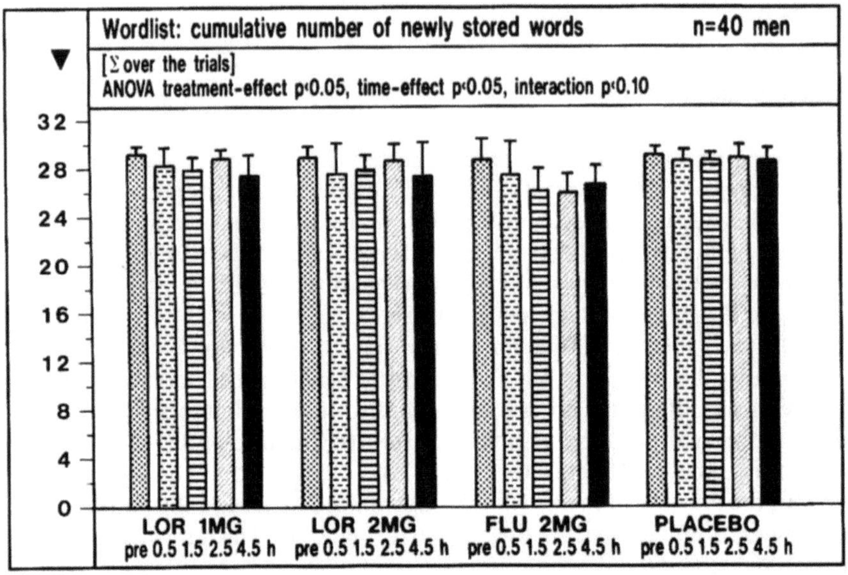

Fig. 3. Means and standard deviations of the cumulative number of stored words for each measurement in the five trial runs ($n=40$ men). Although the mean differences are not very great, the paired comparisons show that fewer new words were stored under 2 mg flunitrazepam ($p<0.05$; LSD tests)

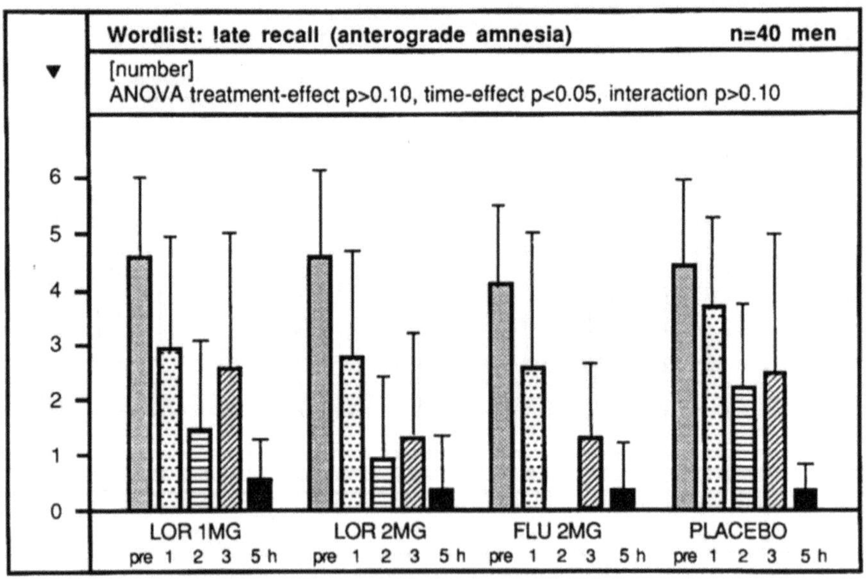

Fig. 4. Means and standard deviations of late recall of items from word lists 30 min after presentation. The anterograde impairment of recall in the benzodiazepine groups is clearly visible. There was a total block of recall 2 h after administration of flunitrazepam

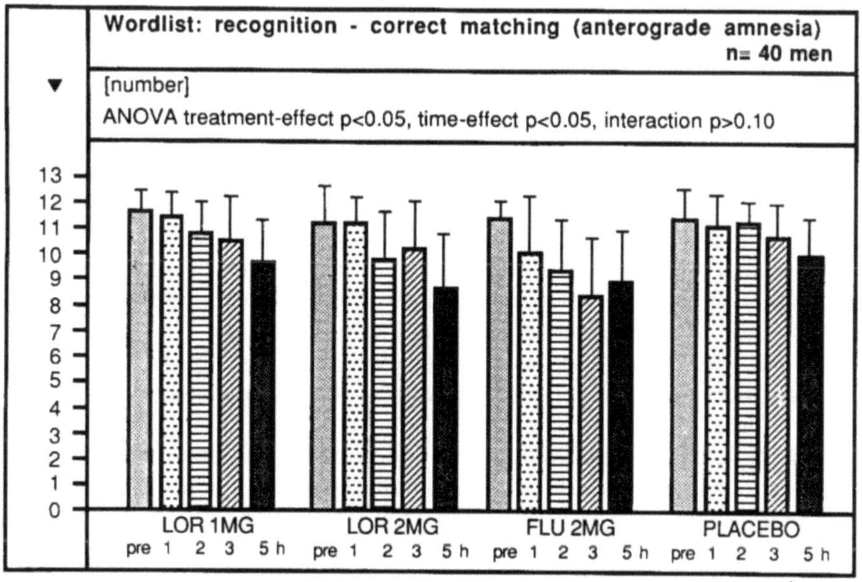

Fig. 5. Means and standard deviations of recognition. After 2 mg flunitrazepam there are significant differences compared with placebo for all postdrug measurements ($p<0.05$; LSD tests). Lormetazepam 2 mg significantly reduced recognition at 3 h and 5 h postadministration. Lormetazepam 1 mg showed no differences in the paired comparison with placebo ($p>0.10$)

the decrease in the cumulative number of stored words. On the other hand, it also inhibits the retrieval of material from long-term memory – evident from the comparison of means. The total number of words reproduced is reduced under flunitrazepam and the number of inconsistently reproduced words is increased. Delayed recall is the measure of the retrieval of information from long-term memory. The subjects were requested to recall the learned word lists 30 min after the acquisition phase of each test version. Figure 4 shows the drop in performance over time in all groups. A particularly striking feature is that 120 min after drug administration not one of the 10 subjects from the flunitrazepam group could recall a single word from a word list learned 30 min previously. The drug led to a total block of recall at this measurement time.

That recall was blocked becomes evident from performance in a recognition task. Immediately after free recall, the subjects were presented 12 words (six learned words and six distractors). Their aim was to decide which words were items from the word list and which were not. Figure 5 shows a reduction of performance under flunitrazepam and 2 mg lormetazepam compared with placebo ($p<0.05$), yet from the means it can be seen that the impairment was less severe than in free recall. We believe that the results indicate that the benzodiazepines affect the transfer from working to long-term memory and the retention of information less strongly than retrieval mechanisms.

3.2 Retrograde Amnesia

In order to investigate possible retrograde impairments of memory, the first version of the word list was tested 90 min, 270 min, and 24 h after drug intake (see Table 2). The aim was to clarify whether benzodiazepines disturb the retrieval from long-term memory of material acquired before drug ingestion. Figure 6 shows the results of free recall. The data show that benzodiazepines do not have retrograde amnesic effects. On the contrary, a "promnesic" profile of action is apparent in that memory was sometimes significantly better in the benzodiazepine groups than in the placebo group. This result is not altogether unexpected (LILJEQUIST et al. 1978) but deserves particular attention.

The following interpretations are put forward as a contribution to more extensive discussion and do not claim to provide a sufficient explanation of the "promnesic" potency of the benzodiazepines.

One should begin by stating that the poor memory performance in the placebo group cannot be interpreted as a chance intergroup difference, since the earlier values did not differ ($p > 0.10$) and the good learning and retrieval of the sample is evident from Figs. 1 and 2. Moreover, the other tests, both the pictures and the sylable pairs, also showed that the benzodiazepine groups were better at recalling information acquired before application, although the initial values were the same in all groups.

The reduction of performance in the placebo group can be explained with the help of the interference theory. According to this theory, which has received wide

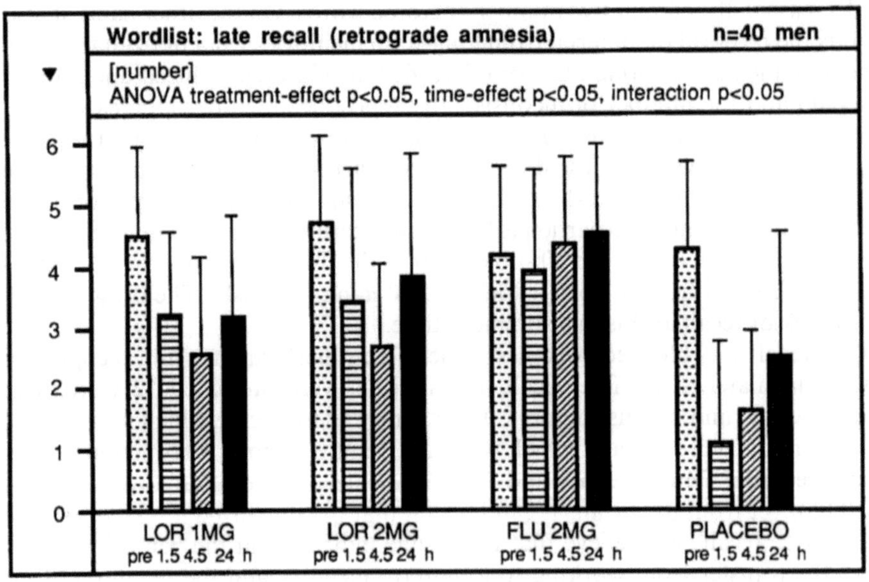

Fig. 6. Means and standard deviations of free recall of items from word list version 1, which was learned before drug administration. The paired comparisons with placebo indicate promnesic effects of the benzodiazepines ($p < 0.05$; LSD tests): flunitrazepam 2 mg at all measurement times; lormetazepam 2 mg at 1.5 h and 24 h postadministration; lormetazepam 1 mg at 1.5 h postadministration

empirical confirmation, the retrieval of learned material is inhibited by the further acquisition of new material, this having a detrimental effect on previously learned information. The greater the similarity between the contents, the greater the performance deficits; this would also apply to the present experiment. Since the disturbance relates back to material acquired earlier, the effect is termed retroactive interference (ZIMBARDO 1983). The question, however, is why the retroactive interference was not as great in the benzodiazepine groups as in the placebo group and what mechanism hindered such interference.

Benzodiazepines have an inhibitory action on the central nervous system, e.g., they relieve tension and anxiety. They are therefore often administered in clinical practice for the treatment of anxiety and tension. In psychological terms, these effects can be characterized as stimulus shielding actions. The patient is protected from pathogenic subjection to external stimuli; relevant internal and external stimuli can thus be processed more appropriately. While the benzodiazepines reduce the functioning of the entire information processing system through their inhibitory actions (anterograde disturbances of acquisition, storage, and retrieval), existing memory contents are, as it were, "protected" by the drug action. This screening effect prevents the formation of retroactive interference: newly acquired material does not disturb the retrieval of existing information in memory.

Another explanation for the "promnesic" effects could be a "reactive increase of tension" (DÜKER 1964) in the benzodiazepine-treated subjects. As the subjects sensed the onset of the drug effect, e.g., through the feeling of sedation (at about 20 min after administration), they may have attempted to counteract the expected performance deficits. One counterregulatory technique could have been increased rehearsal, particularly of the first version of the word list. This strategy would explain the good performance by the benzodiazepine-treated subjects.

Our data suggest that the reduction of memory performance under the effects of benzodiazepines (anterograde amnesia) is inversely proportional to the effect on recall of material learned before drug application (retrograde amnesia). There is a correlation between the potency of benzodiazepines (dependent on dose and substance type) and their anterograde amnesic as well as "promnesic" potency. Lormetazepam at the low dose of 1 mg, for instance, caused very weak anterograde memory deficits and slight "promnesic" effects. Lormetazepam 2 mg, by contrast, caused more marked changes in both directions, and these changes were even more distinct with the more potent compound flunitrazepam.

3.3 Motor Performance and Subjective Mood

Parallel to each test of memory performance, motor performance was measured with the video tracking test (VTT), and subjective mood was assessed with visual analog scales (VAS).

The VTT is a complex test of fine motor performance: two square cursors of different contrasts are generated on a video monitor, the brighter cursor being the target and the other the tracking signal. The subject has to maneuver the tracking signal with a joystick so that it always covers the target. In the present version, the target moves across the screen sinusoidally with uniform amplitude and

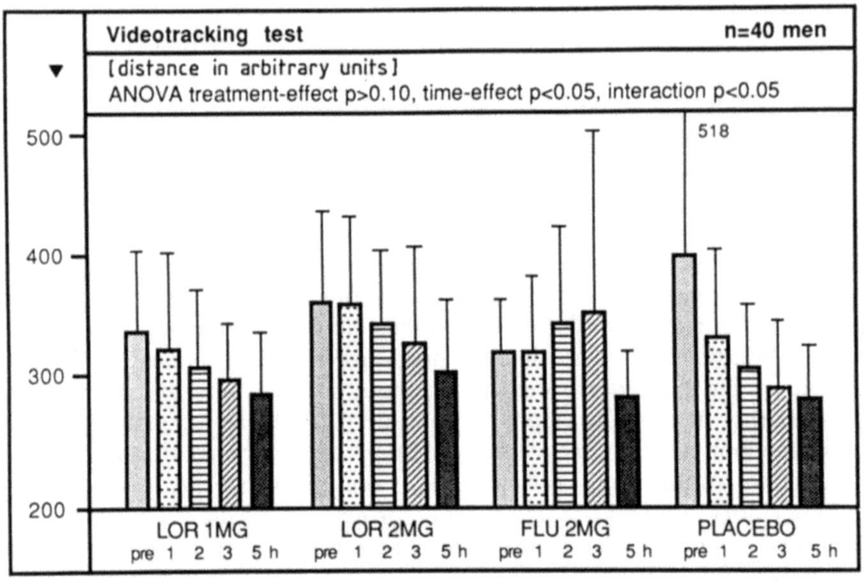

Fig. 7. Means and standard deviations for fine motor performance in the video tracking test. An improvement of performance from measurement to measurement is evident under placebo and lormetazepam 1 mg (learning effect); with lormetazepam 2 mg performance improved at 2 h postadministration while with flunitrazepam 2 mg the values increased at 2 h and 3 h. An improvement of performance under flunitrazepam was not evident until 5 h postadministration

speed. The device calculates the mean radial distance (in arbitrary units) between target and tracking cursor. The higher the value, the poorer the tracking performance of the subject. Figure 7 shows the means from five trial runs at each measurement time. It is evident that the four treatment groups attained different baseline scores. It should, however, be stated that this was the very first time that the VTT – which was developed by the authors and W. Röske – was used in a human pharmacological study and that the subjects were not familiar with the test. To avoid such different baseline values, in future studies subjects will have to be adapted to the test, i.e., the test phase will be preceded by training sessions to ensure that the level of performance is homogeneous across all groups.

The results nevertheless parallel those observed in the memory test: while the group of subjects given 1 mg lormetazepam could improve their performance from measurement to measurement, similarly to the placebo group, the two groups that received 2 mg lormetazepam and flunitrazepam still showed no learning effect 1 h after drug intake. Such an improvement was not seen until 2 h after drug administration in the higher-dose lormetazepam group. Under flunitrazepam, performance poorer at 2 and 3 h after administration. It was at the same measurement times that memory deficits were most pronounced under flunitrazepam.

That the subjects from the flunitrazepam group performed as well as the placebo group 5 h after drug administration could suggest that the drug did not

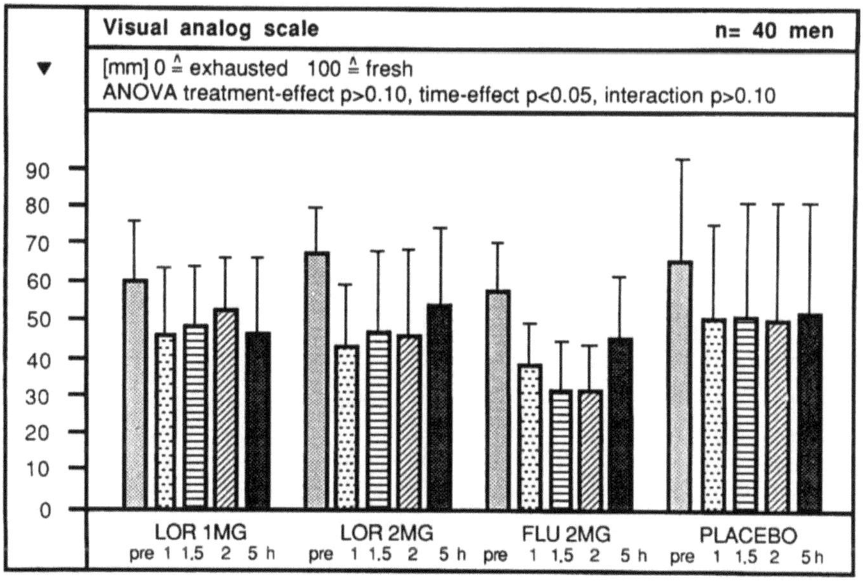

Fig. 8. Means and standard deviations of the visual analog scale for exhaustion. Compared with the other substances tested, with flunitrazepam the feeling of "freshness" tends most markedly towards the pole "exhaustion" after drug administration

altogether prevent psychomotor learning and that mental elements (e.g., perception of the target's characteristic sinusoidal movement and the sensitivity of joystick reactions) were indeed subject to learning. Yet the benzodiazepine-induced inhibition and slowing-down of motor performance prevented an improvement of the overall score. In other words, the benzodiazepines could have a differential influence on mental strategies and motor performance. On the other hand, one should never forget in empirical studies that because of random variation of values, the interpretation of findings not controlled by the design has to remain open, since other random factors could have had an influence.

The subjects completed a VAS to give a subjective rating of drug effects. The VAS is an instrument for recording the subject's momentary mood. The subject marks a 100-mm line with two opposite poles (e.g., refreshed-exhausted) at the point which he thinks best corresponds to his momentary state (Ott et al. 1981). Here too, the results parallel those of the objective methods. The potency of the effects was most evident from the ratings made by subjects from the flunitrazepam group. The feelings of exhaustion (Fig. 8), fatigue (Fig. 9), and distractedness (Fig. 10) were particularly distinct 2 and 3 h after drug administration compared with the other groups. However, the ratings of tenseness and nervousness (results not shown) did not seem to be particular sensitive to effects of benzodiazepines.

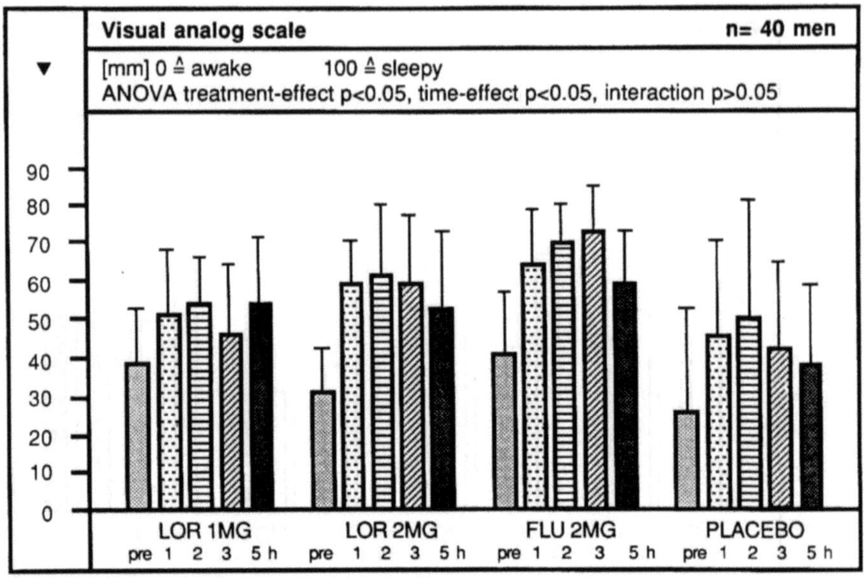

Fig. 9. Means and standard deviations of the visual analog scale for fatigue. In all treatment groups there was an increase in the feeling of tiredness – obviously related to the trial. This effect was particularly evident under flunitrazepam 2 mg

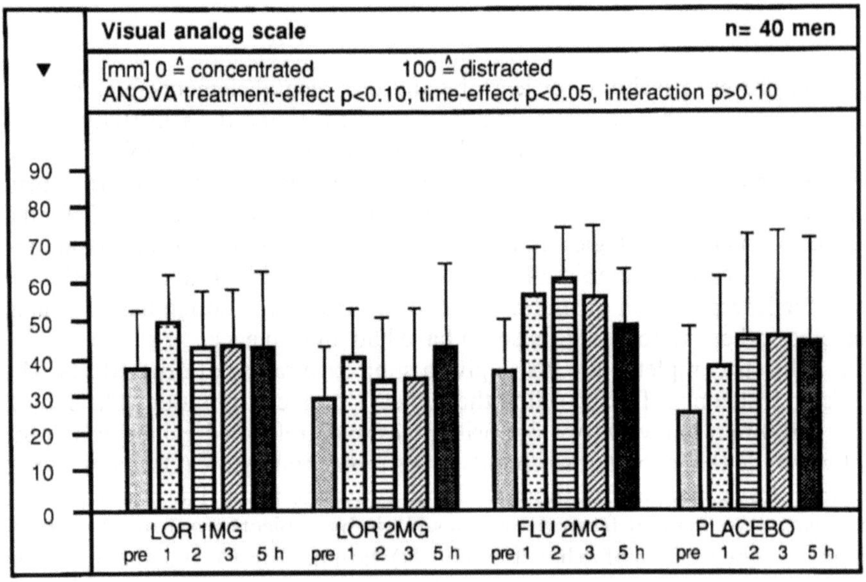

Fig. 10. Means and standard deviations of the visual analog scale for distractedness. An increase in the values is apparent in all treatment groups. The increase in the direction of "distracted" was most evident under flunitrazepam 2 mg

4 Conclusions

The conclusions to be gained from this study can be summarized as follows:

1. Benzodiazepines induce dose- and substance-dependent anterograde amnesia. The interpretation of this is that the amnesia induced is primarily based on the impairment of retrieval mechanisms, whereas the impairment of encoding processes plays a minor role because of controlled vigilance.
2. Benzodiazepines induce dose- and substance-dependent "promnesia" but not retrograde amnesia. This is interpreted as being due to (a) protection of already learned material from the interfering effects of items learned after medication, and (b) a reactive increase of tension (DÜKER 1964).
3. These effects are closely connected to the sedative profile of the benzodiazepines by visual analog scales and a video tracking test.

References

Atkinson RC, Shiffrin RM (1971) The control of short-term memory. Sci Am 224:82–90

Buschke H, Altman Fuld P (1974) Evaluating storage, retention and retrieval in disordered memory and learning. Neurology 24:1019–1025

Düker H (1964) Die reaktive Anspannungssteigerung als Störfaktor bei der Wirkungsprüfung von Schlafmitteln. In: Bradley PB, Flügel F, Hoch PH (eds) Neuro-Psychopharmacology, vol 3. Elsevier, Amsterdam, pp 172–175

Liljequist R, Linnoila M, Mattila MJ (1978) Effect of diazepam and chlorpromazine on memory functions in man. Eur J Clin Pharmacol 13:339–343

Ott H (1984a) Zur Klärung der Konzepte Vigilanz und Aktivierung in Pharmakopsychologie und Elektrophysiologie. Z EEG-EHG 15:190–197

Ott H (1984b) Are electroencephalographic and psychomotor measures sensitive in detecting residual sequelae of benzodiazepine hypnotics? In: Hindmarch I, Ott H, Roth T (eds) Sleep, benzodiazepines and performance. Springer, Berlin Heidelberg New York, pp 133–151

Ott H, Hemmerling KG, Kugler J, Suttmann H, Doenicke A, Tesch C, Straessner G (1980) Amnestische Begleitwirkungen nach i.v.-Gabe von Lormetazepam und Flunitrazepam. In: Doenicke A, Ott H (eds) Anaesthesiologie und Intensivmedizin. Springer, Berlin Heidelberg New York, pp 13.1–13.10

Ott H, Oswald I, Fichte K, Sastre M (1981) Visuelle Analogskalen zur Erfassung von Schlafqualität. VIS–A und VIS–M. In: CIPS Collegium Internationale Psychiatriae Scalarum (eds) Internationale Skalen für Psychiatrie. Beltz, Weinheim

Randt CT, Brown ER, Osborne DP Jr (1980) A memory test for longitudinal measurement of mild to moderate deficits. Clin Neuropsychol II, 4

Schratzer M, Bischoff RC (1984) Anterograde Amnesie nach i. v. Applikation von Benzodiazepinen. Paper at 26th congress ef experimental psychologists, Nuremberg, 1984

Subhan Z (1984) The effects of benzodiazepines on short-term and information processing. In: Hindmarch I, Ott H, Roth T (eds) Sleep, benzodiazepines and performance. Springer, Berlin Heidelberg New York, pp 173–181

Zimbardo PG (ed) (1983) Psychologie. Springer, Berlin Heidelberg New York

Reversal by Caffeine of Triazolam-Induced Impairment of Waking Function

T. ROEHRS[1], A. ZWYGHUIZEN-DOORENBOS, D. SMITH, F. ZORICK and T. ROTH

Abstract

Twelve, healthy normal men aged 21–25 years received each of four treatments (triazolam placebo plus caffeine placebo, triazolam 0.50 mg plus caffeine placebo, triazolam 0.50 mg plus caffeine 4 mg/kg, triazolam 0.50 mg plus caffeine 8 mg/kg), double blind, in a Latin-Square design. Triazolam or placebo was administered at 0830 and caffeine or placebo at 1000 and 1245. On two memory tasks, administered at 1015 with an immediate recall and a delayed recall at 1230 following a 90 min nap (1030–1200), both immediate and delayed recall was impaired by triazolam. Neither caffeine dose reversed the impairments. Sleep latency and sleep efficiency were improved by triazolam and not reversed by caffeine. On a performance battery presented at 1300 most measures of performance were impaired by triazolam; in general the caffeine dose of 4 mg/kg partially reversed the effect while the dose of 8 mg/kg completely restored performance.

1 Introduction

Numerous studies have demonstrated that sedative-hypnotic drugs, specifically the benzodiazepines, impair waking function (JOHNSON and CHERNIK 1982). These impairments, it is hypothesized, are nonspecific and probably related to the hypnotic activity of the benzodiazepines. In fact, hypnotic potency can be directly related to the degree of performance decrement observed. The potency-decrement relation has been shown in studies with daytime administration in subjects remaining awake, with nighttime administration in subjects awakened after 1–3 h of sleep, or with nighttime administration in subjects arising the following day after 8 h of sleep (NICHOLSON 1981; ROTH et al. 1980; ROEHRS et al. 1986). In the latter studies the effects are referred to as residual effects, but the point is that latency to sleep onset during the following day (a direct measure of continued hypnotic activity) is related to the performance decrement observed.

One of the benzodiazepine-induced impairments which has received specific attention is amnesia (LISTER 1985). Amnesia is an effect characteristic of all benzodiazepines, with the magnitude and duration of the effect being a function of the dose, route of administration, and pharmacokinetics of the particular drug. However, whether the amnesia is related to the sedative properties of the benzodiazepines, and hence merely another nonspecific effect, remains an area of controversy (LISTER 1985). In a series of studies assessing the amnesic and sedative effects of benzodiazepines we have argued that the amnesia is in part the result of

[1] Henry Ford Hospital, Sleep Disorders and Research Center, W Grand Boulevard, Detroit, MI 48202, USA.

increased sleepiness and hastened sleep onset which disrupts memory consolidation (ROTH et al. 1984). In addition to showing that there is a high correlation between the morning recall of events occurring during a nighttime awakening and the latency of falling back to sleep after the awakening, we have shown that morning amnesia can be reversed by maintaining wakefulness for 15 min after the memory task and before returning to sleep (ROEHRS et al. 1983).

Caffeine is well known for its stimulant effects. It prolongs sleep latency (i.e., reverses sleepiness) and increases wakefulness during the usual sleep period (BREZINOVA 1974; GOLDSTEIN et al. 1965; KARACAN et al. 1976). Studies of the effects of caffeine on performance and memory, presumably after normal nocturnal sleep, have produced variable results (WEISS and LATIES 1962). But there is no question that caffeine reverses the performance impairments induced by the loss of sleep (BORLAND et al. 1986). There would be both theoretical and practical importance to knowing whether caffeine can reverse the benzodiazepine-induced impairment of waking function.

This study was conducted to assess the capacity of caffeine to reverse the amnesia and psychomotor performance impairment typically seen with benzodiazepines. We hypothesized that, depending on the extent to which caffeine reverses benzodiazepine-induced sleepiness, it should reverse, at least in part, the associated amnesia. The available studies are equivocal as regards the capacity of caffeine to reverse benzodiazepine-induced impairment of memory and psychomotor performance (FILE et al. 1982; GHONEIM et al. 1986; LOKE et al. 1985; MATTILA and NUTTO 1983; MATTILA et al. 1982). More importantly, none of the studies include a concurrent assessment of sedation.

2 Methods

2.1 Subjects

The subjects were 12 healthy, normal-sleeping men aged 21–25 years. All were in good health based on a brief history and physical examination and had normal blood and urine laboratory test results. No subject required concomitant CNS medication, had a history of alcohol or drug abuse, or known hypersensitivity to benzodiazepines. None smoked. They drank no more than one cup of coffee per day. All reported having 7–8 h of sleep each night and none napped during the day. Each subject gave an informed written consent and was paid for participation.

2.2 Design

The study was conducted as a repeated measures design with presentation of treatments in a Latin-Square. Each subject received triazolam 0.5 mg plus caffeine placebo (T0.5+P), triazolam 0.5 mg plus caffeine 4 mg/kg (T0.5+4), triazolam 0.5 mg plus caffeine 8 mg/kg (T0.5+8), or triazolam placebo plus caffeine placebo (P+P). The treatments were administered double blind with at least 2 days between treatments.

2.3 Procedure

After an initial telephone screening regarding sleep habits and health status, subjects were scheduled for a brief physical examination. At that time urine and blood samples were obtained for laboratory analyses. On two other days subjects reported to the laboratory for two practice sessions on the psychomotor performance tests and on the memory tasks.

Those subjects passing the screening were then entered into the study. On the night before each laboratory day subjects slept at home and were asked to maintain a 2400–0600 bedtime. They were asked to refrain from using caffeine or alcohol after 1700 the night before entering the laboratory. Then, each laboratory day they reported at 0800 and were released at 1700. Between laboratory days subjects were asked to maintain their usual sleep-wakefulness schedule.

Each laboratory day subjects received triazolam 0.50 mg or placebo at 0830. Then electrodes were attached at standard placements to obtain standard sleep recordings (Rechtschaffen and Kales 1968). Caffeine or placebo was administered at 1000. The caffeine was administered as a powder in Sanka 97% caffeine-free instant coffee dissolved in 300 ml hot water. The Sanka beverage alone served as the caffeine placebo. A second caffeine or placebo dose was administered at 1245. Since the duration of caffeine's activity at these doses is 2–3 h, the second dose was necessary to cover triazolam's longer duration of activity. Subjects were placed in bed in a dark, quiet room from 1030–1200 and instructed to go to sleep. While they were in bed for the 90 min, continuous sleep recordings were collected.

The memory evaluation consisted of the presentation of two strings of 10 digits and a 16-item memory task (HFH memory task) which has been described in detail previously (Roth et al. 1984). The two memory tasks were presented at 1015 with an immediate recall and a delayed recall at 1230 after the 90 min nap, but before the second caffeine dose. The performance battery, beginning at 1300 and lasting approximately 60 min, consisted of simple and complex reaction time tasks (each about 3–4 min), a divided attention task (15 min), an auditory vigilance task (40 min), and digit-symbol substitution and symbol copying tasks (90 s each) administered in the order listed. The digit-symbol substitution and symbol copying tasks were also administered at 1025 just before the nap.

The sleep recordings were scored manually in 30-s epochs according to the standards of Rechtschaffen and Kales (1968). To assess hypnotic activity, latency to sleep (10 min continuous sleep) and sleep efficiency (minutes of sleep/90 min) were determined. Each dependent measure, including scores on the two memory tests, the various performance measures, and the measures of hypnotic activity, was then analyzed using repeated-measures ANOVAs corrected for possible violations of the assumption of compound symmetry using the Greenhouse-Geisser method. This was followed by post hoc contrasts comparing each treatment to placebo and comparing the two caffeine doses.

3 Results

Means of memory measures and measures of hypnotic activity for each treatment are presented in Table 1. A summary of the statistically significant effects of each treatment on the memory tasks is presented in Table 2. There was a significant effect on the immediate recall of digits ($F = 10.35$; df 3,33; $p < 0.001$). Recall was impaired by triazolam as indicated by the significant P+P vs. T0.5+P contrast (see Table 2). This effect was not reversed by either caffeine dose, which is demonstrated directly by the absence of a difference on the T0.5+P vs. T0.5+4 and

Table 1. Measures of memory and hypnotic activity

Measure and time	Treatment conditions			
	P+P	T 0.5+P	T 0.5+4	T 0.5+8
Memory tasks				
Digit recall 1015	16.9	11.7	11.2	14.0
Digit recall 1230	10.8	0.9	3.7	2.9
HFH memory 1015	15.8	12.7	12.3	12.2
HFH memory 1230	15.5	12.0	11.5	10.7
Hypnotic activity				
Sleep latency (min)	17.4	2.2	4.3	4.2
Sleep efficiency (TST/TIB)	79.0	96.0	95.0	94.0

Recall data are no. correct – 16 possible on HFH memory test, 20 on digit recall.
T 0.5+P, triazolam 0.50 mg+caffeine placebo.
T 0.5+4, triazolam 0.50 mg+ caffeine 4 mg/kg.
T 0.5+4, triazolam 0.50 mg +caffeine 8 mg/kg.
P+P, triazolam placebo+caffeine placebo.
TST/TIB, Total sleep time/time in bed.

Table 2. Effects on measures of memory and hypnotic activity

Measure and time	Contrasts and significance levels					
	P+P vs. T 0.5+P	P+P vs. T 0.5+4	P+P vs. T 0.5+8	T 0.5+P vs. T 0.5+4	T 0.5+P vs. T 0.5+8	T 0.5+4 vs. T 0.5+8
Memory tasks						
Digit recall 1015	0.01	0.01	0.01	NS	NS	0.02
Digit recall 1230	0.01	0.02	0.01	NS	NS	NS
HFH memory 1015	0.02	0.02	0.04	NS	NS	NS
HFH memory 1230	0.01	0.01	0.01	NS	NS	NS
Hypnotic activity						
Sleep latency	0.01	0.01	0.01	NS	NS	NS
Sleep efficiency	0.01	0.01	0.01	NS	NS	NS

For details of treatment conditions see Table 1.

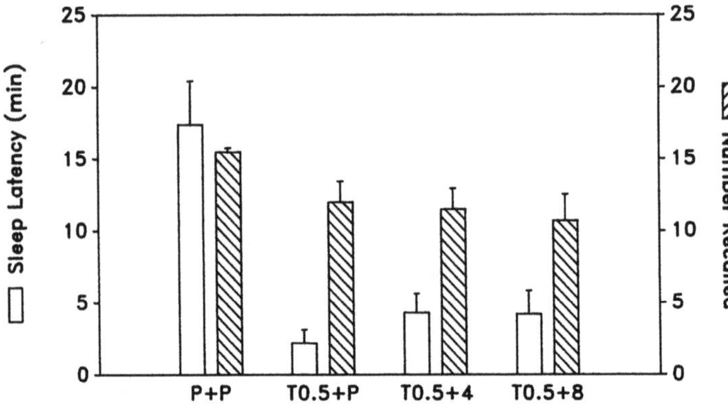

Fig. 1. Number of items recalled on the HFH memory task (▧) and the sleep latency (□) immediately following the memory task in each of the four treatments. *P+P*, triazolam placebo+caffeine placebo; *T0.5+P*, triazolam 0.50 mg+caffeine placebo; *T0.5+4*, triazolam 0.50 mg+caffeine 4 mg/kg; *T0.5+8*, triazolam 0.50 mg+caffeine 8 mg/kg

T0.5+8 contrasts and indirectly by the presence of differences on the P+P vs. T0.5+4 and T0.5+8 contrasts. This significant effect on digit recall remained on the delayed recall test at 1230 ($F=10.67$; df 3,33; $p<0.001$). Again, triazolam impaired recall and caffeine failed to reverse the effect. Performance on the HFH memory task, as with digit recall, was altered significantly on immediate recall ($F=5.20$; df 3,33; $p<0.005$) and on delayed recall ($F=6.08$; df 3,33; $p<0.002$). At both points triazolam impaired recall and neither caffeine dose reversed the effect (see Table 2).

The pattern of significant results for measures of hypnotic activity derived from the sleep recordings are also included in Table 2. Latency to sleep was significantly altered ($F=21.59$; df 3,33; $p<0.001$); triazolam reduced the latency and caffeine failed to reverse the effects of triazolam (shown by no significant difference on T0.5+P vs. T0.5+4 or T0.5+8 contrasts and the significant P+P vs. T0.5+4 or T0.5+8 contrasts). Sleep efficiency was also improved significantly ($F=13.46$; df 3,33; $p<0.001$) with the pattern of significant contrasts similar to those found for sleep latency. The consistency of the pattern of results (see Table 2) for measures of hypnotic activity and those for memory is striking. Figure 1 illustrates this consistency with data for two of the measures. As seen in Fig. 1, with placebo delayed recall on the HFH memory task was nearly perfect (16 correct) and with triazolam, regardless of caffeine dose, at least four items were forgotten. Sleep latency on placebo was 17 min and it was reduced to less than 5 min with triazolam irrespective of whether caffeine was added.

The means for the various performance measures are presented in Table 3 and the pattern of significant drug effects on the various performance measures is summarized in Table 4. No treatment effects were detected on the simple and complex reaction time tasks and thus these measures are not included in the tables. On the divided attention task, tracking error ($F=4.78$; df 3,33; $p<0.01$), central reaction time ($F=8.36$; df 3,33; $p<0.001$), and peripheral reaction time

Table 3. Measures of performance

Measure and time	Treatment conditions			
	P+P	T 0.5+P	T 0.5+4	T 0.5+8
Divided attention				
Tracking error[a]	95.9	139.1	120.3	95.6
Central RT (ms)	254.0	380.0	307.0	271.0
Peripheral RT (ms)	241.0	338.0	298.0	256.0
Vigilance				
Misses (no.)	1.8	11.2	3.0	2.0
Mean RT (ms)	466.0	629.3	538.6	327.9
Block 1 RT (ms)	387.3	470.2	519.3	287.0
Block 2 RT (ms)	447.7	718.9	512.1	355.3
Block 3 RT (ms)	495.7	572.1	560.5	361.3
Block 4 RT (ms)	514.2	752.5	560.3	308.6
Digit – symbol				
Copying (no. cor.) 1025	148.6	121.7	139.6	139.5
Copying (no. cor.) 1400	157.3	138.6	152.0	161.8
Substitution (no. cor.) 1025	69.5	51.0	54.3	53.3
Substitution (no. cor.) 1400	66.0	55.4	65.8	66.6

[a] Tracking error units are CRT pixels.
For details of treatment conditions see Table 1.

Table 4. Effects on measures of performance

	Contrasts and significance levels					
	P+P vs. T 0.5+P	P+P vs. T 0.5+4	P+P vs. T 0.5+8	T 0.5+P vs. T 0.5+4	T 0.5+P vs. T 0.5+8	T 0.5+4 vs. T 0.5+8
Divided attention						
Tracking error[b]	0.02	NS	NS	NS	0.02	NS
Central RT (ms)	0.01	NS	NS	0.05	0.01	NS
Peripheral RT (ms)	0.01	NS	NS	NS	0.01	NS
Vigilance						
Misses	0.02	NS	NS	0.04	0.02	NS
Mean RT (ms)	0.02	NS	0.02[a]	NS	0.01	0.01
Block 1 RT (ms)	NS	0.03	NS	NS	0.01	0.02
Block 2 RT (ms)	0.01	NS	NS	0.01	0.01	0.01
Block 3 RT (ms)	NS	NS	NS	NS	0.02	0.01
Block 4 RT (ms)	0.02	NS	0.01[a]	0.01	0.01	0.03
Digit – symbol						
Copying (no. cor.) 1025	0.01	NS	NS	NS	0.01	NS
Copying (no. cor.) 1400	0.01	NS	NS	NS	0.01	NS
Substitution (no. cor.) 1025	0.01	0.01	0.01	NS	NS	NS
Substitution (no. cor.) 1400	0.01	NS	NS	0.01	0.01	NS

[a] Improvement over placebo; [b] Tracking error units are CRT pixels.
For details of treatment conditions see Table 1.

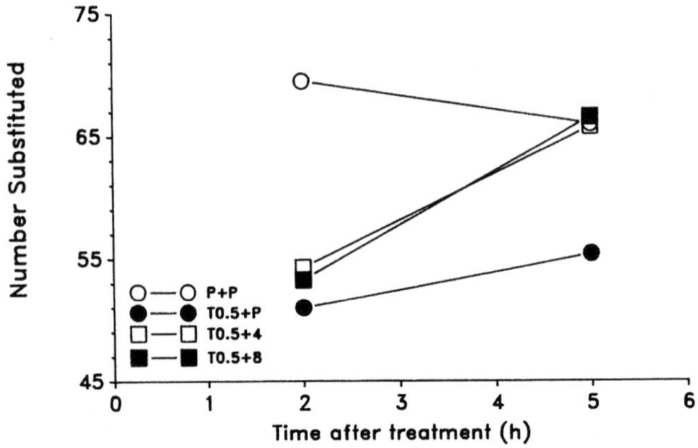

Fig. 2. Number of symbols substituted on the digit-symbol substitution test 2 h and 5.5 h after administration of each of the four treatments. *P+P*, triazolam placebo+caffeine placebo (○); *T0.5+P*, triazolam 0.50 mg+caffeine placebo (●); *T0.5+4*, triazolam 0.50 mg+caffeine 4 mg/kg (□); *T0.5+8*, triazolam 0.50 mg+caffeine 8 mg/kg (■)

($F = +7.06$; df 3,33; $p < 0.001$) all showed treatment effects. On each measure triazolam reduced performance (see Table 4). The impairment was consistently reversed by the high caffeine dose (a significant T0.5+P vs. T0.5+8 contrast) and less consistently by the low dose (only central reaction time showed the reversal). This was also shown indirectly, because after caffeine (both doses) all three divided attention measures no longer differed from placebo.

On the 40-min vigilance test, number of misses ($F = 6.23$; df 3,33; $p < 0.001$) and mean RT ($F = 11.97$; df 3,33; $p < 0.001$) were altered by the treatments. In each case triazolam impaired performance and the high caffeine dose reversed the impairment, while the low dose had a less consistent effect (see Table 4). Interestingly, mean reaction time was improved compared to placebo with the high caffeine dose. Finally, significant caffeine dose differences were also found for mean vigilance reaction time (a significant T0.5+4 vs. T0.5+8 contrast). Analyses of vigilance performance by 10-min blocks yielded a pattern of results comparable to the overall effects and revealed that the caffeine-related improvement in reaction time occurred on the last 10-min block of the vigilance task.

Treatment effects on the paper and pencil tasks are also included in Tables 3 and 4. Symbol copying was altered at 1025 ($F = 3.06$; df 3,33; $p < 0.04$) and at 1400 ($F = 5.21$; df 3,33; $p < 0.005$). At both 1025 and 1400, triazolam impaired performance and the high caffeine dose, but not the low dose, reversed the effect. A different pattern of results was found for symbol substitution. At both 1025 and 1400, treatment effects were found ($F = 7.70$; df 3,33; $p < 0.001$; and $F = 6.60$; df 3,33; $p < 0.001$). However, caffeine only reversed the triazolam-induced impairment at

1400. This apparent interaction was verified with a two-factor repeated-measures ANOVA. A significant time of test by treatment interaction was found ($F = 5.96$; df 3,33; $p < 0.003$). The interaction is illustrated in Fig. 2.

4 Discussion

In this study the previously documented benzodiazepine-induced disruption of memory and psychomotor performance was observed on almost all measures; only the simple and complex reaction time tasks failed to show the effect. The psychomotor impairments produced in this study were reversed by caffeine and in most cases the high dose completely restored performance and the low dose partially restored performance. On the other hand, the memory impairments produced by triazolam were not reversed by caffeine. Likewise, the hypnotic activity of triazolam was also not reversed by caffeine.

Thus, these data do not provide further information as to whether the amnesic effects of benzodiazepines are related to their sedative effects. The data do not rule out that possibility. If caffeine had reversed the sleepiness but not the amnesic effects, one could have ruled out the hypothesis that their amnesic effects are due to the sedative effects. However, caffeine did not reverse the sleepiness and in fact, the data are consistent with our previous work (ROTH et al. 1984). As with our previous studies, nearly perfect recall was associated with sleep latencies of greater than 15 min and impaired recall was associated with shortened sleep latencies, in this case less than 5 min (ROTH et al. 1984).

The failure of caffeine to reverse the memory impairments while successfully reversing the psychomotor impairments is probably related to the different ratios of blood levels of caffeine to triazolam at the times of memory and performance evaluation. The memory tasks were presented 2 h after triazolam administration when triazolam blood levels were reaching their peak, based on what is known of triazolam's pharmacokinetics. Both the initial acquisition and consolidation of the memories for later recall were occurring at this point. Clearly, hypnotic activity was present, probably at its most potent level, and caffeine did not reverse it. On the other hand, the performance testing began 4.5 h after triazolam administration when triazolam blood levels were somewhat lower and furthermore, a second caffeine dose had been administered. While there was sufficient pharmacological activity to produce performance impairment, there was probably a higher ratio of caffeine to triazolam than there was earlier, 2 h after triazolam administration. Unfortunately, a direct assessment of hypnotic activity over hours 4–5 was not included in this study. Consequently, the extent to which caffeine may have reversed hypnotic activity at this point in time is unknown. The hypnotic potency of triazolam relative to that at the time of the memory task also is unknown. Thus, questions remain as to whether higher doses of caffeine would reverse the amnesic effects of triazolam and whether the reversal would be accompanied by a reversal of the hypnotic effect.

References

Borland RG, Rogers AS, Nicholson AN, Pascoe PA, Spencer MB (1986) Performance overnight in shiftworkers operating a day-night schedule. Aviat Space Environ Med 57:241–249
Brezinova V (1974) Effect of caffeine on sleep; EEG study in the late middle age people. Br J Clin Pharmacol 1:203–208
File SE, Bond AJ, Lister RG (1982) Interaction between effects of caffeine and lorazepam in performance tests and self-ratings. J Clin Psychopharmacol 2:102–106
Ghoneim MM, Hinrichs JV, Chiang CK, Loke WH (1986) Pharmacokinetic and pharmacodynamic interactions between caffeine and diazepam. J Clin Psychopharmacol 6:75–80
Goldstein A, Warren R, Kaizer S (1965) Psychomotor effects of caffeine in man. I. Individual differences in sensitivity to caffeine-induced wakefulness. J Pharmacol Exp Ther 149:156–159
Karacan I, Thornby JI, Anch AM, Booth GH, Williams RL, Salis PJ (1976) Dose-related sleep disturbances induced by coffee and caffeine. Clin Pharmacol Ther 20:682–689
Johnson LC, Chernik DA (1982) Sedative-hypnotics and human performance. Psychopharmacology 76:101–113
Loke WH, Hinrichs JV, Ghoneim MM (1985) Caffeine and diazepam: separate and combined effects on mood, memory, and psychomotor performance. Psychopharmacology 87:344–350
Lister RG (1985) The amnesic action of benzodiazepines in man. Neurosci Biobehav Rev 9:1–8
Mattila MJ, Nuotto E (1983) Caffeine and theophylline counteract diazepam effects in man. Med Biol 61:337–343
Mattila MJ, Palva E, Savolainen K (1982) Caffeine antagonizes diazepam effects in man. Med Biol 60:121–123
Nicholson AN (1981) The use of short- and long-acting hypnotics in clinical medicine. Br J Clin Pharmacol 11:615–695
Rechtschaffen A, Kales A (1968) A manual of standardized terminology, techniques and scoring system for sleep stages of human subjects. U.S. Government Printing Office, Washington, DC (National Institute of Health Publication, no 204)
Roehrs T, Zorick F, Sicklesteel J, Wittig R, Hartse K, Roth T (1983) Effects of hypnotics on memory. J Clin Psychopharmacol 3:310–313
Roehrs T, Kribbs N, Zorick F, Roth T (1986) Hypnotic residual effects of benzodiazepines with repeated administration. Sleep 9:309–316
Roth T, Hartse KM, Saab PG, Piccione PM, Kramer M (1980) The effects of flurazepam, lorazepam, and triazolam on sleep and memory. Psychopharmacology 70:231–237
Roth T, Roehrs T, Wittig R, Zorick F (1984) Benzodiazepines and memory. Br J Clin Pharmacol 18:45S–49S
Weiss B, Laties V (1962) Enhancement of human performance by caffeine and amphetamines. Pharmacol Rev 14:1–36

β-Carbolines and Benzodiazepine Antagonist Actions on Memory and Vigilance

Bidirectional Nature of Benzodiazepine Receptor Ligands Extends to Effects on Vigilance

D. N. STEPHENS[1] and M. SARTER

Abstract

The classification of benzodiazepine receptor ligands into agonists, antagonists and inverse agonists is based on biochemical, electrophysiological and behavioural evidence. Agonists potentiate the effects of γ-aminobutyric acid (GABA) and exhibit anxiolytic, anticonvulsant, hypnotic, amnesic and muscle-relaxant properties; inverse agonists show mirror-image effects in that they may be convulsant and anxiogenic and may increase muscle tone. Antagonists antagonise the effects of both agonists and inverse agonists. Some of the most interesting ligands, however, are those substances with actions intermediate between either those of the agonists and the antagonists, or between those of the antagonists and the inverse agonists. These partial agonists and partial inverse agonists possess only some of the properties of the agonists and inverse agonists, respectively.

The present experiments show that the agonist and inverse agonist properties of benzodiazepine receptor ligands can also be revealed in an animal continuous attention task in which rats were required to detect a brief signal during which operation of a lever was rewarded by food. Benzodiazepines and a β-carboline benzodiazepine receptor agonist, ZK 93423, disrupted performance of this task, as did the antimuscarinic substance, scopolamine. Another β-carboline, ZK 91296, which has anxiolytic and anticonvulsant properties like benzodiazepines, did not affect performance of the continuous attention task, demonstrating a separation of anxiolytic and sedative properties of such substances. A partial inverse agonist β-carboline, FG 7142, was able to antagonise the disruptive effects of scopolamine on this task, as was, to a smaller extent, the antagonist ZK 93426.

These results are discussed in terms of vigilance-enhancing properties of the inverse agonist β-carbolines, and the possibility that such vigilance-enhancing effects might contribute to improvement of performance in learning tasks.

1 The Benzodiazepine Receptor

The benzodiazepines appear to achieve their clinically useful effects through an action at specific receptor sites in the brain (BRAESTRUP and SQUIRES 1977; MÖHLER and OKADA 1977). These benzodiazepine receptors are associated with receptors for the major inhibitory transmitter, γ aminobutyric acid (GABA), and benzodiazepine and GABA receptors form parts of a functional unit in the neuronal membrane (OLSEN 1982). The exact nature of the association between GABA and benzodiazepine receptors need not concern us here, but recent work (e.g. MÖHLER et al. 1986) suggests that the two receptor sites are subunits of a single membrane protein.

[1] Research Laboratories, Department of Neuropsychopharmacology, Schering AG, 1000 Berlin 65, FRG.

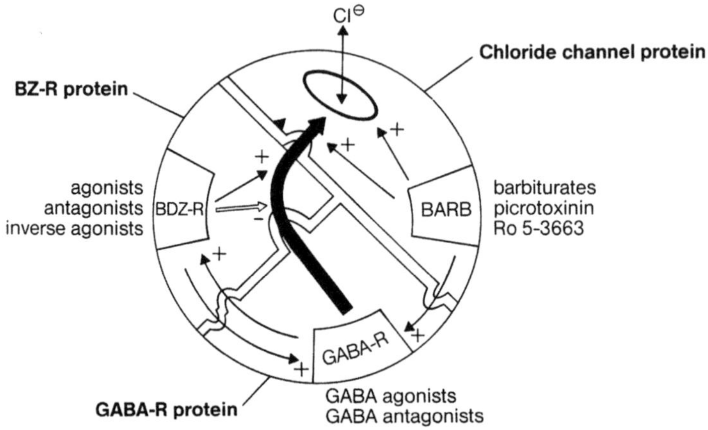

Fig. 1. A schematic diagram of the functional relationships between GABA receptors (*GABA-R*) and benzodiazepine receptors (*BZ-R*) and the GABA-gated chloride ion channel (after POLC et al. 1982). The diagram indicates that the ability of GABA to open chloride ion channels is increased in the presence of benzodiazepines [from POLC et al. 1982 (adapted)]

More importantly, the physiological role of the benzodiazepine receptor seems to be to modulate the effects of GABA on membrane permeability. Electrophysiological studies have shown that benzodiazepines potentiate the inhibitory effects of GABA on membrane excitability, increasing inhibitory post-synaptic potentials at synapses where GABA is the natural transmitter (POLC and HAEFELY 1976). GABAergic inhibition arises predominantly as a result of increased chloride ion permeability (BARKER and RANSOM 1978; BARKER and MATHERS 1981) and this effect is potentiated by benzodiazepines (MACDONALD and BARKER 1978) via an increase in the frequency of chloride channel opening (STUDY and BARKER 1981). These functional interactions are represented diagramatically in Fig. 1, adapted from POLC et al. (1982). It should be emphasised that benzodiazepines themselves appear to have no effects on membrane permeability, but simply modulate the effects of GABA. In this respect the benzodiazepine receptor differs from classical neurotransmitter receptors.

A further difference from classical receptors arises out of the benzodiazepine receptor's function as a modulator. Following the discovery of the benzodiazepine receptor (BRAESTRUP and SQUIRES 1977), a further chemical class was identified which acts at this site. Ethyl β-carboline-3-carboxylate (β-CCE) was isolated as an artefact from human urine, and found to exhibit an affinity for the benzodiazepine receptor as high as that of the benzodiazepines (BRAESTRUP et al. 1980). However, in contrast to the benzodiazepines this substance and some of its related derivatives have been found to depress GABA-mediated responses in electrophysiological studies (JENSEN and LAMBERT 1986). Thus, the benzodiazepine receptor is, under the influence of different ligands, able either to enhance or to reduce the effects of GABA on chloride ion channel conductance; that is, to intensify of minimise the inhibitory effects of GABA on neurotransmission.

2 Pharmacology of Agonists and Inverse Agonists

As might be predicted from these electrophysiological experiments, the two types of benzodiazepine receptor ligand also exert opposing effects on behaviour. The benzodiazepines are used in clinical practice for their anxiolytic, hypnotic, anticonvulsant, muscle relaxant and sedative properties (though for any one of these clinical uses the others often represent undesirable side effects) and each of these properties can be modelled in animal experiments. In contrast to the benzodiazepines, β-CCE was found in such models to exert proconvulsant and anxiogenic properties (BRAESTRUP et al. 1982). Such observations have given rise to the terms benzodiazepine receptor agonist for the benzodiazepine type of ligand and inverse agonist for the β-CCE type of substance.

Figure 2 illustrates the opposing effects of inverse agonists and agonists on one pharmacological measure, the loss of the righting reflex induced by the barbiturate pentobarbital in mice (JENSEN et al. 1986). It can be seen that the benzodiazepines lorazepam and diazepam markedly prolong the time during which the righting reflex is lost. A similar effect is shown by the substance ZK 93423, a β-carboline agonist derivative of β-CCE. In the lower part of the graph, the effects of two β-carboline inverse agonists, dimethoxyethylcarboline carboxylate methyl ester (DMCM) and β-CCM are shown, and these actually reduce the duration of the loss of the righting reflex.

However, the inclusion of four other benzodiazepine receptor ligands illustrates that such ligands are not simply dividable into "agonists" and "inverse ag-

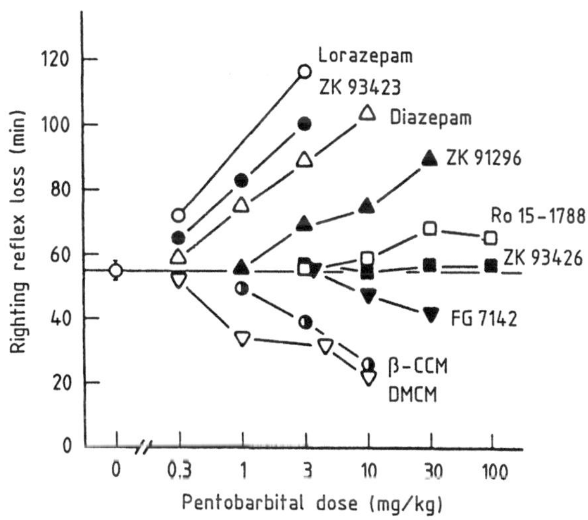

Fig. 2. The effect of several ligands acting at benzodiazepine receptors on the loss of the righting reflex induced by pentobarbital (30 mg/kg, i.p.). Such a procedure is often used as a simple model of hypnotic activity of benzodiazepines. Although the standard benzodiazepines and the agonist β-carbolines potentiate the effects of the barbiturate, the so-called inverse agonist β-carbolines (DMCM, FG 7142) actually reduce righting reflex loss. Between these extremes lies a continuum of substances, including some which have little effect on the barbiturate-induced loss of righting reflex (from JENSEN et al. 1986)

onists", but that a continuum exists with the "agonists" and "inverse agonists" representing two extremes. In the middle of the continuum, compounds like the β-carboline ZK 93426 and the imidazodiazepine Ro 15–1788 have little effect on the loss of righting reflex, even though they also exhibit a high affinity for the benzodiazepine receptor. These compounds, by virtue of their receptor affinity, compete with agonists or inverse agonists for the benzodiazepine receptor and, because they themselves have few and weak pharmacological effects, act to antagonise the agonist and inverse agonist actions.

Figure 2 also includes two further substances, ZK 91296 and FG 7142, which are of special interest for the rest of this article. Their activities lie between those of the agonists and the antagonists, and between those of the inverse agonists and the antagonists, respectively. Pharmacologically, such substances possess only some of the effects of the "full" agonists and "full" inverse agonists, respectively, and are termed partial agonists and partial inverse agonists.

ZK 91296, for instance, has been shown to possess anxiolytic effects in several animal models (STEPHENS and KEHR 1985; STEPHENS et al. 1984; PETERSEN et al. 1984), but unlike the full agonists is non-sedative and non-ataxic (PETERSEN et al. 1984; STEPHENS et al. 1985). On the other hand, FG 7142 does not give rise to convulsions as do the full inverse agonists like DMCM (PETERSEN 1983) but does give rise to anxiety, both in animal models (STEPHENS et al. 1984; STEPHENS and KEHR 1985) and in man (DOROW et al. 1983).

Thus, within the β-carboline series there exists a continuum of substances with activities ranging from benzodiazepine-like to exactly opposite properties, and within this continuum exist series of compounds with rather subtle differences between them.

3 Cognitive Function

An aspect of benzodiazepine receptor pharmacology which is receiving growing attention, and is the basis of this book, is their ability to disrupt cognitive function. This disruption exists at many levels, ranging from sedation (impairment of stimulus input) to frank amnesia (e.g. LISTER 1985). We were therefore interested in investigating the effects of the novel β-carbolines on measures of cognitive function. By a simple pharmacological analogy we wondered whether the partial agonist β-carbolines such as ZK 91296, which show anxiolytic but not locomotor sedative effects, might also lack the amnesic and vigilance-reducing effects of the benzodiazepines.

Furthermore, we speculated that in contrast to the impairment of cognitive performance seen with benzodiazepines, the inverse agonists, at least at non-convulsant and non-anxiogenic doses, might actually improve certain aspects of information processing; in particular we anticipated vigilance-enhancing properties, perhaps leading to enhanced performance in tasks involving learning and memory.

4 Sedation: Impoverished Stimulus Input

Whilst sedation is sometimes recognised as an unwanted effect of benzodiazepines, there is some confusion in the preclinical literature as to what this means. For the most part benzodiazepine-induced sedation in animal studies refers to locomotor sedation. No doubt higher doses of benzodiazepines also lead to frank locomotor sedation in the clinical setting, but this effect occurs at doses which approach those used for the hypnotic properties of benzodiazepines. Much more important in terms of their clinical use is the fact that patients receiving anxiolytic therapy, or on the day following one of the longer-acting benzodiazepine hypnotics, often have problems when they need to operate machinery or drive a car. The animal tests which measure *locomotor* sedation seem to offer an inadequate model for such effects. We were therefore interested to develop an animal test which might more accurately reflect those properties of benzodiazepines which we identify as being associated with reduced vigilance.

5 An Animal Model of Vigilance

In both animal and human experimental psychology the concept of vigilance is related to the ability to detect events of significance occurring in the environment. Classically, changes in vigilance are detectable using signal detection methods and for our experiments we adapted the simple continuous attention task described by WARBURTON and BROWN (1971).

The subjects were eight adult male Wistar rats, 3 months old at the start of training, obtained from the animal breeding department of Schering AG. Each rat was housed individually with water ad libidum and was given enough standard laboratory diet to maintain it at 90% of free-feeding weight. Subjects were trained and tested in standard Campden Instruments operant chambers fitted with two levers (of which only the left was operative), a pellet dispenser, house light and a panel light over the active lever. A 500-kΩ resistor wired in parallel to the panel light reduced its intensity. Illumination of the panel light constituted the discriminative stimulus. During training and drug testing, the stimulus was delivered for 3 s on a variable interstimulus interval schedule with a mean of 15 s and a range of 9–21 s. A lever press during the stimulus was rewarded by the delivery of a standard 45-mg food pellet (Bioserv Ltd.) and terminated the stimulus.

A lever press during the last 9 s of the interstimulus interval postponed the onset of the next stimulus for 9 s. Responses during the last 3 s before stimulus onset and during the 3 s after stimulus onset were used for signal detection analysis. The probability of responding in the 3-s bin which followed light onset was defined as the probability of scoring a hit; the probability of responding in the 3-s bin just prior to the light onset was taken as the probability of a false alarm. Schedule programming and data collection were accomplished using a TRS–80 microcomputer and OPN software (EMMETT-OGLESBY et al. 1984).

Since no assumption could be made about the statistical properties of the sensory events underlying the responses, a non-parametric calculation of response bias (B'') is appropriate (GRIER 1971). The formulae given by GRIER (1971) were used to calculate indices of signal detectability (A') and/or bias (B'').

Following prolonged training (more than 5 weeks at stable performance), the animals were used repeatedly in drug testing, receiving one pharmacological treatment per week. Since further training was given between drug applications, we have treated each drug application as an independent measure and chosen to ignore possible effects of order of treatment. Drugs were given intraperitoneally 30 min before testing.

Each set of data includes a test under treatment with vehicle; this always preceeded a test day with drug treatment. Thus differences between vehicle days and drug days are confounded with order of treatment. This procedure was adopted to prevent possible carry-over effects from drug to vehicle days.

Statistical analysis was carried out using Wilcoxon matched pairs signed-rank test. A target criterion of $p<0.05$ was adopted.

Figures 3 and 4 show the effects of the two β-carbolines, the full agonist ZK 93423 and the partial agonist ZK 91296, on two signal detection parameters, A' a measure of signal detectability, and B'' a measure of perceptual bias in this task. ZK 93423 resembled benzodiazepines (FRANCIS and COOPER 1979) in both reducing A' the level of vigilance, and increasing bias B''. Since there was no dose of ZK 93423 which reduced A' without influencing B'' these results are difficult to interpret, but the clear implication is that ZK 93423 exhibited both locomotor sedative and vigilance-reducing properties.

In contrast to ZK 93423, ZK 91296 exerted no influence on either parameter. The lack of influence on B'' is consistent with the lack of effect of ZK 91296 on motor activity, and the lack of effect on A' indicates that ZK 91296 in addition does not exhibit deleterious effects on vigilance in this model.

This distinction between ZK 93423 and ZK 91296 is interesting for several reasons. Both ZK 91296 and ZK 93423 are anxiolytic in several animal models (STEPHENS and KEHR 1985; STEPHENS et al. 1984) and the difference between the two compounds in this test indicates that anxiolytic effects may be dissociated from vigilance-reducing effects in this class of compounds. Such an observation has important therapeutic implications.

On a more specific level, this result also indicates that anticonflict properties of drugs in animal models of anxiety based on visual discrimination performance are not simply reflections of their disruption of discrimination performance rather than of their anxiolytic effect. In the conventional Geller-Seifter type of task, rats are trained to operate a lever to obtain food. The onset of a stimulus, either visual or auditory, indicates a change in the reinforcement schedule, specifically that further responding will not only be rewarded with food but also punished by electric shock. Such a contingency is usually referred to as a conflict schedule (GELLER and SEIFTER 1960; MARGULES and STEIN 1968).

Although such a procedure has a certain degree of face validity (conflict = anxiety?), its main attraction as a model is that such tests reliably identify benzodiazepine and barbiturate anxiolytics. However, since all such anxiolytics also exhibit vigilance-reducing properties and thus impair discrimination performance, it is by

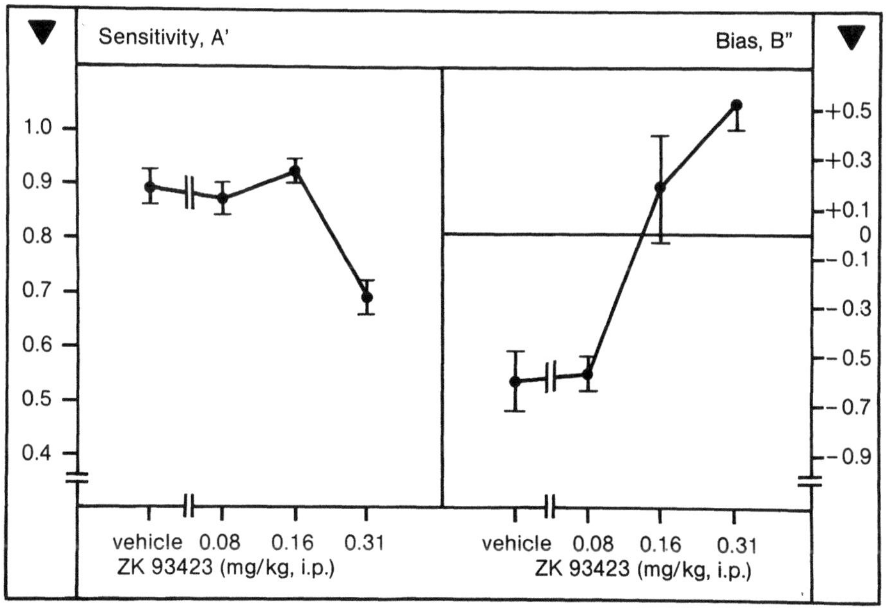

Fig. 3. The effect of the agonist β-carboline ZK 93423 on signal detection indices of performance in a simple successive discrimination task. ZK 93423 reduced stimulus sensitivity *A'*, in a dose-related manner (Friedman $\chi^2 = 12.56$; $p < 0.001$), indicating an impairment of vigilance at this dose. Response bias *B''* was increased at 0.16 mg/kg, reflecting a disruption of lever pressing at this dose ($\chi^2 = 10.75$; $p < 0.01$)

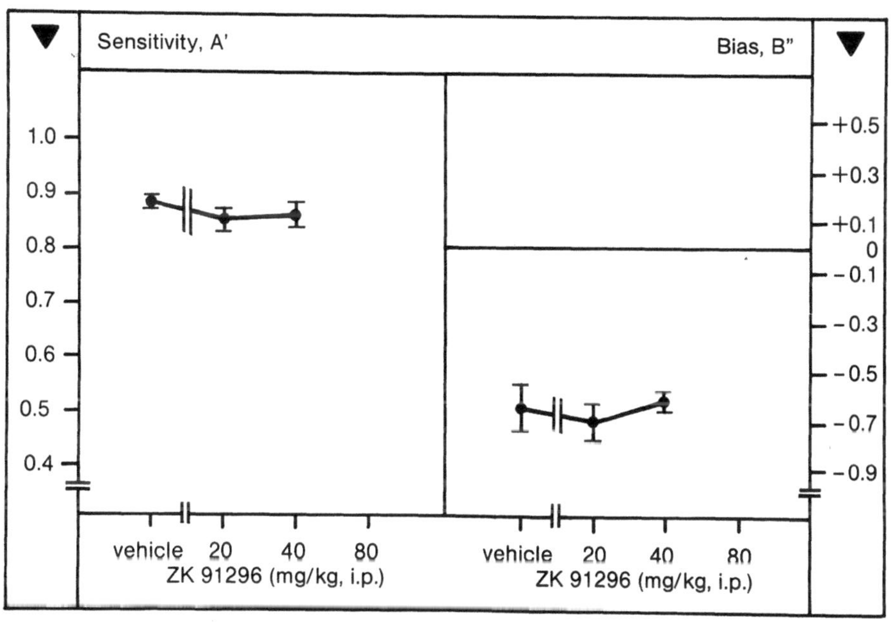

Fig. 4. The effect of the partial agonist β-carboline ZK 91296 on signal detection indices of performance in a successive discrimination task. Neither dose of ZK 91296 affected discrimination performance

no means clear which *aspect* of benzodiazepine or barbiturate pharmacology such discriminations are testing. Our finding with ZK 91296 strengthens the view that such "conflict tests" are indeed true models of anxiolytic activity, and positive effects in such tests do not simply reflect impaired discrimination.

Using the same operant chambers and panel light discriminative stimulus we have trained rats in a conflict schedule in which two reinforcement schedules alternated. In the first, lever pressing was rewarded with a single food pellet at variable intervals of 30 s mean duration (VI 30 s). Illumination of the panel light signalled the onset of the conflict component in which 10 presses of the lever (fixed ratio 10) were reinforced with both a food pellet and a mild foot shock. Both ZK 93423 and ZK 91296 were active in this test (Stephens, unpublished). Furthermore, the failure to impair the discrimination performance with ZK 91296 also suggests a dissociation between vigilance-reducing effects of β-carbolines and their amnesic properties in animal tests. In another paper we (Jensen et al. 1987) demonstrate an amnesic effect of ZK 91296, together with full agonist benzodiazepine receptor ligands, in a simple passive avoidance test (see Sarter and Stephens, this volume). Although this test employed another species (mice), it suggests that, at least in some animal models, amnesic effects may be dissociated from effects on vigilance.

6 Inverse Agonist Effects on Vigilance

As described above, the benzodiazepine receptor possesses not only ligands which enhance the effects of GABA and lead to anxiolytic, sedative and anticonvulsant effects, but also ligands giving rise to convulsant or proconvulsant, anxiogenic and muscle-tone-enhancing properties.

We report elsewhere (Jensen et al. 1987; Sarter and Stephens, this volume) that the amnesic effects of the benzodiazepines find a mirror image in the ability of benzodiazepine receptor inverse agonists to improve performance in tasks requiring learning and memory (see also Venault et al. 1986). Tentative evidence that a weak inverse agonist benzodiazepine receptor ligand also improves memory in man has been reported by Duka et al. (this volume). Although the basis of these improvements in performance in cognitive tasks is not yet clear (but for discussion see Sarter and Stephens, this volume) we were also interested in investigating the effects of inverse agonist β-carbolines in the visual discrimination task described above.

Since these experiments took place in the context of attempting to enhance performance of aged animals it is necessary first to describe the effect of senescence on vigilance in our animals.

7 Vigilance in Senescent Rats

The rats used in this study were 24 months old at the beginning of training. Especial care was taken to ensure that the animals were healthy, and ones with obvious tumours or other signs of disease or degeneration were excluded. The schedule described above was also modified to control for possible locomotor deficits of the senescent animals. Thus the signal lasted 6 s instead of 3 s to allow for reduced locomotor activity preventing the rats from responding promptly. However, to allow comparison with the data reported above, only the first 3-s bin following signal onset was used for calculation of signal detection parameters. In practice, the extension of the signal duration to 6 s proved unnecessary since less than 1% of the signal-appropriate responses of the old animals were emitted in the second 3-s bin and these results will not be considered further. In the initial part of this experiment, a further eight 3-month-old rats were included for comparison purposes.

Figure 5 compares the stable performance of the senescent and young rats in this modified task. The probability of the senescent rats responding during the first 3 s of the signal [probability of a hit, *p*(hit)] was less than that of the young animals, whilst the probability of responding in the absence of the signal, *p*(false), was non-significantly greater in the old animals. These results are reflected in the signal detection parameters, A' and B''. The senescent rats showed a reduced signal sensitivity index, A', as well as a lower tendency to respond (bias, B'').

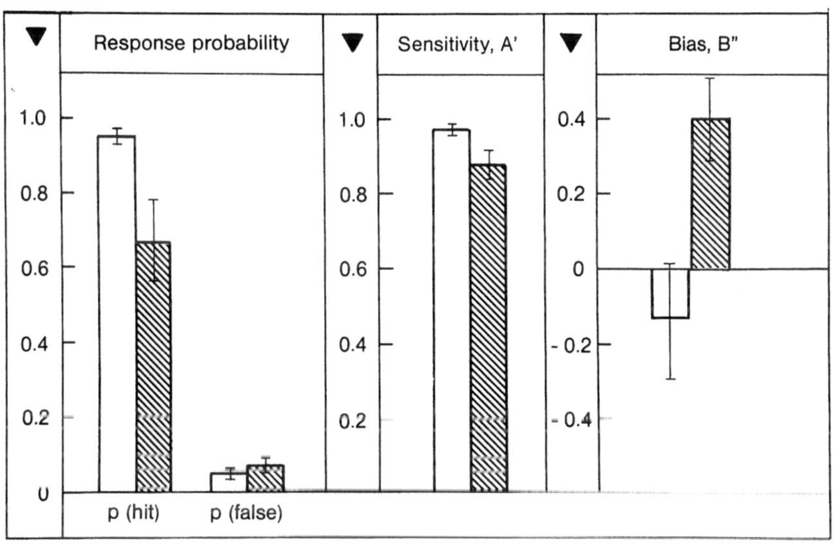

Fig. 5. Comparison of young (3 months old, □) and senescent (24 months old, ▨) male rats performing a simple successive discrimination task. The senescent rats were less likely to operate the lever in the presence of a light signal indicating the availability of food, (Mann-Whitney $U=13$; $n_1=n_2=8$; $p<0.05$) and non-significantly more likely to operate the lever in the absence of the signal. This poorer discrimination performance of the senescent animals is reflected in the lower value of signal sensitivity (A'; Mann-Whitney $U=14$; $p<0.05$), a measure of vigilance, and an increased bias B'' (Mann-Whitney $U=9$; $p<0.01$)

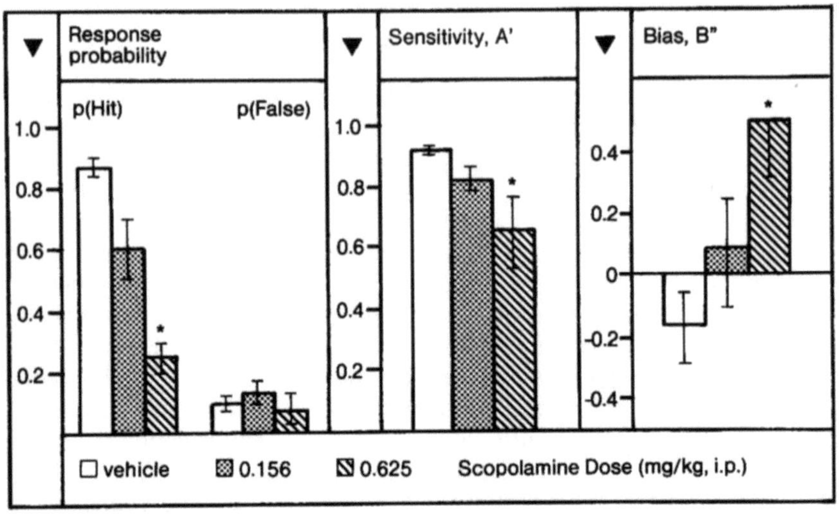

Fig. 6. The effect of two doses of scopolamine on discrimination performance of senescent rats. Scopolamine worsened performance in this task, reducing the probability of an appropriate response while leaving inappropriate responses unchanged. This is reflected in a dose-related effect of scopolamine on signal sensitivity *A'*, and on response bias *B''* (significantly different from vehicle, $p<0.05$)

Although the aged rats showed a reduced level of vigilance (sensitivity index, *A'*), compared with the young animals, in order to test the possible vigilance-enhancing effects of inverse agonist *β*-carbolines we felt it desirable to reduce performance in this task yet further. The means which we chose to impair performance were based on the observations of WARBURTON and BROWN (1971) who demonstrated that the muscarinic antagonist scopolamine reduces performance in such tasks by reducing signal detectability.

Figure 6 shows that scopolamine dose-dependently reduced signal detectability, *A'*, and at higher doses also increased bias, *B''*. These results are consistent with those of WARBURTON and BROWN (1971) and confirm that our adaptation of their methods is sensitive to the effects of muscarinic antagonists.

8 Antagonism of Scopolamine by *β*-Carboline Inverse Agonists

Figure 7 shows that the *β*-carboline inverse agonist FG 7142 antagonised the effects of scopolamine on the index of signal detectability, *A'*, in these senescent rats (JENSEN et al. 1987). The *β*-carboline exerted no effects on signal detectability in non-scopolamine-treated rats, and in contrast to the antagonism of scopolamine's effects on signal detectability, tended to increase the disruption of responding seen following scopolamine. This result is thus consistent with an ability of FG 7142 to enhance vigilance, at least when it has been reduced by scopolamine.

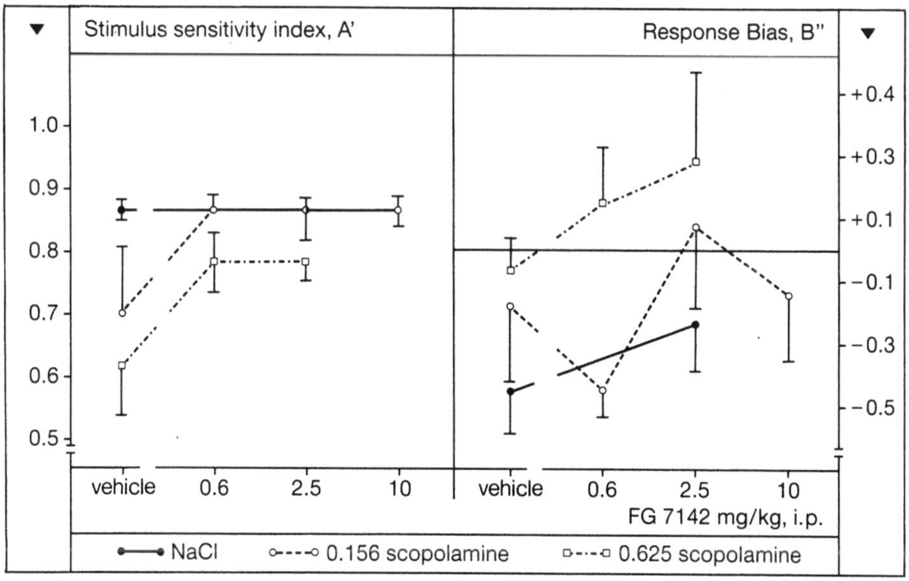

Fig. 7. The effect of the partial inverse agonist FG 7142 on signal detection indices of performance in a discrimination task following its disruption by scopolamine. FG 7142 reversed the impairment of stimulus sensitivity *A'*, induced by scopolamine while further disrupting responding, as reflected by a tendency to increase *B''*

JENSEN et al. (1987) also report that a similar weak ability to antagonise the effects of scopolamine on signal detectability was seen with two further β-carbolines, ZK 90886, a partial inverse agonist with anxiogenic properties, and ZK 93426, a substance characterised as a benzodiazepine receptor antagonist (JENSEN et al. 1984), though with some weak anxiogenic-like activity in certain animal models (JENSEN et al. 1984; FILE et al. 1986).

Thus, inverse agonist β-carbolines acting at the benzodiazepine receptor appear, at least in particular circumstances, to be able to increase vigilance. ZK 93426 also increases vigilance in human volunteers, both as assessed by changes in EEG (DUKA et al. in press) and in behavioural experiments employing visual and auditory continuous attention tasks (DUKA et al. 1987; DUKA, personal communication).

These observations extend the bidirectional nature of benzodiazepine receptor ligands to their effects on vigilance. Since vigilance is an important factor in learning, it seems possible that weak inverse agonist benzodiazepine receptor ligands, or even antagonists like ZK 93426, might give rise to improved performance in learning tasks. This possibility is explored in an accompanying paper (SARTER and STEPHENS, this volume).

9 Conclusion

The benzodiazepine receptor is unusual in that its ligands are able to exert pharmacological effects in opposite directions. Thus, different substances binding to the benzodiazepine site may be anxiolytic or anxiogenic, anti- or proconvulsant, etc. The present paper indicates that such bidirectionality extends to effects on vigilance. Furthermore, the existence of partial agonists and partial inverse agonists at the receptor offers the possibility of identifying substances exhibiting only some of the effects of, on the one hand the benzodiazepines, and on the other, the inverse agonists. Thus, it has proved possible to synthesise β-carbolines which, in animal models, exhibit anxiolytic effects without inducing sedation. On the other hand, β-carbolines have been identified which exhibit vigilance-enhancing effects without being anxiogenic. Such compounds may exert beneficial effects in tasks of learning and memory in animal models, and perhaps even in man.

Acknowledgements. We thank Heidemarie Schäfer for technical assistance, Bettina Fogel for preparation of the manuscript and John Andrews for helpful discussion.

References

Barker JL, Mathers DA (1981) GABA analogues activate channels of different duration on cultured mouse spinal neurones. Science 212:358–361

Barker JL, Ransom BR (1978) Amino acid pharmacology of mammalian central neurones grown in tissue culture. J Physiol 280:331–354

Braestrup C, Squires RF (1977) Specific benzodiazepine binding receptors in rat brain characterised by high affinity ^{3}H-diazepam. Proc Natl Acad Sci USA 74:3805–3809

Braestrup C, Nielsen M, Olsen CE (1980) Urinary and brain β-carboline-3-carboxylates as potent inhibitors of brain benzodiazepine receptors. Proc Natl Acad Sci USA 77:2288–2292

Braestrup C, Schmiechen R, Nielsen M, Petersen EN (1982) Interaction of convulsive ligands with benzodiazepine receptors. Science 216:1241–1243

Dorow R, Horowski R, Paschelke G, Amin M, Braestrup C (1983) Severe anxiety induced by FG 7142, a β-carboline ligand for benzodiazepine receptors. Lancet 2:98–99

Duka T, Stephens DN, Krause W, Dorow R (1987) Human studies on the benzodiazepine receptor antagonist β-carboline ZK 93426: preliminary observations on psychotropic activity. Psychopharmacology 93:421–427

Duka T, Goerke D, Dorow R, Höller L, Fichte K (1988) Human studies on the benzodiazepine receptor antagonist β-carboline ZK 93426: antagonism of lormetazepam's psychotropic effects. Psychopharmacology (in press)

Emmett-Oglesby MW, Spencer DGH, Arnoult DE (1982) A TRS–80 based system for the control of behavioral experiments. Pharmacol Biochem Behav 17:583–587

File SE, Pellow S, Jensen LH (1986) Actions of the β-carboline ZK 93426 in animal tests of anxiety and the holeboard: interaction with Ro 15–1788. J Neural Transm 65:103–114

Francis RL, Cooper S (1979) Chlordiazepoxide-induced disruption of discrimination behavior: a signal detection analysis. Psychopharmacology 63:307–310

Geller I, Seifter J (1960) The effects of meprobamate, barbiturates, d-amphetamine and promazine on experimentally induced conflict in the rat. Psychopharmacologia 1:482–492

Grier JB (1971) Non parametric indexes for sensitivity and bias: computing formulas. Psychol Bull 75:424–429

Jensen LH, Petersen EN, Braestrup C, Honore T, Kehr W, Stephens DN, Schneider HH, Seidelmann D, Schmiechen R (1984) Evaluation of the β-carboline ZK 93426 as a benzodiazepine receptor antagonist. Psychopharmacology 83:249–256

Jensen LH, Petersen EN, Honore T, Dreijer J (1986) Bidirectional modulation of GABA function by β-carbolines. Adv Biochem Psychopharmacol 41:79–89, Raven Press, New York

Jensen LH, Stephens DN, Sarter M, Petersen EN (1987) Bidirectional effects of β-carbolines and benzodiazepines on memory processes. Brain Res Bull 19:359–364

Jensen M, Lambert J (1986) Electrophysiological studies in cultured mouse CNS neurones of the actions of an agonist and an inverse agonist at the benzodiazepine receptor. Br J Pharmacol 88:717–731

Lister RG (1985) The amnesic action of benzodiazepines in man. Neurosci Biobehav Rev 9:87–93

MacDonald RC, Barker JL (1978) Benzodiazepines specifically modulate GABA-mediated postsynaptic inhibition in cultured mammalian neurones. Nature 271:563–564

Margules D, Stein C (1968) Increase of antianxiety activity and tolerance to behavioural depression during chronic administration of oxazepam. Psychopharmacology 13:74–80

Möhler H, Okada T (1977) Properties of ^{3}H-diazepam binding to benzodiazepine receptors in rat cerebral cortex. Life Sci 20:2101–2110

Möhler H, Schoch P, Richards JG, Häring P, Takacs B, Stähli C (1986) Monoclonal antibodies: probes for structure and location of the GABA receptor/benzodiazepine receptor/chloride channel complex. In: Olsen RW, Venteri JC (eds) Benzodiazepine/GABA receptors and chloride channels. Liss, New York, pp 285–297

Olsen R (1982) Drug interactions at the GABA receptor ionophore complex. Annu Rev Pharmacol Toxicol 22:245–277

Petersen EN (1983) DMCM: a potent convulsive benzodiazepine receptor ligand. Eur J Pharmacol 94:117–124

Petersen EN, Jensen LH, Honore T, Braestrup C, Kehr W, Stephens DN, Wachtel H, Seidelmann D, Schmiechen R (1984) ZK 91296, a partial agonist at benzodiazepine receptors. Psychopharmacology 83:240–248

Polc P, Haefely W (1976) Effects of two benzodiazepines, phenobarbitone and baclofen on synaptic transmission in the cat cuneate nucleus. Naunyn-Schmiedebergs Arch Pharmacol 294:121–131

Polc P, Bonetti EP, Schaffner R, Haefely W (1982) A three-state model of the benzodiazepine receptor explains the interactions between the benzodiazepine antagonist Ro 15–1788, benzodiazepine tranquillizers, β-carbolines, and phenobarbitone. Naunyn-Schmiederbergs Arch Pharmacol 321:260–264

Stephens DN, Kehr W (1985) β-carbolines can enhance or antagonise the effects of punishment in mice. Psychopharmacology 85:143–147

Stephens DN, Shearman G, Kehr W (1984) Discriminative stimulus properties of β-carbolines characterised as agonists and inverse agonists at central benzodiazepine receptors. Psychopharmacology 83:233–239

Stephens DN, Kehr W, Wachtel H, Schmiechen R (1985) The anxiolytic activity of β-carboline derivatives in mice, and its separation from ataxic properties. Pharmacopsychiatry 18:167–170

Study RE, Barker JL (1981) Diazepam and (−)-pentobarbital: fluctuation analysis reveals different mechanisms for potentiation of GABA-mediated inhibition after chronic exposure of spinal cord cultures to diazepam. Brain Res 268:171–176

Venault P, Charpouthier G, de Carvalho CP, Simiand J, Morre M, Dodd RH, Rossier J (1986) Benzodiazepine impairs and β-carboline enhances performance in learning and memory tasks. Nature 321:364–366

Warburton DM, Brown K (1971) Attenuation of stimulus sensitivity induced by scopolamine. Nature 230:120–127

Animal Model Studies of Benzodiazepine-Induced Amnesia

E. R. GAMZU[1]

Abstract

The development of a mouse passive avoidance test as a model for amnesia produced by benzodiazepines is described. The model appropriately classifies the amnesic potential of a wide range of psychoactive drugs as validated by clinical findings. Control experiments indicate that the effect is best described as anterograde amnesia resulting from a failure of consolidation. γ-Aminobutyric acid (GABA) antagonists had almost no effect on benzodiazepine-induced amnesia, whereas the benzodiazepine-receptor antagonist Ro 15–1788 completely and specifically reversed it. This clinically confirmed finding suggests that benzodiazepine-induced amnesia is mediated through the benzodiazepine-receptor. However, in vivo inhibition of benzodiazepine binding does not correlate well with amnesia in the mouse, and some benzodiazepine-receptor agonists with potent CNS effects in other in vitro models do not produce amnesia. Additional work is needed to clarify what aspects of benzodiazepine receptor occupancy mediate amnesia.

Benzodiazepine-induced amnesia is an example of a phenomenon that was first described as a result of clinical experience. This led to research directed at delineating the clinical manifestations, including comparisons of various therapeutic agents. Only then were pharmacological and mechanistic studies in animals employed. Despite great advances in our understanding of the neurobiology of benzodiazepine action only limited effort has been focused directly on the effects of benzodiazepines on memory.

As early as 1965, NUTTER and MASSUMI described the use of diazepam in cardioversion. Prior to that point this rather drastic therapy was carried out under general anesthesia. The use of diazepam resulted in a much more benign pharmacological manipulation with the additional advantage of pronounced muscle relaxation. An unexpected bonus was the fact that almost all of the patients had no recollection of the procedure. The failure to recall an unpleasant medical intervention was soon to become a boon in dental surgery, endoscopies, and various other mildly invasive outpatient surgical procedures. Indeed, the vast majority of the early studies were undertaken by anesthesiologists, especially DUNDEE and colleagues (HASLETT and DUNDEE 1968; PANDIT and DUNDEE 1970; PANDIT et al. 1971; DUNDEE and PANDIT 1972). For quite some time the focus was on intravenously delivered benzodiazepines. It was to be a number of years until the first reports (BAIRD and HAILEY 1972; MCKAY and DUNDEE 1980) and double-blind demonstration (JONES et al. 1978) of the amnesic effects of orally administered

---

[1] Department of Pharmacology, Hoffmann-La Roche Inc., Nutley, NJ 07110, USA.
Current address: Clinical Research Department, Warner Lambert/Parke-Davis, 2800 Plymouth Rd., Ann Arbor, MI 48105, USA.

benzodiazepines. More recently a number of investigators have confirmed the earlier findings and expanded the number of benzodiazepines which can be shown to have this property. A few anecdotes by noted investigators (SHADER and GREENBLATT 1983) have refocused scientific attention on the amnesic effect.

In the interim, our understanding of the mechanism of benzodiazepine pharmacology was advanced by demonstration that some of their effects could be explained by interactions with the most common CNS inhibitory transmitter, γ-aminobutyric acid (GABA; HAEFELY et al. 1975; COSTA et al. 1975). Subsequently the simultaneous description of high-affinity stereoselective benzodiazepine binding sites by two sets of investigators (MOHLER and OKADA 1977; SQUIRES and BRAESTRUP 1977) led to an explosion in benzodiazepine research. The synthesis of an imidazo-benzodiazepine (Ro 15–1788) that blocked all the known actions of benzodiazepines (HUNKELER et al. 1981; BONETTI et al. 1982) heralded the first step in the identification of the binding site as a receptor. The final criterion, isolation of an endogenous ligand from brain tissue has been claimed (SKOLNICK et al. 1983; GUIDOTTI et al. 1983). The receptor is part of a macromolecular protein complex. Activation of the benzodiazepine receptor component of the protein by compounds such as diazepam enhances the coupling of the GABA receptor and ion channel components such that normal GABA stimuli are more effective in opening the ion channel. These receptors have a distinct topographical distribution in the CNS including high densities in the hippocampus, an area known to play a major role in memory function. In addition, there are suggestions of the existence of at least two distinct subtypes of receptor. These concepts have permitted considerable resolution of apparently conflicting data in the areas of behavior, electrophysiology, and neurochemistry of benzodiazepines, as has been shown in recent reviews (SEPINWALL 1983; HAEFELY and POLC 1983; MOHLER et al. 1983).

The first animal laboratory demonstration of the amnesic properties of the benzodiazepines by SOUBRIE and colleagues occurred only in 1976, 11 years after the first clinical description. Despite a tremendous volume of behavioral literature on benzodiazepines and related anxiolytic compounds (excellently reviewed by SEPINWALL, 1984), only a handful of studies have focused on amnesia. THIEBOT's (1985) recent review of these shows considerable agreement and concordance between amnesia in animal tests and clinical findings, as is further demonstrated by our own data.

In order to characterize both the beneficial effects of potential preanesthetic agents as well as the potential liability of anxiolytic or hypnotic medications, we at Hoffmann-La Roche, started in the mid-1970s to study systematically the amnesic effects of benzodiazepines in animals (GAMZU et al. 1979). We used the deceptively simple one-trial passive avoidance procedure. Each mouse was placed in one side of a two-compartment chamber, from which it would wander into the second compartment where it experienced a brief foot-shock. Testing usually occurred 24 h later. Animals not exposed to the shock will enter the second compartment in less then 30 s, whereas control (placebo- or more correctly, vehicle-treated) mice will stay in the start box for as long as the experimenter is willing to wait (in our case 2 min). The essence of the procedure is captured by the common expression "once bitten, twice shy." In fact, considerable parametric work (especially around the response-shock interval, the shock duration, and the shock intensity) was conducted to ascertain those conditions under which the phenomenon could be reliably replicated, and experiments were only conducted between 1300 and 1700 each day. The final parameters that were employed are described in an earlier publication (BONETTI et al. 1982). In order to maximize our confidence in the data, we employed a number of conservative measures. The first was the use of a light-to-dark task (opposite to the most common form of passive avoidance) to use the nocturnal mouse's preference for the dark as a bias against amnesia. Similarly, the drug data are described primarily in terms of whether or not the mouse entered the second box, regardless of latency. Mice that experienced shock on the training day and entered the second box on the test day were considered to have "forgotten" the shock experience. Considered together with the control experiments described below, this phenomenon can reasonably be interpreted as anterograde amnesia.

There were two distinct sets of theoretical issues to be clarified. One set focuses on the pharmacology and mechanism of action and has been reported in some detail (GAMZU 1987), and so will only be summarized here. The second set of issues can be classified as cognitive or psychological.

The first task was to identify those compounds that produced the phenomenon. Certain benzodiazepines and other compounds administered orally prior to training reliably produced an anterograde amnesia when the animals were tested the next day, although immediate recall was not impaired. Based on the variability seen in many experiments, we decided that a meaningful effect was obtained if 50% of the mice "forgot." Those compounds that produced this effect at two or more doses separated by half a log unit were considered to be "active." For these compounds the dose that would produce an amnesic effect in half of the mice (the ED_{50}) was computed in order to compare the potency of different compounds. Active compounds also differed in the maximum percentage of mice "forgetting," which is captured as the peak percentage effect. These parameters are shown for active known compounds in Table 1. In subsequent experiments on mechanisms of action, or on comparisons between compounds, these data were used to choose equally effective doses, either the ED_{50} or the dose producing the maximum effect. Posttrial electroconvulsive shock was also effective in producing amnesia in 90% of mice so treated.

Based on these experiments with mice, we were willing to predict that the compounds listed in Table 1 would have amnesic potential in the clinic. Indeed, recall failure had already been, or would subsequently be, demonstrated in humans for all of the compounds in Table 1 except estazolam (GHONEIM and MEWALDT 1975; DUNDEE and GEORGE 1976; ROTH et al. 1980; DUNDEE and WILSON 1980; BLOCK and BERCHOU 1984; SUBHAN and HINDMARCH 1984; LISTER 1985; KOEPPEN et al. 1985). In the case of the latter compound, there are no reports of any such study, and the possibility of such an effect is unresolved.

A second group of compounds was clearly inactive in the mouse passive avoidance test. Included in this group were amobarbital, amphetamine, chlorpheniramine, chlorpromazine, clonidine, imipramine, methyl-scopolamine, pentobarbital, physostigmine, and protriptyline. Also in the inactive class were two phenylquinolines (PK 8156 and PK 9084) with claimed anxiolytic effects (LE FUR et al.

Table 1. Active compounds in the mouse passive avoidance amnesia test

Generic name	Pretreatment (min)[a]	ED_{50} (mg/kg)	Peak % effect
Triazolam	30	0.30	88
	60	0.30	75
Lorazepam	60	0.47	60
Alprazolam	60	0.51	89
Flunitrazepam	60	0.87	69
Diazepam	60	8.39	58
	1 i. v.	2.80	50
Estazolam	60	12.90	86
Midazolam	60	13.59	85
	30	6.23	82
	1 i. v.	0.15	88
Clobazam	60	29.34	73
Scopolamine	15 i. p.	0.39	82
Morphine	60	44.90	100
Zopiclone	60	65.30	65

[a] Route of administration p. o. except where otherwise noted.

1981) and the benzodiazepine antagonist Ro 15–1788. It is immediately obvious that neither CNS depression nor general sedation was sufficient to produce amnesia in this model. To the best of our knowledge, none of these compounds has ever been reported to produce a reliable clinical amnesia.

A third group of compounds that produced a 50% amnesia at only one dose was more difficult to evaluate. Of these, many, but not all, were active only at the highest testable dose. In this general class were four benzodiazepines – chlordiazepoxide, flurazepam, nitrazepam, and oxazepam. All have been in clinical use for many years, and there are only sporadic (mostly anecdotal) reports of their causing amnesia. In fact, at least one clinical study failed to obtain amnesic effects with oxazepam (LILJEQUIST et al. 1979). Also in this class were amitriptyline, atropine, meprobamate, and CL 218 872. The latter is a triazolopyridazine that has anxiolytic-like effects without sedation in animals and has played an important role in the study of the biology of the benzodiazepine receptor because of the possibility that it binds to a subset of the receptors (KLEPNER et al. 1979). We would consider compounds in this class as having, at best, limited potential for producing clinical amnesia.

In subsequent experiments we were able to demonstrate that the specific benzodiazepine receptor antagonist Ro 15–1788, although inactive by itself, reversed the amnesic effects of triazolam in a dose-dependent (BONETTI et al. 1982) and relatively specific (GAMZU et al. 1982) manner when administered just prior to the triazolam. At the clinical level it is clear that Ro 15–1788 can reverse benzodiazepine-induced amnesia (O'BOYLE et al. 1983). Despite the clarity of the findings with Ro 15–1788, there are a number of benzodiazepines that bind to the benzodiazepine receptor, are clinically active anxiolytics, and yet have minimal amnesic effects. Moreover, using only compounds that had in vivo activity predictive of centrally mediated antianxiety effects we were unable to show any clear correlation between activity in the mouse passive avoidance amnesia test and in vitro inhibition of labelled benzodiazepine receptor binding (GAMZU 1987). This is somewhat surprising, since bioassays of other benzodiazepine effects do correlate highly with in vitro binding (MOHLER and RICHARDS 1983). Although benzodiazepine-induced amnesia is most likely mediated through the benzodiazepine receptor, receptor occupancy alone, without invoking either regional or receptor subtype differences, is inadequate to explain the complexities of the phenomenon.

All of the conclusions above are based on the assumption that the observed behavior can be interpreted as reflecting anterograde amnesia. Before these conclusions could be reached careful testing and elimination of other confounding variables was necessary.

That the phenomenon is not dependent on the use of shock was demonstrated in two ways. First, in separate experiments rats treated with flunitrazepam, lorazepam, or triazolam prior to training in an automated Y-maze discrimination for food showed no evidence of retention when tested 24 h later. Secondly, in an experimental design attributable to SOUBRIE et al. (1976), mice given benzodiazepines prior to an extinction session (during which the mouse was left in the second box for 5 min in order to disassociate the shock from that specific location) subsequently avoid the second box (i.e., have amnesia for the extinction session). In contrast, mice treated with vehicle just prior to the extinction session now enter the second box (i.e., "forget" the initial experience). Thus, the behavioral outcome is completely reversed from that in the standard experiment. Finally, although it might be tempting to ascribe the phenomenon to an anxiolytic effect of the benzodiazepines during the testing session, it is also the case that anxiolytic compounds that are effective clinically and in other animal models employing shock-suppressed behavior (such as the barbiturates, meprobamate, and some benzodiazepines) are inactive in this test.

The question of whether the effect is anterograde (affecting events occurring after medication) or retrograde (affecting events occurring prior to medication) was easier to resolve. In a relatively extensive series of experiments, we never reliably obtained differences in 24-h recall between mice treated with vehicle or benzodiazepines when the drugs were injected immediately after training. This is consistent with the absence of any claims of retrograde amnesia resulting from the use of benzodiazepines in humans.

From the cognitive and psychological perspective one of the major issues is whether the benzodiazepines actually produce "amnesia" or whether the failure of recall is secondary to sleepiness and sedation. The implication of the latter perspective is that the information to be learned is simply not registered. As is demonstrated in the other articles in this book, this question is not simply answered by asking whether people or animals appear to be asleep shortly after taking benzodiazepines. At least three different lines of research addressed this issue. In the most simple case, mice that were highly sedated simply were unable to complete the training session. More compelling are the data from the training sessions, an exemplar of which is shown in Fig. 1 from an early experiment with lorazepam. The data on the left-hand side of the figure indicate that the groups of lorazepam-treated mice were equivalent to the vehicle-treated group in latency to exit the start box shortly after receiving the drug. In contrast, 24 h later the groups differ considerably in retention of the task, as can be seen on the right-hand side of the

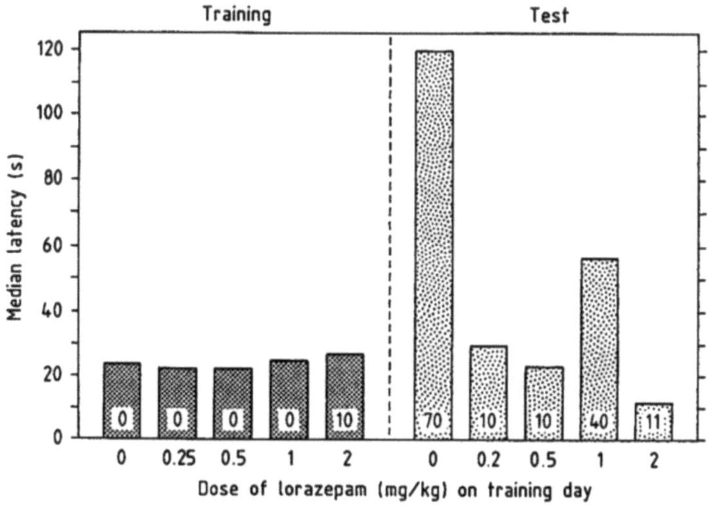

Fig. 1. The effect of various oral doses of lorazepam given 1 h prior to a training session on the median latency of groups of 10 mice to enter the shock compartment of a two-compartment passive avoidance apparatus. Latencies in the *left-hand* half of the figure are for the training session and indicate that the doses were equivalent to vehicle. Latencies in the *right-hand* portion of the figure are from the test session conducted 24 h later and indicate that whereas vehicle-treated mice "remember" the shock, most lorazepam-treated mice do not. *Numbers* in bars indicate the percentage of mice that did not enter the shock compartment

figure. While it is undoubtedly true that sedation can interfere with learning, the vast majority of our data were consistent with those shown in Fig. 1 in that at appropriate doses of the "amnesic" benzodiazepine compounds, treated mice that appeared perfectly normal and were no different behaviorally from vehicle-treated mice would subsequently show no evidence of the prior experience.

More compelling evidence against an explanation of the data based on the failure to learn were obtained in an experiment in which vehicle or triazolam 1 mg/kg were given orally to four groups of mice prior to training. Two of the groups followed the procedures described above with the anticipated results. The other two groups were tested within 10–20 min of the training session. Under these conditions there were no differences between vehicle- or triazolam-treated mice, indicating that both groups had adequately learned the task! On the other hand, mice treated equivalently with triazolam at the same time but tested 24 h later showed no evidence of retention of the prior shock experience.

Finally, it is relatively well established that tolerance develops to the sedative effects of benzodiazepines (Cook and Sepinwall 1975). Consequently, we treated groups of mice for 3 consecutive days with either vehicle or lorazepam 0.5 mg/kg p.o. to see if a prior treatment regimen known to eliminate the sedative and other (Gamzu 1977) effects of benzodiazepines would eliminate the amnesic effects. Subsequently, the groups were subdivided and some treated with vehicle and some with the same dose of lorazepam immediately before training in the passive avoidance test described above. When tested 24 h later the results were en-

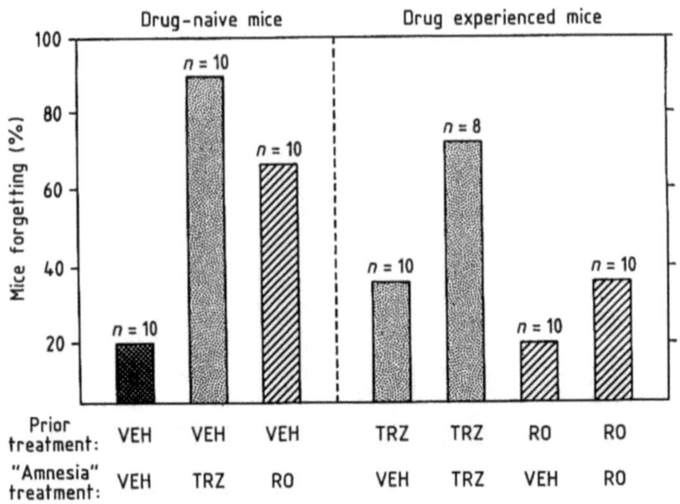

Fig. 2. The effect of prior drug experience on the acute effects of oral doses of triazolam and Ro 23–1590 (a non-benzodiazepine anxiolytic that binds to the benzodiazepine receptor) on the retention of a passive avoidance task in mice (groups of 10 each). All mice were given three prior exposures on three separate days to vehicle, triazolam 1 mg/kg, or Ro 23–1590 30 mg/kg. On the fourth day groups were exposed acutely to one of the treatments immediately prior to training in the passive avoidance task. Data are presented as percentage of mice forgetting on the test session that occurred 24 h later. In the *left-hand* panel, it can be seen that both active compounds produced amnesia compared to vehicle-treated mice. In the *right-hand* panel, it is obvious that triazolam continues to produce amnesia in triazolam-experienced mice. In contrast, prior exposure to Ro 23–1590 eliminated the amnesic effects of that compound. *VEH*, vehicle; *TRZ*, triazolam; *RO*, Ro 23–1590

tirely independent of prior experience. Mice treated with vehicle immediately before training avoided the shock compartment, while those treated with lorazepam did not. This absence of tolerance to the amnesic effect was true for the "active" benzodiazepines that were so tested. Indeed, tolerance to the "amnesic" effect was seen for only one compound – Ro 23–1590. This is a newly reported phenylquinolone that binds to the benzodiazepine receptor (BAUTZ et al. 1986) and has anxiolytic properties in neuropharmacologic, anticonflict (SULLIVAN et al. 1986), and anticonvulsant tests in animals (ANDERSON et al. 1986). This compound was compared to triazolam in an experiment identical to the one described above. The amnesic effects of these two compounds after acute administration can be seen in the left-hand panel of Fig. 2. The right-hand panel demonstrates that the triazolam effect is still manifest after prior exposure to triazolam, but prior exposure to Ro 23–1590 apparently abolishes the ability of the compound to interfere with retrieval of the task. Despite the data with Ro 23–1590, the series of experiments again demonstrates that anterograde amnesia resulting from administration of benzodiazepines can occur without any obvious evidence of sedation or interference with learning.

Another major cognitive or psychological issue is whether the phenomenon can be subsumed under the category of state dependency. In essence, this concept sug-

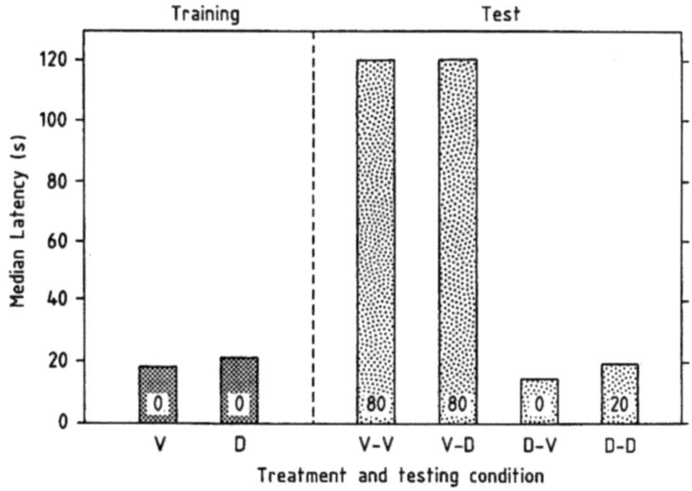

Fig. 3. Results of lorazepam 0.5 mg/kg p.o. on median latency in a 2×2 amnesia versus state-dependency paradigm employing the passive avoidance task in mice ($n = 10$ per group). All animals were treated 1 h prior to both training and testing with either vehicle (*V*) or drug (*D*). In the *left-hand* panel one can see that the dose of lorazepam had no effect on the median training latency. Each of these two groups of 20 mice were split into two groups of 10. In the *right-hand* portion of the figure the median latency is given for the resultant four groups (designated by *two initials* indicating the training and testing treatments respectively). Mice retain the task requirements if they received vehicle prior to training and forget it if they received lorazepam. The testing treatment had essentially no effect, indicating an amnesic effect. *Numbers* in bars indicate the percentage of mice that did not enter the shock compartment

gests that material learned is coded by internal as well as external stimuli and that all, or most, of these must be present for recall. Thus, information learned in a drugged state might only be totally retrievable in the same state. Consequently, any change from drugged to undrugged state or vice versa should result in decreased recall. It should be noted that this is not a memory deficit, since the information can theoretically be retrieved by reinduction of the drugged state. Indeed, this is the basis for the first full-length detective novel in the English language (COLLINS 1981/1868). However, demonstration of state dependence requires rigorous experimentation (OVERTON 1974).

To study this phenomenon we employed a 2×2 design in which mice were treated prior to both training and testing sessions with either vehicle (V) or drug (D). This results in four groups that are referred to as V-V, V-D, D-V, or D-D. An example of the type of data that were obtained using lorazepam 0.5 mg/kg p.o. are shown in Fig. 3. It is obvious that retention of the task depends on what was administered prior to training and is completely independent of the treatment prior to testing. This indicates that the phenomenon is most reasonably described as amnesia. Interestingly, all the benzodiazepines that were tested in this paradigm showed identical results. The other compounds in question (and the doses used) were: diazepam (10 mg/kg p.o.), flunitrazepam (1 mg/kg p.o.), and triazolam (1 mg/kg p.o.). This was also true for scopolamine (3 mg/kg i.p.).

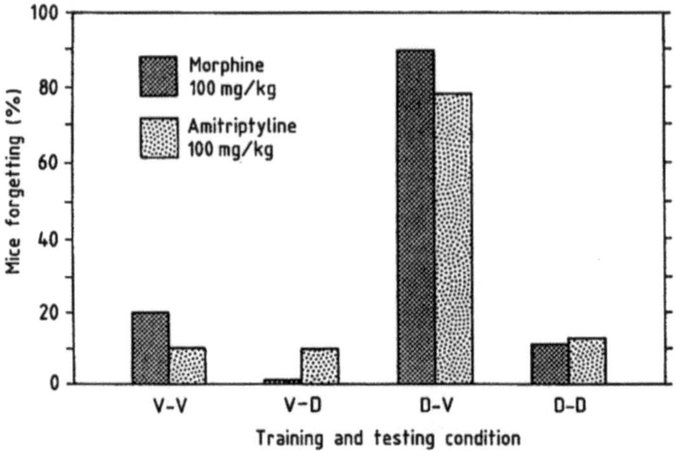

Fig. 4. Asymmetric state-dependency effects of oral doses of morphine 100 mg/kg and amitriptyline 100 mg/kg on percentage of mice forgetting in a 2 × 2 amnesia versus state-dependency paradigm employing the passive avoidance task in mice ($n = 10$ per group). Procedural details are equivalent to those described for Fig. 3. Only test data are presented

Pure state dependency would involve no recall in the D-V and V-D conditions and almost perfect recall when the training and testing treatments are identical. This is extremely rare. More common are asymmetric situations in which recall is impaired only when the drug is present at recall but not at learning (or vice versa), or in which recall is not complete when the drug state is reintroduced. In fact, asymmetric state dependency was obtained with morphine and amitriptyline, as is shown in Fig. 4. Clinically, data have been reported that clearly show no state dependency, but some asymmetries have been noted.

In summary, at certain doses it is possible to demonstrate that some benzodiazepines given shortly before a passive avoidance task will have no effect on either learning or performance, but will render mice unable to recall the task 24 h later. The animal model parallels clinical experience in terms of active and inactive compounds, duration of action, and pharmacologic manipulations. As such, it is a useful model for predicting clinical effects and for analysis of the phenomenon. The results of our investigations provided evidence, also available at the clinical level, that led to the conclusion that the active benzodiazepines interfere with the consolidation of short-term memory into long-term memory.

As a final comment, it is worth noting that the exciting research on the GABA-benzodiazepine-chloride channel macromolecule suggests some intriguing possibilities. It has been suggested that in addition to antagonists, there are compounds called inverse agonists that bind to the benzodiazepine receptor but produce effects that are directly opposite to those produced by the classical agonists such as diazepam (POLC et al. 1982). Should this be borne out for all the functional effects of the benzodiazepines, it is conceivable that inverse agonists might enhance memory rather than interfere with it (CHAPOUTHIER et al. 1984; VENAULT et al. 1986). Other research suggests the possibility of partial agonists that are specific to one but not all of the effects of benzodiazepines (Sepinwall et al. 1986); for example antianxiety without sedation or muscle relaxation. Should this be true for inverse agonists as well, it raises the possibility of a compound that would specifically improve memory through an interaction at the benzodiazepine receptor, with obvious therapeutic applications (Olton et al. 1985).

Acknowledgments. Thanks are due to the many people who have contributed to this work: E. Boff, L. Cook, A. Davidson, L. Lee-Perrone, D. Poonian, B. Salsitz, A. Verderese, and G. Vincent. R. A. 0'Brien, G. Bautz, N. Spirt, and M. Zanko provided the binding data.

References

Anderson C, Sullivan JW, Boff E, Horst D, Furman S, Pietrusiak N, Zavatsky E, Keim K, Gold L, Sepinwall J (1986) Preclinical evaluation of a novel anxiolytic phenylquinoline (Ro 23-1590) in anticonvulsant and CNS sedative/depressant test procedures. Soc Neurosci Abstr 12:661

Baird ES, Hailey DM (1972) Delayed recovery from a sedative, correlation of the plasma levels of diazepam with clinical effects after oral and intravenous administration. Br J Anaesth 44:803–808

Bautz G, Spirt NM, Mangano RM, O'Brien RA, Horst WD (1986) Ro 23-1590: just another ligand for the benzodiazepine receptor. Soc Neurosci Abstr 12:662

Block RI, Berchou R (1984) Alprazolam and lorazepam effects on memory acquisition and retrieval. Pharmacol Biochem Behav 20:233–241

Bonetti EP, Pieri L, Cumin R, Schaffner R, Pieri M, Gamzu ER, Muller RKM, Haefely W (1982) Benzodiazepine antagonist RO 15–1788: neurological and behavioral effects. Psychopharmacology 78:8–18

Chapouthier G, Venault P, Prado de Carvalho L, Simiand J, Rossier J (1984) Possible effects of β-carbolines on memory. Soc Neurosci Abstr 10:647

Collins W (1981) The moonstone. Penguin, New York (first published 1868)

Cook L, Sepinwall J (1975) Parameters of emotion: psychopharmacological parameters and methods. In: Levi L (ed) Emotions – The parameters and methods. Raven, New York, pp 379–404

Costa E, Guidotti A, Mao CC (1975) Evidence for involvement of GABA in the action of benzodiazepine. Studies on rat cerebellum. In: Costa E, Greengard P (eds) Mechanism of action of benzodiazepines. Raven, New York, pp 113–130

Dundee JW, George KA (1976) The amnesic action of diazepam, flunitrazepam and lorazepam in man. Acta Anaesthesiol Belg (Suppl) 27:3–11

Dundee JW, Pandit SK (1972) Anterograde amnesic effects of pethidine, hyoscine and diazepam in adults. Br J Pharmacol 44:140–144

Dundee JW, Wilson DB (1980) Amnestic action of midazolam. Anaesthesia 35:459–461

Gamzu E (1977) The multifaceted nature of the taste-aversion inducing agent: is there a single common factor? In: Barker LM, Best M, Domjan M (eds) Learning mechanisms in food selection. Baylor University Press, Waco, Texas, pp 477–509

Gamzu ER (1987) The role of benzodiazepine receptors in amnesia; laboratory predictors. J Clin Psychiatry Monogr 5:8–13

Gamzu ER, Perrone L, Salsitz B (1979) An animal model for drug induced amnesia. Bull Psychon Soc 12:253

Gamzu ER, Perrone L, Keim K, Smart T, Davidson AB, Cook L (1982) A mouse passive avoidance model of anterograde amnesia: comparison of benzodiazepine and scopolamine induced amnesias and EEG effects. Fed Proc 41:1067

Ghoneim MM, Mewaldt SP (1975) Effects of diazepam and scopolamine on storage, retrieval and organisational processes in memory. Psychopharmacologia 44:257–262

Guidotti A, Forchetti CM, Corda MG, Konkel D, Bennett CD, Costa E (1983) Isolation characterization, and purification to homogeneity of an endogenous polypeptide with agonistic action on benzodiazepine receptors. Proc Natl Acad Sci USA 80:3531–3535
Haefely W, Polc P (1983) Electrophysiological studies on the interaction of anxiolytic drugs with GABAergic mechanisms. In: Malick JB, Enna SJ, Yamamura HI (eds) Anxiolytics: neurochemical, behavioral and clinical perspectives. Raven, New York, pp 113–146
Haefely W, Kulcsar A, Mohler H, Pieri L, Polc P, Schaffner R (1975) Possible involvement of GABA in the central actions of benzodiazepines. In: Costa E, Greengard P (eds) Mechanism of action of benzodiazepines. Raven, New York, pp 131–151
Haslett WHK, Dundee JW (1968) Studies of drugs given before anaesthesia. XIV: Two benzodiazepine derivatives, chlordiazepoxide and diazepam. Br J Anaesth 40:250–258
Hunkeler W, Mohler H, Pieri L, Polc P, Bonetti EP, Cumin R, Schaffner R, Haefely W (1981) Selective antagonists of benzodiazepines. Nature 290:514–516
Jones DM, Lew MJ, Spriggs TLB (1978) The effects of low doses of diazepam on human performance in group administered tasks. Br J Clin Pharmacol 6:333–337
Klepner CA, Lippa AS, Benson DI, Sano MC, Beer B (1979) Resolution of two biochemically and pharmacologically distinct benzodiazepine receptors. Pharmacol Biochem Behav 11:457–462
Koeppen D, Netter P, Fischer C (1985) Individual differences in the effects of clobazam on memory functions – clobazam and memory. Pharmacopsychiatry 18:12–14
Le Fur G, Mizoule J, Burgevin MC, Ferris O, Heaulme M, Gauthier A, Gueremy C, Uzan A (1981) Multiple benzodiazepine receptors: evidence of a dissociation between anticonflict and anticonvulsant properties by PK 8165 and PK 9084 (two quinoline derivatives). Life Sci 28:1439–1448
Liljequist R, Palva E, Linnoila M (1979) Effects on learning and memory of 2-week treatments with chloridiazepoxide lactam, n-desmethyldiazepam, oxazepam and methyloxazepam, alone or on combination with alcohol. Int Pharmacopsychiat 14:190–198
Lister RG (1985) The amnesic action of benzodiazepines in man. Neurosci Behav Rev 9:87–96
McKay AC, Dundee JW (1980) Effect of oral benzodiazepines on memory. Br J Anaesth 52:1247–1257
Mohler H, Okada T (1977) Demonstration of benzodiazepine receptors in the central nervous system. Science 198:849–851
Mohler H, Richards JG (1983) Receptors for anxiolytic drugs. In: Malick JB, Enna SJ, Yamamura HI (eds) Anxiolytics: neurochemical, behavioral, and clinical perspectives. Raven, New York, pp 15–40
Nutter DO, Massumi RA (1965) Diazepam in cardioversion. N Engl J Med 273:650–651
O'Boyle C, Lambe R, Darragh A, Taffe W, Brick I, Kenny M (1983) RO 15–1788 antagonizes the effects of diazepam in man without affecting its bioavailability. Br J Anaesth 55:349–350
Olton D, Gamzu E, Corkin S (eds) (1985) Memory dysfunctions: An integration of animal and human research from preclinical and clinical perspectives. New York Academy of Sciences (Ann NY Acad Sci vol 444)
Overton DA (1974) Experimental methods for the study of state-dependent learning. Fed Proc 33:1800–1813
Pandit SK, Dundee JW (1970) Preoperative amnesia, the incidence following the intramuscular injection of commonly used premedicants. Anaesthesia 25:493–499
Pandit SK, Dundee JW, Keilty SR (1971) Amnesia studies with intravenous premedication. Anaesthesia 26:421–428
Polc P, Bonetti EP, Schaffner R, Haefely W (1982) A three-state model of the benzodiazepine receptor explains the interactions between the benzodiazepine antagonist RO 15–1788, benzodiazepine tranquilizers, β-carbolines, and phenobarbitone. Naunyn-Schmiedeberg's Arch Pharmacol 321:260–264
Roth T, Hartse KM, Saab PG, Piccione PM, Kramer M (1980) The effects of flurazepam, lorazepam, and triazolam on sleep and memory. Psychopharmacology 70:231–237
Sepinwall J (1983) Behavioral studies related to the neurochemical mechanisms of action of anxiolytics. In: Malick JB, Enna SJ, Yamamura HI (eds) Anxiolytics: neurochemical, behavioral, and clinical perspectives. Raven, New York, pp 147–171

Sepinwall J (1984) Behavioral effects of antianxiety agents: Possible mechanisms of action. In: Seiden LS, Balster RL (eds) Behavioral pharmacology: The current status. Liss, New York, pp 181–203

Sepinwall J, Sullivan JW, Glinka S, Gold L, Boff E, Gamzu E, Keim K, Pietrusiak N, Smart T (1986) Assessment of the anxiolytic properties of a novel benzodiazepine derivative (Ro 23–0364) with a mixed agonist/antagonist profile. Soc Neurosci Abstr 12:661

Shader RI, Greenblatt DJ (1983) Triazolam and anterograde amnesia. All is not well in the Z-zone. J Clin Psychopharmacol 3:273

Skolnick P, Marangos PJ, Paul SM (1983) Endogenous ligands of the benzodiazepine receptor. In: Malick JB, Enna SJ, Yamamura HI (eds) Anxiolytics: neurochemical, behavioral, and clinical perspectives. Raven, New York, pp 41–56

Soubrie P, Simon P, Boissier JR (1976) An amnesic effect of benzodiazepines in rats? Experientia 32:359–360

Squires RF, Braestrup C (1977) Benzodiazepine receptors in rat brain. Nature 266:732–734

Subhan Z, Hindmarch I (1984) Effects of zopiclone and benzodiazepine hypnotics on search in short-term memory. Neuropsychobiology 12:244–249

Sullivan JW, Gold L, Cumin R, Keim K, Smart T, Vincent G, Verderese A, Gamzu E, MacNeil D, D'Amico J, Sepinwall J (1986) Preclinical evaluation of a novel phenylquinoline anxiolytic (Ro 23–1590) in neuropharmacologic and anticonflict test procedures. Soc Neurosci Abstr 12:661

Thiebot MH (1985) Some evidence for amnesic-like effects of benzodiazepines in animals. Neurosci Behav Rev 9:95–100

Venault P, Chapouthier G, Prado de Carvalho L, Simiand J, Morre M, Dodd RH, Rossier J (1986) Benzodiazepine impairs and β-carboline enhances performance in learning and memory tasks. Nature 321:864–866

β-Carbolines as Tools in Memory Research: Animal Data and Speculations

M. Sarter[1] and D. N. Stephens

Abstract

Benzodiazepines induce in animals, as in humans, almost exclusively anterograde amnesia. The mechanism of this effect is still unsettled; however, explanations like state dependency which may be based on sedative or emotional properties of benzodiazepines are usually favoured in contrast to an interpretation in terms of true amnesia. It is proposed that by the use of β-carbolines with agonist, partial agonist, antagonist and partial inverse agonist properties, the nature of the amnesia induced by benzodiazepine receptor agonists may be characterised. From a series of experiments it is concluded that the major reason for benzodiazepine-induced amnesia might be an impaired ability to filter interfering stimuli; that is, an attentional deficit. Since the antagonist β-carbolines may play a key role in providing evidence as to the GABAergic involvement in cognitive processes, the pharmacological profile of ZK 93426 is presented. The results of the interaction of β-carbolines with scopolamine will provide a basis on which to speculate on the GABAergic control of cholinergic neurotransmission and its therapeutic implications.

1 Drugs as Tools

There are at least two different aims in studying the psychopharmacology of cognitive processes. First, drugs, neurotransmitters, neurohormones or neuromodulators (for definition see Osborne 1981) are examined in order to characterise the contribution of specific brain systems to cognitive abilities (e.g. McGaugh 1983; Mason 1983). Secondly, there is a still small but increasing interest in evaluating psychological concepts by the use of pharmacological research strategies, i.e. using drugs as tools in psychology (Warburton and Wesnes 1984). In the past the latter approach was not paid much attention by those behavioural pharmacologists interested in cognitive functions. This might have been related to a situation where, with the exception of McGaugh's early work on strychnine and pentylenetetrazol (McGaugh 1968) and Drachman's studies on physostigmine (Drachman and Leavitt 1974), specific tools to impair and enhance particular cognitive functions were not available.

This paper focuses on the use of β-carbolines as tools in animal research on learning and memory in general and on the analysis of the mechanisms of benzodiazepine-induced amnesia in particular. This new class of benzodiazepine receptor ligands allows the manipulation of the behavioural effects mediated via this receptor complex in a bidirectional way (Stephens and Sarter, this volume) and

[1] Research Laboratories, Department of Neuropsychopharmacology, Schering AG, 1000 Berlin 65, FRG.

offers new tools for characterising the nature of benzodiazepine-induced amnesia. Furthermore, some of these new ligands may offer a new approach to developing nootropic drugs based on receptor antagonists and inverse agonists.

Lastly, by studying the neurochemical effects of benzodiazepine receptor ligands, both agonists and inverse agonists, it may be possible to derive information on the neurochemical processes underlying aspects of cognitive function. One possible process, GABAergic modulation of cholinergic transmission, will be explored in the last part of this article.

2 Benzodiazepine-Induced Amnesia

The amnesic effects of benzodiazepines form the central theme of this book and do not need to be further commented on here. Suffice it to say that in both humans and animals benzodiazepines appear to impair memory by effects on acquisition rather than on performance or retrieval (e.g. COLE 1986), and that the existence of a good correlation between the sedative and amnesic potencies of benzodiazepines (CLARK et al. 1983; ROEHRS et al. 1983) suggests that the benzodiazepines' amnesic properties may be due to an effect on stimulus filtering.

Of the novel benzodiazepine receptor ligands, certain compounds, like the β-carboline ZK 91296, possess the anxiolytic and anticonvulsant properties of the benzodiazepines but appear to lack their sedative and hypnotic effects (PETERSEN et al. 1984; STEPHENS et al. 1985; PELLOW and FILE 1986). It is of both theoretical and clinical interest to know whether such compounds also lack the amnesic effects of the benzodiazepines.

In the preliminary experiment we found that ZK 91296 was as effective as benzodiazepines and full agonist β-carbolines in inducing amnesia (JENSEN et al. 1987). In a simple passive avoidance task, using mice, the benzodiazepines lorazepam and diazepam as well as the β-carboline full agonist ZK 93423 and the partial agonist ZK 91296 given before acquisition induced amnesia for the task when the animals were tested 24 h following acquisition.

The interpretation of this finding, however, is not straightforward. It is not clear, for instance, whether the anxiolytic properties of these compounds contributed to a weaker acquisition since, although the animals were all trained to the same criterion of avoidance, the number of trials required to reach criterion was not recorded. Furthermore, no attempt was made to control for effects of state dependency (OVERTON 1974) on performance of this task. Thus, the results are at best only tentative evidence that the amnesic and sedative effects of benzodiazepine receptor ligands are separable, and further evidence from animal and human studies are urgently needed.

3 Behavioural Effects of the Benzodiazepine Receptor Antagonist ZK 93426

ZK 93426 (5-isopropoxy-4-methyl-β-carboline-3-carboxylic acid ethylester) was found to have no effect on passive avoidance learning in the experiment of JENSEN et al. (1987), and also lacked overt effects in most of the more common tests of anxiolytic or sedative effects (JENSEN et al. 1984). From these results, and if the hypothesis were true that state-dependency might have been responsible for the results with the agonists, ZK 93426 should not show state-dependent effects when tested in an appropriate experiment (OVERTON 1974).

3.1 State-Dependent Learning

The concept of state-dependent learning is based on the assumption that a psychotropic drug induces a certain "brain state" which, if it exists both during training and retest, will lead to improved retrieval in comparison to a situation where different "states" are present during acquisition and retrieval. On the understanding that retrieval of memories is more efficient when similar cues are present as during acquisition, the phenomenon is better termed "state-dependent retrieval".

The 2×2 design which we used to test for possible state dependency follows OVERTON's recommendation (OVERTON 1974). The experiment was carried out with young (24 weeks; $n = 40$) and senescent rats (120 weeks; $n = 40$) and a one-trial passive avoidance task using a step-down paradigm. The apparatus consisted of a platform (20×20 cm) which was elevated (5 cm) from a grid (1.5 cm distance between bars). The grid was charged by a separate power supply. Shock was administered by a manually operated switch. The animals were injected with ZK 93426 (5 mg/kg i.p.) or Cremofor EL (BASF, Ludwigshafen, FRG; 10% in saline) as vehicle 30 min before they were placed on the platform. When they stepped down on the grid, they were immediately shocked (a single shock, 2 mA for 0.5 s). Thereafter, the animals were replaced in their home cage. Step-down latency was measured manually using a stopwatch. The animals were treated 24 h later with ZK 93426 or Cremofor EL according to the 2×2 design and 30 min later placed on the platform. Step-down latency was measured without shocking the animal following step-down. The test was terminated after 200 s if the animal completely avoided stepping down.

Statistical evaluation was based on a three-factor ANOVA (age, pretreatment, pre-retest treatment). In addition, the effects of age and pretreatment on the step-down latency in the acquisition phase were tested by a two-factor ANOVA.

The results are summarised in Fig. 1. The step-down latencies in the acquisition phase were somewhat influenced by age ($p < 0.1$), i.e. the senescent rats stepped down with a shorter latency than the young ones. No pretreatment effect was found and no interaction between age and pretreatment.

There was a clear effect of age on retest step-down latency (the senescent rats being faster than the young ones), and the interaction between pretreatment (be-

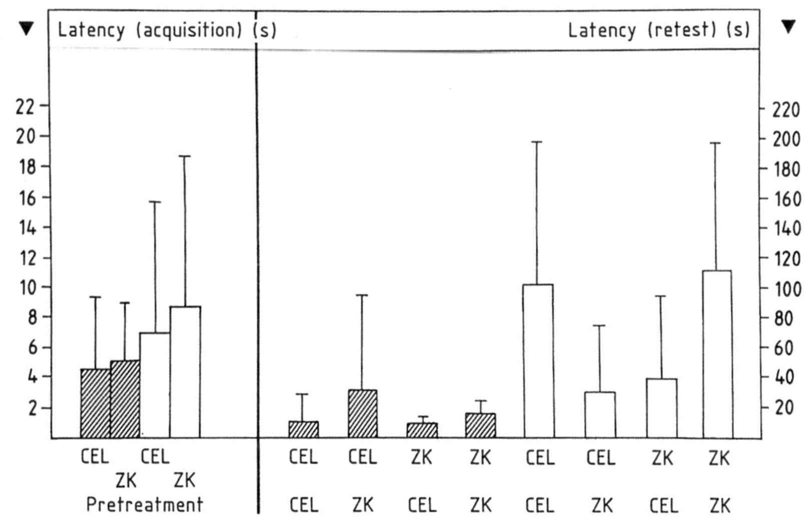

Fig. 1. *Left*, the acquisition latencies in senescent and young animals injected with vehicle (*CEL*) or ZK 93426 (*ZK*). *Right*, retest latencies. The young animals (*open bars*) generally showed a better memory for the one-trial step-down experience than the old ones (*shaded bars*); the longest retest latencies were shown by the young animals which were injected with the same substance at both acquisition and retest. *Bars* show mean + SD

fore acquisition) and treatment (before retest) as well as between those two and age reached significance. Figure 1 shows that in young animals the retest latencies were increased in those two groups of animals which received the same treatment before acquisition and retest. The interaction of all three factors in this experiment results from this age-specific effect. The increase in the latencies did not differ between groups treated with either Cremofor EL on both occasions or with ZK 93426 on both occasions.

Thus, this experiment demonstrated a clear state-dependent effect in the young rats, suggesting that ZK 93426 induced a different brain state from vehicle. As the senescent animals did not show any difference between acquisition and retest latencies, there may be no evidence for state-dependent effects, because they never learned the task.

In the experiment of Jensen et al. (1987), ZK 93426 had no amnesic effect even though the animals were treated only before acquisition and therefore state dependency could have occurred. It seems possible that the one-trial paradigm is more sensitive to state-dependent effects and that the acquisition training to a criterion (as carried out in Jensen's experiment) hindered the appearance of state-dependent effects. However, the nature of the brain state induced by the benzodiazepine receptor antagonist ZK 93426 remains unclear from this experiment, and it may not resemble that produced by the other two β-carbolines.

3.2 Spatial Delayed Alternation Learning

A simple T-maze was used (total length 110 cm; width of the cross-path 50 cm in each direction) with retractable food holes. Male senescent rats (27 months; body weight 499 ± 28 g at the beginning of the experiment; $n = 10$) and mature young animals (5 months; 342 ± 94 g; $n = 10$) were trained. Both groups were handled extensively and body weight was reduced to 75% (senescent animals) and 85% (young animals) before training was started. After shaping the animals to enter the arms of the maze to receive food reinforcement, they were trained to alternate between the two arms with the minimum possible delay between trials. Ten trials per day were given and in all stages of the experiment, the learning criterion was defined 80% correct responses on three consecutive days ($p = 0.04$ for 8 correct responses out of 10 trials when the probability of the event is 0.5). After the acquisition criterion was reached, a 10-s delay was introduced (during which time the animal was kept within the start-box). Delays of 20 s and 40 s were then introduced for animals that reached the criteria of the previous stages.

Administration of ZK 93426 (5 mg/kg) or Cremofor EL (1 ml/kg) was introduced at the beginning of the stage with a 20-s delay. Intraperitoneal injections were made 30 min before the session was started. Statistical analysis was performed non-parametrically using the Mann-Whitney *U* test. The results of this experiment are shown in Fig. 2.

Following introduction of the 10-s delay, the old animals performed worse than the young ones. Following the application of ZK 93426 at the stage with a 20-s delay, the young animals treated with ZK 93426 needed more trials and

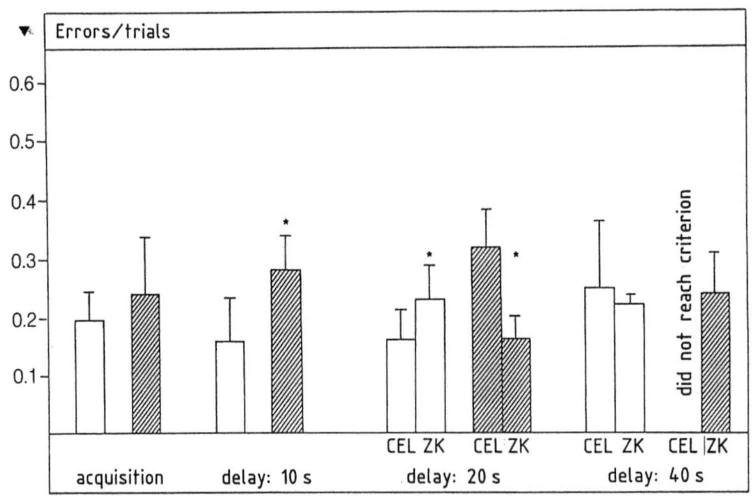

Fig. 2. Effects of ZK 93426 (*ZK*) on delayed alternation learning of young and senescent animals. At the 10-s delay, the old rats showed impaired learning compared to the young ones. ZK 93426 was injected from the beginning of the stage with 20-s delay and improved the performance of senescent animals only. None of the vehicle treated (*CEL*) senescent animals reached the criterion in the stage with 40-s delay. Performance is indicated by the ratio between errors and trials (mean + SD)

made more errors than vehicle-treated animals before they reached the criterion. The senescent animals, however, were found to benefit from the compound. Whereas only one of the young animals did not reach the criterion at the stage with a 40-s delay, none of the vehicle-treated senescent animals reached this criterion, but 40% of the ZK 93426-treated senescent rats did.

The main result from this experiment suggests that senescent animals are impaired in learning a delayed alternation task and, at an intermediate degree of task difficulty, performance is affected by ZK 93426 negatively in young animals but positively in senescent animals.

We do not assume that the impaired performance of the senescent animals during the stage with a 10-s delay necessarily reflects impaired working memory. Since the old animals spent more time in exploratory activities such as sniffing and rearing, differences in selective attentional processes might have been the basis for the age differences. This would be consistent with experiments demonstrating that senescent animals are not impaired compared to young ones in tasks which are controlled for the slower locomotor activity of senescent animals as well as for their different exploratory behaviour (SARTER 1986; SARTER and MARKOWITSCH 1983; STEPHENS et al. 1985).

In order to test whether ZK 93426 may have caused opposite effects in young and senescent animals by differentially altering exploratory or attentional processes, the level of locomotor activity, or emotional (anxiolytic, anxiogenic) states, the effects of the compound in young and senescent animals on locomotor and exploratory behaviour were studied.

3.3 Locomotor and Exploratory Behaviour

This experiment was carried out using a hole board apparatus. The hole board is a widely used apparatus although not fully characterized as a tool in behavioural pharmacology. In different modifications of the hole board, only a few substances have been reported to selectively increase the open field counts (amphetamine) or to selectively decrease the "hole" counts (dopamine agonists; LJUNGBERG and UNGERSTEDT 1976; MAKANJUOLA et al. 1977). The parameters measured with this test include head dipping, which is sometimes regarded as a measure of exploratory activity (e.g. FILE 1985; CRAWLEY 1985; FILE and WARDILL 1975). Due to our lack of knowledge, the validity of certain parameters is rather unclear; nevertheless, it is assumed that the hole board allows observations concerning general locomotor activity (total horizontal activity), exploratory activity (hole counts) and anxiety (ratio between activity spent in the centre and the total activity).

The hole board apparatus used in this study consists of a square open field (65 × 65 cm) with 16 holes (3 cm diameter) arranged in four parallel rows of four holes each. The field is surrounded by vertical Plexiglas walls 30 cm high. Automation of the hole board is provided by photobeam devices which also monitor the holes. Using an automated counter, data were collected and printed out every second minute. The following parameters were recorded: total horizontal activity, activity within the central area, number of head dips, and time spent in head dip-

Table 1. Summary of the results of the hole board experiment (four-factor ANOVA)

Variables	Factors			Interactions		
	Age (A)	Dose (D)	Intra-session habituation (IH)	A × D	A × IH	D × IH
Total activity	*		*		*	
Centre activity	*		*		*	
Number of head dips	*	*	*		*	
Time spent head dipping	*	(*)		(*)		*
Time per head dip		*		*		*

* $p<0.05$
(*) $p<0.1$
Hypotheses concerning different observed variables were tested independently of each other although the variables are related. Among the significant effects, the interaction between ZK 93426 dose and age for the head-dipping behaviour and between ZK 93426 dose and the intrasession habituation are considered to be the most important results. They suggest that ZK 93426 increased head-dipping behaviour in aged animals only and attenuated their habituation to exploring the holes.

ping. Male senescent (30 months; 495 ± 65 g; $n=40$) and mature young rats (6 weeks; 235 ± 20 g; $n=40$) were injected with physiological saline containing 10% Cremophor EL (BASF, Ludwigshafen, FRG; 1 ml/kg) or ZK 93426 suspended in such a solution (1.56, 6.25 or 25 mg/ml kg, i.p.) 30 min before the test. Each animal was tested for 14 min. Data were analysed in a four-factor ANOVA with age, dose and time (seven levels) as fixed factors and animals as a random factor. The results are summarised in Table 1.

There was an effect of age on all behavioural parameters except the time per head dip. The effects of age on total activity and activity in the central area indicate generally reduced locomotor activity in the aged rats. The decrease in locomotor activity occurred faster in aged rats than in the young ones, this being represented statistically by the interaction between the factors age and intra-session habituation.

There was an effect of drug dose on the number of head dips and the time per head dip; the former was reduced with increasing dose of ZK 93426 in both young and senescent animals. The senescent animals showed an increase in the total time spent in head dipping when injected with the compound; this is indicated in Table 1 by the interaction of dose with age.

Of all the parameters observed or calculated from the hole board, ZK 93426 affected only those that were related to head dipping behaviour. Whereas the number of head dips decreased with increasing dose in both young and senescent animals, the time per head dip increased drastically in old animals at the lowest dose of ZK 93426. The data also suggest that the exploratory activity of old rats normally habituates faster than that of young ones and that ZK 93426 retarded habituation in old animals only.

Thus, ZK 93426 did not alter locomotor activity but enhanced exploratory activity in senescent animals only. Inasmuch as exploratory activity may reflect attentional processes (HALLIDAY 1968; STEPHENS et al. 1982), this result suggests that the findings from the delayed alternation experiment might have been based on changes in attentional processes rather than changes in locomotion because the compound did not change the reduced locomotor activity in senescent rats.

3.4 Age-Dependent Effects

The difference between old and young animals in their responses to ZK 93426 is of some interest. Senescent rats differ from young animals in a number of neurotransmitter systems and the difference in the effect of the drug presumably reflects these changes. In behavioural changes reflecting memory or vigilance, a primary candidate would be the cholinergic system (LIPPA et al. 1980; 1985; STRONG et al. 1980). In order to test an interaction between the β-carboline antagonist ZK 93426 and the cholinergic system, we examined the potency of ZK 93426 to antagonise the disruptive effects of scopolamine on the spontaneous alternation behaviour of mice (SARTER et al., 1988).

3.5 Scopolamine Antagonism

The apparatus used to investigate spontaneous alternation behaviour was a Y-maze, automated using photo beams (Fig. 3). Two variables were recorded during an 8-min session: number of arm entries and alternation performance, that is the percentage of successive arm entries which are organised in a systematic way (triplets like ABC or CBA, but not for example ACA; for details see SARTER et al., 1988).

As illustrated in Fig. 4, ZK 93426 neither affected the activity of the animals nor reduced the scopolamine-induced enhancement of activity. Alternation performance was not influenced by the compound on its own, but the scopolamine-induced impairment of spontaneous alternation was significantly antagonised at 6.25 mg/kg.

Alternation behaviour requires the use of a rudimentary kind of working memory; working memory in general is sensitive to scopolamine (e.g. SPENCER et al. 1985) and ZK 93426 appears to posses anti-amnesic activity for the amnesia caused by the anticholinergic treatment. This effect of ZK 93426 seems not to be a task-specific one (see STEPHENS and SARTER, this volume). As lesions of the basal nucleus of Meynert impair alternation behaviour in rats (BENINGER et al. 1986) and GABAergic compounds injected into this area have been shown to influence cortical acetylcholine turnover, this part of the cholinergic brain system may be critically involved in the anti-amnesic effects of ZK 93426 in the case of scopolamine-induced amnesia. Thus the benzodiazepine receptor antagonist ZK 93426 might have disinhibited acetylcholine turnover and consequently resulted in displacement of scopolamine from the receptor. This indirect effect on acetylcholine receptor occupation would explain the antiscopolaminergic effects of ZK 93426.

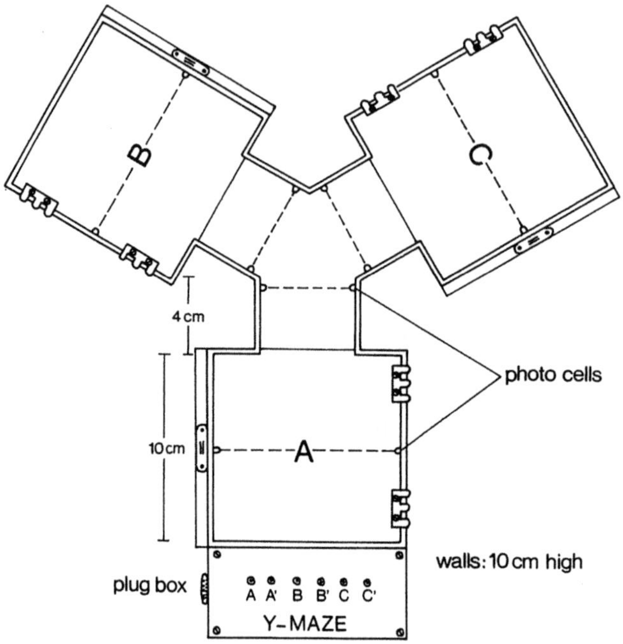

Fig. 3. The Y-maze used in the spontaneous alternation test. It consists of black Plexiglas walls with transparent ceilings. The size of the Y-maze and the location of the photo beams (*broken lines*) are indicated

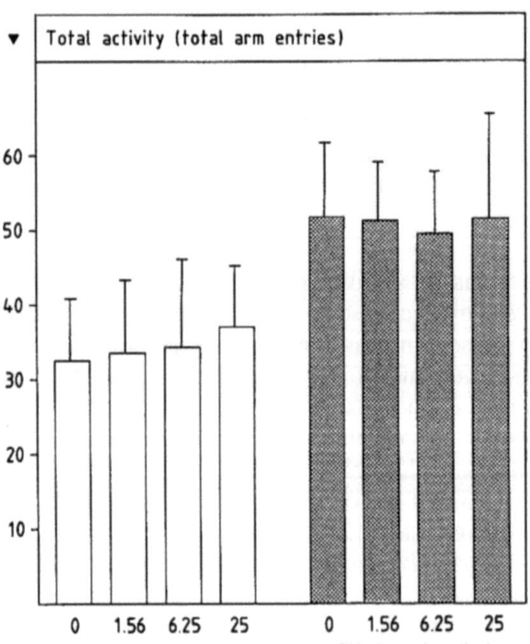

Fig. 4. The effects of ZK 93426 (*ZK*) on locomotor activity (total arm entries) in the spontaneous alternation test in vehicle-treated (CEL 10 ml/kg; *open bars*) and scopolamine-treated (1 mg/kg; *shaded bars*) mice. ZK 93426 affected neither locomotor activity in vehicle-treated animals nor the scopolamine-induced increase of locomotor activity (mean + SD)

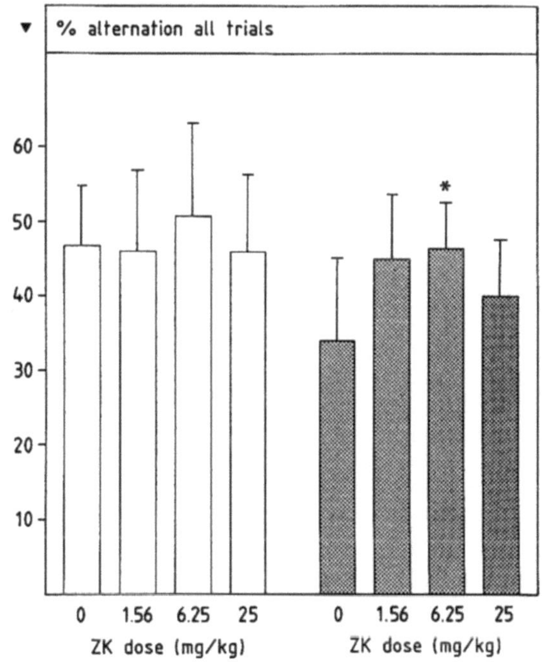

Fig. 5. Effects of ZK 93426 (*ZK*) on the alternation performance of vehicle (CEL)-treated (*open bars*) and scopolamine-treated mice (*shaded bars*). The alternation performance is given as the percentage of total arm entries (see text; mean + SD). ZK 93426 did not affect alternation performance when co-injected with vehicle but at a dose of 6.25 mg/kg it antagonised the decrease induced by scopolamine (*asterisk*)

3.6 Effects on Acquisition Versus Retrieval

The effects of ZK 93426 shown in the experiments described above are related exclusively to acquisition processes. Thus, the question remains of whether ZK 93426 affects learning exclusively or whether retrieval processes are also influenced. Furthermore, it remains unsettled whether the antagonism of the effects of scopolamine is restricted to antagonism of scopolamine-induced impairments of acquisition. For this reason we carried out an experiment in which we administered ZK 93426 at different times, before and during acquisition and before retention testing of a learning task.

Mice were treated with ZK 93426 (10 mg/kg) or vehicle (10% Cremofor EL) 30 min before acquisition in a one-trial passive avoidance task (step-down passive avoidance). Immediately afterwards, the animals were treated with scopolamine (3 mg/kg) or saline and, 24 h later, treated pre-retest with ZK 93426 or vehicle. Additionally, the effects of chronic pretreatment with this β-carboline on its acute effects and its interaction with post-trial scopolamine were tested. The design of this experiment is summarised in Table 2.

The step-down latencies during acquisition were not found to differ between the four groups, indicating that in this experiment neither chronic nor acute ZK 93426 influenced activity. The retest latencies were statistically compared by the use of χ^2 tests on the basis of the number of animals which reached the 200-s criterion. The main statistical results are summarised in Table 3.

There was only one significant effect, namely that acute pretreatment with ZK 93426 increased the number of animals which reached the retest criterion.

Table 2. Interaction of chronic and acute pretreatment with ZK 93426, post-trial scopolamine and pre-retest treatment with ZK 93426. (a 2×2×2×2 design)

Chronic pretreatment	Acute pretreatment	One-trial passive avoidance	Post-trial treatment	Pre-retest treatment
ZK 93426	ZK 93426	×	Scopo	ZK 93426
(10 mg/kg),		×	(3 mg/kg)	CEL
$n=80$		×	NaCl	ZK 93426
		×		CEL
	CEL	×	Scopo	ZK 93426
		×		CEL
		×	NaCl	ZK 93426
		×		CEL
CEL	ZK 93426	×	Scopo	ZK 93426
(1 ml/kg),		×		CEL
$n=80$		×	NaCl	ZK 93426
		×		CEL
	CEL	×	Scopo	ZK 93426
		×		CEL
		×	NaCl	ZK 93426
		×		CEL

Chronic pretreatment; 5 days; CEL, Cremophor EL; Scopo, Scopolamine.

Table 3. Results of the experiment based on the design summarised in Table 2

Components of variation	χ^2	df	p
Chronic pretreatment	0.36	1	0.54
Acute pretreatment	10.92	1	0.001
Differences between all combinations of pretreatment	13.04	3	0.0046
Post-trial treatment	0	1	1.0
Differences between all 16 groups	21.9	15	0.1

χ^2 test; $\alpha=0.1$; $\alpha'=0.0125$.
All further comparisons were non-significant. There is only one important result: acute (30 min before) treatment with ZK 93426 before learning of the step-down avoidance task yields enhanced retest performance 24 h later.

Neither chronic pretreatment, post-trial scopolamine nor pre-retest ZK 93426 affected retest performance.

The main result of this experiment suggests that ZK 93426 is predominantly active in improving learning, fails to enhance retrieval of a one-trial passive avoidance experience, but does not affect acquisition. If it is assumed that the effects of ZK 93426 are based on improved attentional processes, than this property would be specifically related to the evaluation of *external* stimuli since the attentional processes during retrieval might be predominantly related to the screening of internal associations or cues in order to reactivate long-term stored information (see Sect. 4).

4 General Discussion

From the experiments described above it can be concluded that:

1. Sedation did not seem to be the cause of benzodiazepine-induced amnesia inasmuch as a non-sedative β-carboline also induced amnesia in a passive avoidance test
2. The benzodiazepine receptor antagonist ZK 93426 exerted state-dependent effects
3. ZK 93426 improved performance of a delayed alternation task in senescent rats but impaired performance in young rats
4. ZK 93426 enhanced exploratory activity only in senescent rats and did not affect locomotor activity in general
5. ZK 93426 antagonised the scopolamine-induced impairment of alternation performance in mice
6. ZK 93426 seemed to act on acquisition processes rather than on retrieval

4.1 Benzodiazepine-Induced Amnesia

In both human and animal research, benzodiazepines appear to impair acquisition rather than performance or retrieval (e.g. COLE 1986). Although the degree to which amnesia may be due to state-dependent effects has been assumed to be relatively small (LISTER 1985; THIEBOT 1985; COLE 1986), the number of studies using the classic 2×2 design still seems too small to draw a definite conclusion. Most frequently, acquisition and retest have been performed at different "brain states" induced by a benzodiazepine or vehicle, respectively. If state dependency is a reasonable explanation of the mechanism of benzodiazepine-induced amnesia, what kind of pharmacological property could function as the cue for state-dependent effects?

From the amnesic effects of the non-sedative β-carboline ZK 91296 in the passive avoidance test cited above it seems unlikely that a lowered state of arousal plays a necessary role in the benzodiazepine-induced amnesia in such a task. This hypothesis, however, stands in contrast to numerous findings suggesting that there is a relationship between the sedative and amnesic potencies of benzodiazepines (CLARK et al. 1983; ROEHRS et al. 1983). Mood as a cue for benzodiazepine-induced state-dependent amnesia seems more difficult to test, but mood state is unlikely to account for the finding that the benzodiazepine receptor antagonist ZK 93426, which largely lacks emotional effects, also induced state-dependent effects (see above).

Possibly, the cue subserving state-dependent effects for the agonist, antagonist, inverse agonist β-carbolines is related to their effects on cognition. From GRAY's (1982) analysis of the behavioural effects of anti-anxiety drugs, it could be suggested that benzodiazepines act on cognition by impairing the screening of interfering stimuli. This idea is deduced primarily from the effects of such drugs on discrimination learning: GRAY concluded that impairments in discrimination learning are only found in cases when the discriminanda are presented successively (Gray 1982). Performance in such tasks depends on the ability to process interfering stimuli, which may be impaired by benzodiazepines. The hypothesis is supported by observations on the effects of benzodiazepines on discrimination tasks which have been analysed according to signal detection theory (see Stephens and Sarter, this volume). Such tasks indicate that benzodiazepines interfere with the evaluation of environmental stimuli. This would be consistent with the view that benzodiazepines selectively affect acquisition but not retrieval.

4.2 Improved Filtering of Interfering Stimuli Following ZK 93426

In keeping with this hypothesis it might be assumed that the effects of ZK 93426 are due to an improved filtering of interfering stimuli. First of all, the effects of this compound in the signal detection task (see Stephens and Sarter, this volume) can be interpreted in terms of this hypothesis. It might be speculated that the differences between senescent and young animals found in the delayed alternation task might be related to impaired selective attention in the senescent animals. If the antagonist β-carboline ZK 93426 focuses attention then we might expect improvement of performance in the working memory-related task (delayed alternation). It cannot be excluded that the senescent animals were simply less anxious and the compound induced an emotional shift which secondarily affected attention. Such an explanation is rendered somewhat less likely by the observation that, in the hole board, ZK 93426 did not decrease either the number of head dips, or the time spent in the centre of the area.

4.3 GABAergic-Cholinergic Interaction and Attentional Processes

It has been demonstrated that the antagonist β-carboline ZK 93426 can antagonise the effects of scopolamine on spontaneous alternation performance in mice (see Sarter et al., 1988). This property has also been demonstrated in other behavioural paradigms (avoidance tasks, signal detection tasks; see Stephens and Sarter, this volume). The effects of scopolamine seem generally related to an impairment of working memory-related abilities (e.g. Spencer et al. 1985) and quite specifically related to acquisition rather than to other stages of learning and memory (Ghoneim et al. 1984). These similarities between the effects of scopolamine and benzodiazepine receptor agonists and the assumption that scopolamine models the loss of cholinergic neurons in senile dementia suggest that benzodiazepine receptor agonists may also be used to model the cognitive decline of senile dementia (Block et al. 1985). Thus, the psychopharmacological similarities between benzodiazepine receptor agonists and muscarinic receptor blockers suggest that there may also be a direct neuropharmacological interaction between (parts of) both systems in the brain.

There are some arguments from different fields of research suggesting that such an interaction takes place in the basal nucleus of Meynert and its cholinergic projections to the cortex. Recently, a GABAergic input to the cholinergic cell bodies of the basal nucleus has been demonstrated anatomically (Zaborsky et al. 1986).

Pharmacologically, there are several reports suggesting that cortical acetylcholine turnover is controlled by a GABAergic input to the basal nucleus and can be altered by benzodiazepines administered systematically or into the basal nucleus (WOOD 1986; ZSILLA et al. 1976; TANGANELLI et al. 1985; WENK 1984; CASAMENTI et al. 1986).

On that basis it could be speculated that the benzodiazepine receptor antagonist ZK 93426 acts to reduce the inhibitory effects of GABA on acetylcholinergic activity and hence on acetylcholine release in cortical areas, leading to displacement of scopolamine from the receptor. As such an effect has not been observed on the scopolamine-induced increase of locomotor activity, it is necessary to postulate that this interaction between ZK 93426 and acetylcholinergic neurones is not general but restricted to, possibly, the basal forebrain-cortical cholinergic system.

4.4 Therapeutic Possibilities

Among the variety of neuropathological effects in the brains of patients with senile dementia, the loss of neurons arising from the basal nucleus of Meynert and terminating in several cortical areas is closely related to the degree of dementia, and may be seen as the most important pathological feature of this disease (COLLERTON 1986). If benzodiazepine receptor antagonists like ZK 93426 disinhibit the release of acetylcholine by the remaining neurons without accelerating the ongoing degenerative processes, such compounds would be very attractive for the treatment of senile dementia. There is, moreover, a possibility that such disinhibition also takes place at the cortical cholinergic terminals by GABAergic interneurons (BEANI et al. 1986). At least, the vigilance-enhancing properties shown in aged animals in different tasks suggest that such compounds may be valuable for cognitive impairments associated with normal ageing. These impairments are characterised primarily by attentional deficits and reduction of speed of performance in cognitive tasks.

Acknowledgements. We wish to thank G. Bodewitz, A. Neumann, G. Rose and T. Steckler for their excellent technical assistance, A. van der Linde for the statistical work, R. Kollberg and D. Niemeyer for help in automating several methods and Bettina Fogel for typing the manuscript with her usual friendly efficiency. As always, John Andrews, Herbert Schneider and Theodora Duka have provided useful discussions.

References

Beani L, Tanganelli S, Antonelli T, Bianchi C (1986) Noradrenergic modulation of cortical acetylcholine release is both direct and γ-aminobutyric acid-mediated. J Pharmacol Exp Ther 236:230–236

Beninger RJ, Jhamandas K, Boegman RJ, El-Defrawy SR (1986) Effects of scopolamine and unilateral lesions of the basal forebrain on T-maze spatial discrimination and alternation in rats. Pharmacol Biochem Behav 24:1353–1360

Block RI, DeVoe M, Stanley B, Stanley M, Pomara N (1985) Memory performance in individuals with primary degernative dementia: its similarity to diazepam-induced impairments. Exp Aging Res 11:151–155

Casamenti F, Deffenu G, Abbamondi AL, Pepeu G (1986) Changes in cortical acetylcholine output induced by modulation of the nucleus basalis. Brain Res Bull 16:689–695

Clark MS, Milberg S, Ross J (1983) Arousal cues arousal related material in memory: implications for understanding effects of mood on memory. J Verb Learn Verb Behav 22:633–649

Cole SO (1986) Effects of benzodiazepines on acquisition and performance: A critical assessment. Neurosci Biobehav Rev 10:265–272

Collerton D (1986) Cholinergic function and intellectual decline in Alzheimer's disease. Neuroscience 19:1–28

Drachman DA, Leavitt J (1974) Human memory and thecholinergic system. Arch Neurol 30:113–121

File SE (1985) What can be learned from the effects of benzodiazepines on exploratory behavior? Neurosci Biobehav Rev 9:45–54

Crawley JN (1985) Exploratory behavior models of anxiety in mice. Neurosci Biobehav Rev 9:34–44

File SE, Wardill AG (1975) The reliability of the hole board apparatus. Psychopharmacology 44:47–51

Ghoneim MM, Hinrichs JV, Mewaldt SP (1984) Dose-response analysis of the behavioral effects of diazepam: I. Learning and memory. Psychopharmacology 82:291–295

Gray JA (1982) The neuropsychology of anxiety: an enquiry into the functions of the septo-hippocampal system. Oxford University Press, Oxford

Halliday MS (1968) Exploratory behavior. In: Weiskrantz L (ed) Analysis of behavioral change. Harper and Row, New York

Jensen LH, Petersen EN, Braestrup C, Honore T, Kehr W, Stephens DN, Schneider HH, Seidelmann D, Schmiechen R (1984) Evaluation of the β-carboline ZK 93426 as a benzodiazepine receptor antagonist. Psychopharmacology 83:249–256

Jensen LH, Stephens DN, Sarter M, Petersen EN (1987) Bidirectional effects of β-carbolines and benzodiazepines on memory processes. Brain Res Bull 19:359–364

Lippa AS, Loullis CC, Rotrosen J, Cordasco DN, Critchett DJ, Joseph JA (1985) Conformational changes in muscarinic receptors may produce diminished cholinergic neurotransmission and memory deficits in aged rats. Neurobiol Aging 6:317–323

Lippa AS, Pelham RW, Beer B, Critchet DJ, Dean RL, Bartus RT (1980) Brain cholinergic dysfunction and memory in aged rats. Neurobiol Aging 1:13–19

Lister RG (1985) the amnesic action of benzodiazepines in man. Neurosci Biobehav Rev 9:87–94

Ljungberg T, Ungerstedt U (1976) Automatic registration of behavior related to dopamine and noradrenaline transmission. Eur J Pharmacol 36:181–188

Makanjvola ROH, Hill G, Mabne I, Dow RC, Ashcroft GW (1977) An anatomical method for studying exploratory and stereotyped behavior in rats. Psychopharmacology 52:271–277

Mason ST (1983) The neurochemistry and pharmacology of extinction behavior. Neurosci Biobehav Rev 7:325–347

McGaugh JL (1968) Drug facilitation of memory and learning. In: Efron DH, Cole JO, Levine J, Wittenborn JR (eds) Psychopharmacology: a review of progress 1957–1967. U.S. Government Printing Office, Washington DC, pp 891–904

McGaugh JL (1983) Hormonal influcences on memory. Annu Rev Psychol 34:297–323

Pellow S, File SE (1986) Evidence that the β-carboline, ZK 91296, can reduce anxiety in animals at doses well below those causing sedation. Brain Res 363:174–177

Osborne NN (1981) Communication between neurons: current concepts. Neurochem Int 3:3–16
Overton DA (1974) Experimental methods for the study of the state dependent learning. Fed Proc 33:1800–1813
Petersen EN, Jensen LH, Honore T, Braestrup C, Kehr W, Stephens DN, Wachtel H, Seidelmann D, Schmiechen R (1984) ZK 91296, a partial agonist at benzodiazepine receptors. Psychopharmacology 83:240–248
Roehrs T, Zorick FJ, Siclesteel JM, Wittig RM, Hartse KH, Roth T (1983) Effects of hypnotics on memory. J Clin Psychopharmacol 3:310–313
Sarter M (1986) Some considerations of different modes of action of nootropic drugs. Neuropsychobiology 15:192–200
Sarter M (1987) Measurement of cognitive abilities in senescent animals. Int J Neurosci 32:765–774
Sarter M, Markowitsch HJ (1983) Reduced resistance to progressive extinction in senescent rats: a neuroanatomical and behavioral study. Neurobiology of Aging 4:203–216
Sarter M, Bodewitz G, Stephens DN (1988) Attenuation of scopolamine-induced impairments of spontaneous alternation behavior by antagonist but not inverse agonist and agonist β-carbolines. Psychopharmacology (in press)
Spencer DG Jr, Pontecorvo MJ, Heise GA (1985) Central cholinergic involvement in working memory: effects of scopolamine on continuous nonmatching and discrimination performance in the rat. Behav Neurosci 99:1049–1065
Stephens DN, Kehr W (1985) β-carbolines can enhance or antagonize the effects of punishment in mice. Psychopharmacology 85:142–147
Stephens DN, Tonkiss J, Wearden JH (1982) Differences between rats undernourished preweaning, and controls in learning about a redundant stimulus during acquisition of a conditioned emotional response. Physiol Behav 28:95–101
Stephens DN, Weidmann R, Quartermain D, Sarter M (1985) Reversal learning in aged rats. Behav Brain Res 17:193–202
Strong R, Hicks P, Hsh L, Bartus RT, Enna SJ (1980) Age-related alteration in the rodent brain cholinergic system and behavior. Neurobiol Aging 1:59–63
Tanganelli S, Bianchi C, Beani L (1985) The modulation of cortical acetylcholine release by GABA, GABA-like drugs and benzodiazepines in freely moving guinea pigs. Neuropharmacology 24:291–299
Thiebot MH (1985) Some evidence for amnesic-like effects of benzodiazepines in animals. Neurosci Biobehav Rev 9:95–100
Warburton DM, Wesnes K (1984) Drugs as research tools in psychology: cholinergic drugs and information processing. Neuropsychobiology 11:121–132
Wenk GL (1984) Pharmacological manipulations of the substantia innominata-cortical cholinergic pathway. Neurosci Lett 51:99–103
Wood PL (1986) Pharmacological evaluation of GABAergic and glutamatergic input to the nucleus basalis-cortical and the septal-hippocampal cholinergic projections. Can J Pharmacol 64:325–328
Zaborsky L, Heimer L, Eckenstein F, Leranth C (1986) GABAergic input to cholinergic forebrain neuros: An ultrastructural study using retrograde tracing of HRP and double immunolabeling. J Comp Neurol 250:282–295
Zsilla G, Cheney DL, Costa E (1976) Regional chenages in the rate of turnover of acetylcholine in rat brain following diazepam or muscimol. Arch Pharmacol 294:251–255

β-Carbolines as Tools in Memory Research: Human Data with the β-Carboline ZK 93426

T. DUKA[1], V. EDELMANN, B. SCHÜTT and R. DOROW

Abstract

The discovery of substances which bind with high affinity to benzodiazepine receptors but have no pharmacological effects (antagonists) or effects opposite to those of benzodiazepines (inverse agonists) have introduced a new approach to elucidating mechanisms underlying memory and other cognitive processes. Since benzodiazepines induce anterograde amnesia and sedation, these substances should show an opposite effect and so enhance memory and/or increase vigilance.

In the present report we present data obtained in humans with a benzodiazepine receptor antagonist with weak inverse agonist properties, ZK 93426. The drug was given intravenously to human volunteers in double-blind, placebo-controlled designs and performance on several psychometric tests was evaluated. In a general estimation of behavioural changes volunteers experienced a stimulatory, activating effect of the drug. An improvement was observed in two cognitive tasks, the logical reasoning task and the pictures difference task, which estimate concentration and attentional processes respectively. No effects were found in a letter cancellation test or in time estimation. In another study utilizing EEG recording, we demonstrated that ZK 93426 increased wakefulness (vigilance) in healthy subjects during the daytime. The effect of ZK 93426 upon memory processes was also investigated utilizing a visual memory test and word lists. A slight improvement in some memory processes, especially long-term retrieval, was found. The present data suggest that benzodiazepine receptor antagonists with weak inverse intrinsic activity possess some effects opposite to those of benzodiazepines.

1 Introduction

It is generally agreed that memory is not a unitary process but rather comprises several psychobiologically distinct aspects which may be differentially affected by drugs. A better understanding of the mechanism of action of these drugs, as well as an analysis of their effects upon different behavioural aspects which may influence memory, would offer new insights into understanding how memory works on the one hand, and into treating memory deficits on the other. Benzodiazepines are one group of drugs which have been shown to cause memory impairment. Although the magnitude of the memory disruption induced by benzodiazepines appears to be dependent upon the dose and the route of administration used (PANDIT et al. 1976; GHONEIM et al. 1984), the patterns of amnesia produced by the different benzodiazepines seem to be qualitatively the same (for review see LISTER 1985). Recently, substances have become available which, although they act via the benzodiazepine receptor, exert no intrinsic effects (antagonists; HUNKELER et

[1] Research Laboratories, Schering AG, 1000 Berlin 65, FRG.

al. 1981; DUKA et al. 1986a) or have effects which are opposite to those of benzodiazepines (inverse agonists; BRAESTRUP et al. 1982). Accordingly, since benzodiazepines induce sedation and impair memory (GHONEIM and MEWALDT 1977; for review see LISTER 1985; CURRAN 1986), it has been suggested that inverse agonists might increase vigilance and alertness and improve memory. Several recent reports from studies using animals as models to test learning and memory processes have demonstrated that inverse agonists improve learning and memory (VENAULT et al. 1986; JENSEN et al. 1987; SARTER and STEPHENS, this volume) and increase vigilance (STEPHENS and SARTER, this volume). Substances which oppose the effects of benzodiazepines and are classified as inverse agonists may be convulsants (BRAESTRUP et al. 1982), and substances classified as partial inverse agonists which reveal selected parts of the whole pharmacological spectrum of inverse agonists may be proconvulsant or anxiogenic (STEPHENS and KEHR 1985). For example, one such substance, FG 7142, has been reported to induce anxiety when administered in humans (DOROW et al. 1983).

For human studies it is essential that putative vigilance enhancers from amongst benzodiazepine receptor ligands should be restricted to weak partial inverse agonists which ideally should increase vigilance and improve memory without possessing proconvulsant or anxiogenic activity. Weak vigilance-enhancing properties have been described for the neutral benzodiazepine antagonist Ro 15–1788 (SCHÖPF et al. 1984; ZIEGLER et al. 1986; HIGGIT et al. 1986; but see EMRICH et al. 1984). The weak inverse intrinsic activity of Ro 15–1788, especially at high doses (DUKA et al. 1986a), may underly these effects (for review see PELLOW and FILE 1986).

A substance from the chemical group of β-carbolines, ZK 93426 (ethyl-5-isopropoxy-4-methyl-6-carboline-3-carboxylate), which according to its biochemical and pharmacological characteristics can be classified as a neutral antagonist with weak inverse intrinsic activity, has been recently synthesized (JENSEN et al. 1984; STEPHENS et al. 1984; STEPHENS et al. 1986). Specifically, ZK 93426 exhibited "proconflict" activity in the lick suppression test (JENSEN et al. 1984) and produced effects in a social interaction test which were interpreted as being anxiogenic (FILE et al. 1986).

Recent animal studies which were carried out to estimate the effects of the substance on memory and vigilance processes demonstrated that it had a beneficial effect on learning and memory (JENSEN et al. 1987; SARTER and STEPHENS, this volume) and enhanced vigilance in a visual discrimination paradigm (JENSEN et al. 1987; STEPHENS and SARTER, this volume). We were therefore interested in investigating the effects of ZK 93426 in humans, from the point of view of estimating changes in various psychobiological aspects which may contribute to performance in different memory tasks (i.e. vigilance, mood stages). Since a significant number of investigations in this direction have been performed using full agonists at benzodiazepine receptors (for review see LISTER 1985) and the neutral benzodiazepine antagonist Ro 15–1788 (SCHÖPF et al. 1984; ZIEGLER et al. 1986; EMRICH et al. 1984; DOROW et al. 1987), direct comparisons may further contribute to conclusions drawn in the present review.

Data will be presented here which were obtained from three different studies. All the studies were performed with healthy male subjects, mostly students, aged between 18 and 40 years, who had given their informed consent. The studies were performed according to a double-blind, placebo-controlled design. The number of subjects per group was always 10. Because of its low bioavailability, ZK 93426 was administered intravenously as an Intralipid solution (0.2 mg/ml).

2 Psychotropic Activity of ZK 93426

In the first study, designed to establish an appropriate dose at which ZK 93426 affected mood and cognitive function (mostly as indicated by measurements on concentration tests), two doses of ZK 93426 (0.01 and 0.04 mg/kg) were administered and compared to placebo (for details see DUKA et al. 1987). One of these doses was then used for further testing of vigilance and memory. A test battery was applied which included:

1. Visual analogue scales (bipolar), ranging from 0 to 100 mm, which were designed to estimate a stimulatory effect of the substance (low drive–high drive, 0–100) and an effect which would imply "excitement" or "nervousness" (agitated–quiet, 0–100).
2. The logical reasoning task first described by BADDELEY (1968), which involves higher mental processes, requires high memory load and also estimates the ability to concentrate. Sixty-four logical statements relating two letters accompanied by the pair of letters (e.g., "A follows B", AB), are presented and the subject has to decide whether the statement is correct or not. The time needed and the percentage of errors were evaluated.
3. Time estimation (subjects' approximation of 15 s) as a control of the internal clock (ELSASS et al. 1979; STEPHAN and DOROW 1986) in order to detect changes in activation of the subjects.

In addition, a novel test was applied which involved detection of small differences between two otherwise similar coloured pictures. This test was introduced in response to reports from an uncontrolled pilot experiment of improvements in vision, such as increased brightness of colours. All these tests, except the picture difference task which was evaluated 45 min after injection, were evaluated before, 30–45 and 120–135 min after injection.

In their subjective ratings using the visual analogue scale subjects who received the higher dose (0.04 mg/kg) showed an increase in drive when compared to subjects who received either placebo or the lower dose of ZK 93426 (0.01 mg/kg; Fig. 1). The same subjects reported a feeling of "relaxation" when compared with the subjects from the other groups (placebo or ZK 93426 0.01 mg/kg), who remained unchanged according to their subjective ratings (Fig. 1). That the time courses of these two subjective ratings were different may indicate that we measured two independent effects of the drug. These findings also demonstrated that the activation induced by the drug is not accompanied by nervousness or agitation.

How these subjective responses to the drug parallel performance on other tasks may give some additional insight into the drug's effect. In the logical reasoning

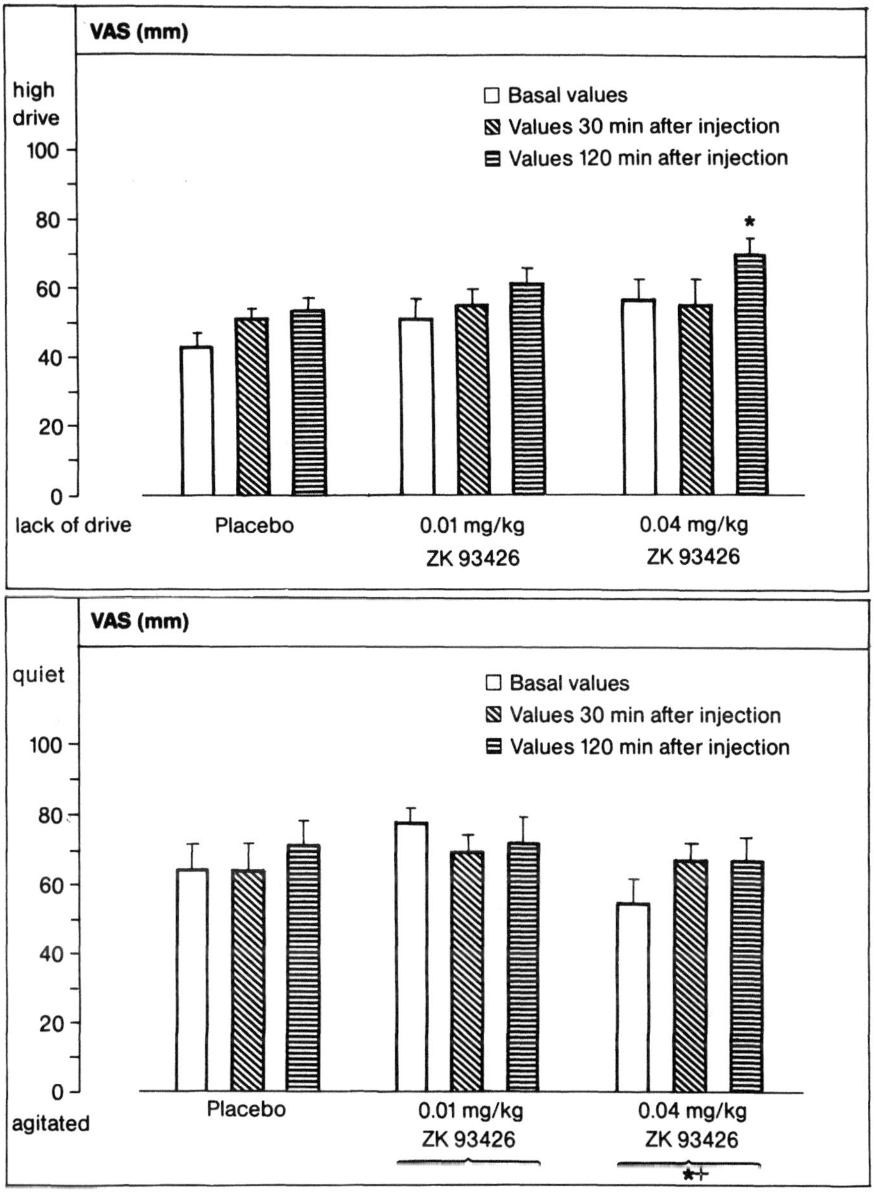

Fig. 1. Visual analogue scores (*VAS*) indicative of "activation" (low drive–high drive, 0–100 mm) and "nervousness" (agitated–quiet, 0–100 mm) following treatment with ZK 93426 0.01 mg/kg and 0.04 mg/kg. *$p<0.05$, 0.04 mg/kg ZK 93426 versus placebo, ANOVA; *+$p<0.05$, 0.01 mg/kg versus 0.04 mg/kg ZK 93426, MANOVA

Table 1. Subjects' performance on the cognitive tasks

	Logical reasoning		
	Time (min) needed for completion (mean ± SD)		
Time after injection (min)	0	45	120
Placebo	5 ± 1	5 ± 1	5 ± 2
ZK 93426 0.01 mg/kg	5 ± 1	5 ± 1	5 ± 1
ZK 93426 0.04 mg/kg*	7 ± 2	6 ± 2	5.5 ± 2
	No. of right/wrong answers (mean ± SD)		
Time after injection (min)	0	45	120
Placebo	60 ± 6/4 ± 6	59 ± 3/ 5 ± 2	61 ± 3/3 ± 2
ZK 93426 0.01 mg/kg	62 ± 1/2 ± 1	61 ± 1/ 4 ± 1	62 ± 1/2 ± 1
ZK 93426 0.04 mg/kg	54 ± 14/9 ± 14	53 ± 12/10 ± 12	57 ± 12/7 ± 12
	Time estimation (s; mean ± SD)		
Time after injection (min)	0	45	120
Placebo	17.0 ± 2	16.0 ± 3	14.0 ± 2
Zk 93426 0.01 mg/kg	15.5 ± 6	16.4 ± 5	16.6 ± 4
ZK 93426 0.04 mg/kg	16.0 ± 5	16.4 ± 4	14.8 ± 2

Performance for logical reasoning is estimated by the time needed to complete the test and by the number of right and wrong answers. Time estimation simply gives the approximation of 15 s calculated by the subjects. * $p<0.1$, ZK 93426 0.04 mg/kg versus placebo, MANOVA.

Table 2. Performance of the subjects on picture difference test, evaluated on the basis of the number of correct answers and errors (median) in a fixed time (5 min)

	Placebo	ZK 93426 0.01 mg/kg	ZK 93426 0.04 mg/kg
No. of correct answers (median, range)	5.5 (3–7)	5.5 (1–9)	4.5 (1–10)
No. of errors (median, range)	0 (0–3)	0 (0–0)*	0 (0–0)*

* $p<0.05$ compared to placebo (Kruskal – Wallis).

test an improvement in performance was observed, i.e. the subjects needed less time to complete the task (Table 1). Although no correlation was found between the two measurements, the self-ratings indicative of a "relaxing" effect of the drug closely paralleled the effect of the drug on logical reasoning. This observation may imply that the effect of the drug in relaxing the subjects benefited their performance in the logical reasoning test. The stimulatory effect of the drug indicated by the subjective ratings (low drive–high drive) may also underly the improvement of performance in logical reasoning. Time estimation measurements did not indicate any effect of the drug (Table 1).

In the picture difference task, subjects under the influence of the drug made significantly less mistakes than those receiving placebo (Table 2). This finding further suggests that ZK 93426 improves performance in concentration-demanding tasks. An improvement in visual perception by ZK 93426 may also contribute to this effect, since a small number of volunteers spontaneously reported "crystal clear vision" under the drug (DUKA et al. 1987).

Taking all these findings together it is suggested that the higher dose (0.04 mg/kg) demonstrated a weak inverse agonist effect of ZK 93426. In particular, its effect in improving performance in the logical reasoning task is the opposite of the effect of the benzodiazepine diazepam (BROSAN et al. 1986). Studies investigating possible weak inverse agonist features of Ro 15–1788 on tests which evaluate the ability to concentrate failed to demonstrate any effect (EMRICH et al. 1984).

3 Effects on EEG

In a study designed primarily to investigate antagonism of the effects of benzodiazepines by ZK 93426 (DUKA et al. 1986b; DUKA et al. 1988) we also evaluated the effects of ZK 93426 on its own; vigilance and daytime alertness were estimated using EEG measurements. Such measurements would allow us to draw further conclusions about ZK 93426's ability to improve performance in tasks which involve attentional or other cognitive processes.

EEG recordings in the form of vigilosomnograms (RECHTSCHAFFEN and KALES 1968) were made for 1 h, during which time drug administration took place. After 5 min EEG recording, placebo was administered intravenously and 30 min later either ZK 93426 (0.04 mg/kg) or a further volume of placebo.

A multiple sleep latency test (MSLT; RICHARDSON 1978) was also used. MSLT is based on EEG measurements which last for 20 min and are performed repeatedly throughout the day at 2-h intervals. During this 20 min subjects were asked to be quiet, close their eyes, and try to sleep. The time required for the subjects to reach sleep stage 1 was evaluated. The test is generally used to estimate the influence of drugs on daytime sleepiness. The evaluation of the vigilosomnograms revealed that sleep stages 1 and 2 were equally distributed in both groups before the injections of ZK 93426 or placebo. After the injection (up to 30 min after application), however, subjects who received ZK 93426 reached only sleep stages 1 and 2, while the placebo group attained deeper stages (Fig. 2; $p<0.01$, placebo + placebo versus placebo + ZK 93426, ANOVA). This effect indicated a vigilance-enhancing property of ZK 93426 which did not allow the subjects to doze. The follow-up of their daytime sleepiness as measured by MSLT demonstrated that the ZK 93426 group did not experience the midday dip in alertness (Fig. 3), but this difference between the two groups was not significant. On account of the pharmacokinetic characteristics of ZK 93426 ($t_{1/2}$ 60 min) a direct influence on daytime sleepiness at that point in time can be ruled out; on the other hand, the drug may have shifted the subjects internal clock with regard to alertness or sleepiness during the day. Similar effects have been demonstrated for the benzodiazepine antagonist Ro 15–1788 (SCHÖPF et al. 1984; ZIEGLER et al. 1986; HIGGIT

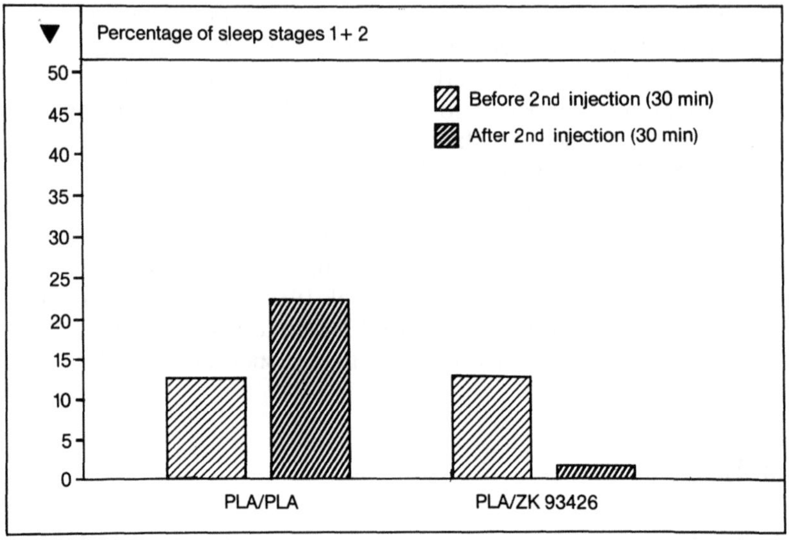

Fig. 2. Percentages of different sleep stages of subjects under different treatment combinations as evaluated from the EEG recordings (1st injection placebo, *PLA*; 2nd injection placebo or ZK 93426)

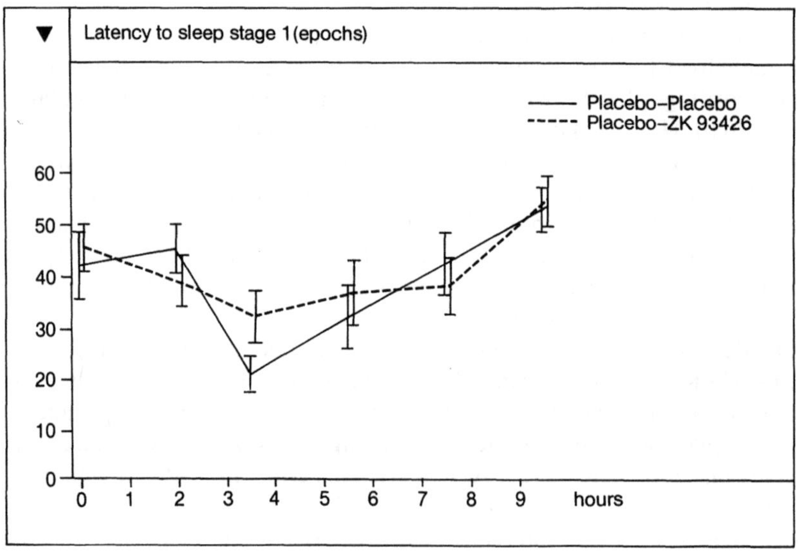

Fig. 3. Multiple sleep latency test (MSLT) measurements (time required to reach sleep stage 1) after different treatment combinations. Test duration was 20 min = 60 epochs

et al. 1986), which has been shown to retain some weak inverse intrinsic activity in some tests (for review see PELLOW and FILE 1986).

The data obtained from the EEG recordings indicate that ZK 93426 induces central activation and thus provide further support for the hypothesis that weak inverse agonists give rise to effects opposing those of benzodiazepines such as sedation.

4 Studies on Memory

Although much has been learned recently about the relation between memory and brain structures, the large majority of memory disorders remain difficult to explain on an anatomical, physiological or biochemical basis. The availability of drugs that selectively block or enhance the functions of certain neuronal systems, offers a means of exploring neuronal-behavioural relationships in intact human subjects and gaining better insight into the pathology of memory. Such studies will give a direction to therapeutic efforts. Benzodiazepines, as discussed at the start of this review, influence memory in a way which appears to be specific for this group of drugs; they cause an anterograde amnesia which is mainly a result of decreased acquisition of new information (for review see LISTER 1985). As a result of reduced interference from interpolated learning, anterograde amnesia is accompanied by an increased ability to recall information acquired before benzodiazepine administration (GHONEIM et al. 1984).

Both findings suggest that benzodiazepines decrease the ability to acquire new information and thus anterograde amnesia occurs under their influence. ZK 93426, like benzodiazepines, acts via the benzodiazepine-γ-aminobutyric acid (GABA) receptor complex but induces effects partially opposite to those of benzodiazepines.

Investigation of the effects of ZK 93426 on memory processes would offer a pharmacological tool for looking at memory using a different aspect of benzodiazepine receptor pharmacology. Thus, we studied the effects of ZK 93426 on performance on memory tests which have also been used to demonstrate effects of benzodiazepine receptor agonists and antagonists (KUBICKI et al. 1986; OTT et al. this volume; BERENBERG et al., this volume). Another objective of the study was to investigate whether ZK 93426 would antagonize the amnesic effects of scopolamine (data will be presented elsewhere). Thus, subjects were first given a subcutaneous injection of scopolamine 0.5 mg or saline followed 1.5 h later by either ZK 93426 (0.04 mg/kg) or placebo (Intralipid) intravenously.

In the current paper, however, only data obtained from subjects who received placebo as their first injection will be presented. Auditory and visual memory were tested using word lists (OTT et al., this volume) and a series of slides (BERENBERG et al., this volume) which were presented 1 h after the first injection. The six words which comprised a list were read out at 2-s intervals and one word list (the whole battery included five word lists) was presented every 15 min for 1 h (Fig. 4). Immediately after stimulus presentation, subjects were required to perform simple arithmetic tasks for 15 s in order to prevent rehearsal. A restrictive remind-

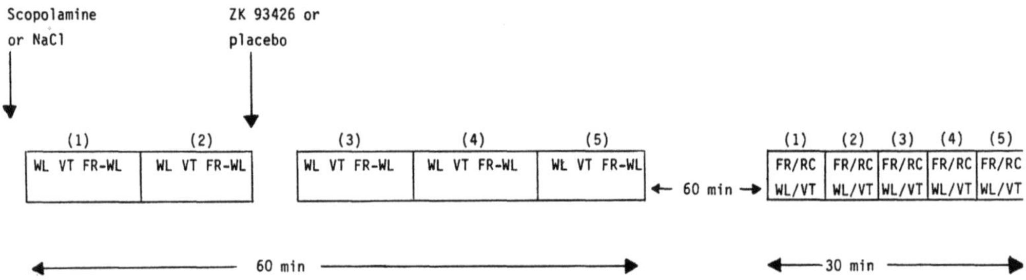

Fig. 4. Experimental design of the study on the effects of ZK 93426 on memory processes. NaCl (saline) was injected subcutaneously 1 h before the memory test started. *WL*, Word list; *VT*, Visual test; *FR*, Free recall; *RC*, Recognition; *number* in parentheses indicate sessions

ing procedure (as described by BUSCHKE and ALTMANN FULD 1974) whereby subjects had to obtain a predetermined criterion, in this case recall of the whole list, was adopted to ensure uniform acquisition.

Visual memory was tested by projecting pairs of simple pictures onto a screen for 5 s, following which subjects were again required to perform simple arithmetic tasks. Four pairs of pictures were presented every 15 min for 1 h following presentation of the word lists (Fig. 4). Free recall of auditory material was tested 10 min later (before the next word list was presented) and 1.5 h after the last word list (Fig. 4).

Free recall of visual material was tested 1 h after presentation of the last slide pair (Fig. 4). Recognition of both visual and auditory material was tested at the end of free recall.

Measurement of acquisition (acquisition score) and learning (total number of words recalled) showed marginal time-dependent effects, especially in the performance of the last session; this was presumably due to interference from newly acquired information (Table 3). This time effect was also evident in the evaluation of free recall 10 min after stimulus presentation (Table 3). Treatment with ZK 93426 did not induce any change of these parameters (Table 3). Delayed recall of word lists was generally poor. It is noteworthy that the recall of word lists read out in the later session was better (Table 3), either indicating a recency effect in this memory measurement or reflecting an increase in attention as a result of the injection.

Basal performance in the immediate recall of visual material was too high to detect any improvement under drug treatment. In the tests of delayed recall, ZK 93426, compared with placebo, impaired performance for the pictures shown before drug application but improved performance for the pictures shown after drug application (Fig. 5; $p < 0.05$, placebo versus ZK 93426, MANOVA). It is assumed that the improvement of recall for material acquired after ZK 93426 administration is due to an effect of the drug on acquisition whereas the decrease in the number of predrug pictures recalled is probably best explained as a result of interference from increased interpolated learning which did not take place in the placebo group. Both these effects are opposite to the effects of benzodiaz-

Table 3. Scores (mean ±SD) obtained by the different treatment groups on acquisition, learning, short term and long term recall. ZK 93426 (0.04 mg/kg) or placebo was injected at the end of session 2. Numbers indicate sessions, each of which lasted 12 min

Treatment	Word acquisition score (acquisition)					Total sum of words recalled (learning)				
	1	2	3	4	5	1	2	3	4	5
NaCl + placebo	27±3	26±2	26±3	28±2	24±4	28±1	27±2	27±2	29±1	27±2
NaCl + ZK 93426	27±2	28±1	27±1	25±3	25±3	28±2	28±1	28±1	27±3	27±1
Treatment	**Free recall (10 min) (no. of words)**					**Delayed recall (1.5 h) (no. of words)**				
	1	2	3	4	5	1	2	3	4	5
NaCl + placebo	5.7±0.4	5.3±0.9	4.8±1.3	5.2±0.7	2.9±1.7	1.7±1	2.5±1.4	2.7±1.3	4.3±11.4	2.7±1.4
NaCl + ZK 93426	5.7±0.4	5.0±4.2	4.7±1.6	4.7±1.4	3.1±2	1.7±1	3.3±1	3.3±1.5	3.9±1.6	2.9±1.7

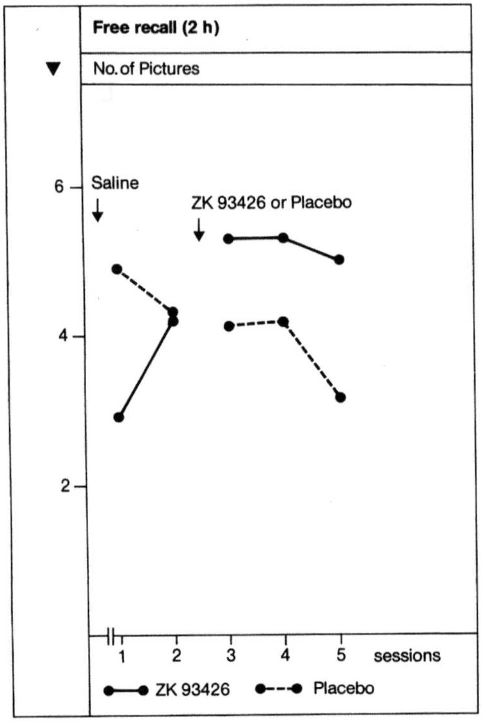

Fig. 5. Effects of ZK 93426 on free recall 1 h after picture presentation. Number of pictures being recalled are referred to the sessions during which they were presented (max. 8 pictures)

epines on similar memory tests (CLARKE et al. 1970; GHONEIM et al. 1984; KUBICKI et al. 1986; BERENBERG et al., this issue).

The test of recognition showed that subjects' performance reached a ceiling, especially for material shown in the first sessions (Fig. 6). Recognition was slightly impaired for material shown immediately after injection and during the last two sessions. This impairment proved to be more clear in the placebo group (Fig. 6; $p<0.01$, placebo versus ZK 93426, ANOVA). It is assumed that ZK 93426, in contrast to benzodiazepines (BROWN et al. 1982), improves recognition, but this effect of ZK 93426 could only be detected at one time point when performance did not reach ceiling levels, probably as a result of decreased attention due to the experimental situation (i.e. injection, expectation of drug effects).

In the present study ZK 93426 was shown to have effects on memory mechanisms. The experimental procedure allowed us to separate effects on immediate and delayed recall. The requirement to recall information presented after a 10- to 15-s period of distraction allowed the premise that the information was retrieved from storage in working memory. In addition, in the case of word lists, the introduction of a restrictive reminding technique and the requirement for complete list recall allowed uniform acquisition and controlled the retrieval conditions (BUSCHKE and ALTMANN FULD 1974). Although no effects of ZK 93426 were observed in the measurement of immediate recall, the fact that ceiling effects were noticed in some parameters could indicate that the tests were not sensitive

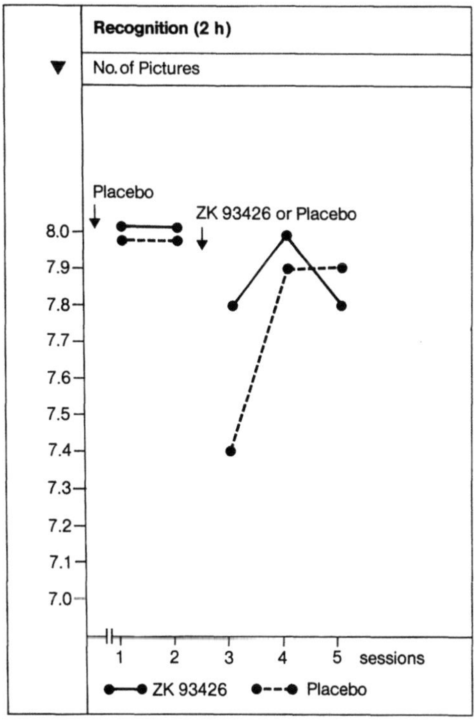

Fig. 6. Effects of ZK 93426 on recognition (visual memory test). Number of pictures correctly recognized are referred to the sessions during which they were presented

enough to detect such effects. The improved performance induced by ZK 93426 in delayed recall of visual material indicated that information acquired during the sessions which followed drug administration could be better transformed into long-term memory when ZK 93426 was given instead of placebo. Lack of this effect of ZK 93426 in the case of verbal material could be due to differences in acquisition conditions between the two tests. In the case of word lists, the repeated retrieval during acquisition until perfect list recall was obtained may have resulted in better encoding of the new material, which did not allow any drug effect to appear. Consequently, in the case of the visual memory test, if incomplete encoding takes place as a result of the lack of additional retrieval procedures, a drug effect may have occurred at any stage of the encoding process (e.g. focusing of attention or during decay; BROADBENT, 1958). This also applies for the effects on recognition of visual material.

An alternative explanation for these differences between visual and verbal tests could be a specific effect of the drug on visual memory. It is not known whether benzodiazepines influence visual and auditory memory in different ways, and no attempt was made here to resolve this issue. Taking into account the effects of ZK 93426 on the picture difference test (Table 2) and the data from free recall and recognition of visual material, it is possible that ZK 93426 has an effect on visual perception which may further contribute to the improvement of memory processes related to visual information.

Recently, the inhibitory neurotransmitter GABA was identified in at least one type of photoreceptor terminal in a primate retina (NISHIMURA et al. 1986). Assuming that benzodiazepine receptors are associated with this GABAergic system, it seems possible that benzodiazepine receptor ligands may have an effect on an early stage of visual information processing. This might account for the effects of ZK 93426 on the visual memory task, especially since we also saw effects in the picture difference task as reported above.

That ZK 93426 increases the ability of the subjects to recall only visual material presented after its administration indicates that its effects were not on retrieval; if they were, then recall of all shown pictures would have improved.

5 General Remarks

The effects of ZK 93426 on visual memory represent a mirror image of the effects of benzodiazepines on memory. While benzodiazepines impair memory anterogradely and improve memory retrogradely, ZK 93426 improved performance anterogradely but impaired it retrogradely. The terms "impair" or "improve" for the effects observed retrogradely may be misused in this context since, as discussed above, what we measure may simply be an interference process due to additionally acquired information which is either suppressed, as in the case of benzodiazepines (HINRICHS et al. 1984), or potentiated, as in the case of ZK 93426. This effect of ZK 93426 on memory may be due to its stimulatory, activating central action. Such an effect on memory could result from the influence of ZK 93426 on vigilance and alertness as indicated by EEG measurements and self-rating scales, which may have improved attentional processes. It is also possible that an influence of the drug on the visual system underlies the effects on visual tasks.

On the other hand, the effect of the drug on the reasoning task could be due to both an improvement of memory processes and/or an increase in attention. These findings further support the hypothesis that weak partial inverse agonists like ZK 93426 exert effects opposite to those of benzodiazepines, such as the effects shown here on sedation and anterograde amnesia. They may also help to introduce new pharmacological tools to study memory and give rise to new therapeutic ideas.

Acknowledgements. We would like to thank I. Columbus, R. Herrmann and P. Schulte for their valuable assistance in carrying out the study, L. Höller, H. Riemer and M. Knabe-Rühl for the application and evaluation of the EEG, and A. Dahrmann for preparing the manuscript. We also want to thank A. Rohloff and R. Kiefer for their help with the memory tasks and K. Fichte for his helpful advice on statistics.

References

Baddeley AD (1968) A 3 min reasoning test based on grammatical transformation. Psychonom Sci 10:341–342

Braestrup C, Schmiechen R, Neff G, Nielsen M, Petersen EN (1982) Interaction of convulsive ligands with benzodiazepine receptors. Science 216:1241–1243

Brown J, Lewis N, Brown MW, Horn G, Bowes JB (1982) A comparison between transient amnesias induced by two drugs (diazepam or lorazepam) and amnesia of organic origin. Neuropsychologica 20:55–70

Broadbent D (1958) Perception and communication. Pergamon, Oxford

Brosan L, Broadbent D, Nutt D, Broadbent M (1986) Performance effects of diazepam during and after prolonged administration. Psychol Med 16:510–522

Buschke H, Altman Fuld P (1974) Evaluating storage, retention and retrieval in disordered memory and learning. Neurology 24:1019–1025

Clarke PRF, Eccersley PS, Frisby JP, Thornton JA (1970) The amnesic effect of diazepam (valium). Br J Anaesth 42:690–697

Curran HU (1986) Tranquilizing memories: a review of the effects of benzodiazepines on human memory. Biol Psychol 23:179–213

Dorow R, Horowski R, Paschelke G, Amin M, Braestrup C (1983) Severe anxiety induced by FG 7142, a β-carboline ligand for benzodiazepine receptors. Lancet ii:98

Dorow R, Berenberg D, Duka T, Sauerbrey N (1987) Amnesic effects of lormetazepam and their reversal by the benzodiazepine antagonist Ro 15–1788. Psychopharmacology 93:507–514

Duka T, Ackenheil M, Noderer J, Doenicke A, Dorow R (1986a) Changes in noradrenaline plasma levels and behavioural responses induced by benzodiazepine agonists with the benzodiazepine antagonist Ro 15–1788. Psychopharmacology 90:351–357

Duka T, Höller L, Obeng-Gyan R, Dorow R (1986b) Initial human pharmacology of the neutral benzodiazepine receptor antagonist β-carboline ZK 93426. Br J Clin Pharmacol 22:228

Duka T, Stephens DN, Krause W, Dorow R (1987) Human studies on the benzodiazepine receptor antagonist β-carboline ZK 93426: preliminary observations on psychotropic activity. Psychopharmacology 93:421–427

Duka T, Goerke D, Dorow R, Höller L, Fichte K (1988) Human studies on the benzodiazepine receptor antagonist β-carboline ZK 93426: antagonism of lormetazepam's psychotropic effects. Psychopharmacology (in press)

Elsass P, Mellrup ET, Rafaelsen OJ, Theilgaard A (1979) Lithium effects on time estimation and mood in manic-melancholic patients. A study of diurnal variations. Acta Psychiatr Scand 60:263–271

Emrich HM, Sonderegger P, Mai N (1984) Action of the benzodiazepine antagonist Ro 15–1788 in humans after sleep withdrawal. Neurosci Lett 47:369–373

File SE, Pellow S, Jensen LH (1986) Actions of the β-carboline ZK 93426 in animal test anxiety and the holebord: interactions with Ro 15–1788. J Neural Transm 65:103–114

Ghoneim MM, Mewaldt SP (1977) Studies on human memory: the interactions of diazepam, scopolamine and physostigmine. Psychopharmacology 52:1–6

Ghoneim MM, Hinrichs JV, Mewaldt SP (1984) Dose-response analysis of the behavioural effects of diazepam: learning and memory. Psychopharmacology 82:291–295

Higgit A, Lacer M, Fonagy P (1986) The effect of the benzodiazepine antagonist Ro 15–1788 on psychophysiological performance and subjective measures in normal subjects. Psychopharmacology 89:395–403

Hinrichs JV, Ghoneim MM, Mewaldt SP (1984) Diazepam and memory: retrograde facilitation reduction. Psychopharmacology 82:158–162

Hunkeler W, Möhler H, Pieri L, Polc P, Bonetti EP, Cumin R, Schaffner R, Haefely W (1981) Selective antagonists of benzodiazepines. Nature 290:514–516

Jensen LH, Petersen EN, Braestrup C, Honore T, Kehr W, Stephens DN, Schneider H, Seidelmann D, Schmiechen R (1984) Evaluations of the β-carboline ZK 93426 as a benzodiazepine receptor antagonist. Psychopharmacology 83:249–256

Jensen LH, Stephens DN, Sarter M, Petersen EV (1987) Bidirectional effects of β-carbolines and benzodiazepines on memory processes. Brain Res Bull 19

Kubicki S, Rohloff A, Ott H, Fichte K (1986) Vigilanz und Amnesie nach Benzodiazepingabe. Neurophysiologische und pharmakopsychologische Aspekte. In: Schulte am Esch J (ed) Benzodiazepine in Anaesthesie und Intensivmedizin. Roche, Basel

Lister RG (1985) The amnesic action of benzodiazepines in man. Neurosci Biobehav Rev 9:87–93

Nishimura Y, Schwartz ML, Rakic P (1986) GABA and GAD immunoreactivity of photoreceptor terminals in primate retina. Nature 320:753–756

Pandit SK, Heiserkamp DU, Cohen PJ (1976) Further studies of the anti-recall effect of lorazepam: a dose-time effect relationship. Anaesthesiologie 45:495–500

Pellow S, File SE (1986) Intrinsic actions of the benzodiazepine receptor antagonist Ro 15–1788. Psychopharmacology 88:1–11

Rechtschaffen A, Kales A (1968) A manual of standardized terminology, techniques and scoring system for sleep stages of human subjects. NIH, Bethesda

Richardson G, Carskadon MA, Flagg W, van den Hoed J, Dement WC, Mitler MM (1978) Excessive daytime sleepiness in man, multiple sleep latency measurement in narcoleptic and control subjects. Electroencephalogr Clin Neurophysiol 45:621–627

Schöpf J, Laurian S, Le PK, Gaillard JM (1984) Intrinsic activity of the benzodiazepine antagonist Ro 15–1788 in man: an electrophysiological investigation. Pharmacopsychiatry 17:79–83

Stephan K, Dorow R (1986) The influence of the new antidepressant rolipram on alertness, performance and mood in healthy volunteers varied with time of day and diurnal type of subject. In: Labrecque G (ed) Biological rhythms and medications. Pergamon, Oxford 437–440 (Annual review of chronopharmacology, vol 3)

Stephens DN, Shearman GT, Kehr W (1984) Discriminative stimulus properties of β-carbolines characterized as agonists and inverse agonists at central benzodiazepine receptors. Psychopharmacology 83:233–239

Stephens DN, Kehr W (1985) β-carbolines can enhance or antagonize the effects of propunishment in mice. Psychopharmacology 85:143–147

Stephens DN, Kehr W, Duka T (1986) Anxiolytic and anxiogenic β-carbolines: Tools for the study of anxiety mechanisms. In: Biggio G, Costa E (eds) GABAergic transmission and anxiety. Raven, New York

Venault P, Chapoutier G, de Carvalho LP, Simiand J, Morre M, Dodd RH, Rossier J (1986) Benzodiazepine impairs and β-carboline enhances performance in learning and memory. Nature 321:864–866

Ziegler G, Ludwig L, Fritz G (1986) Effects of the specific benzodiazepine antagonist Ro 15–1788 on sleep. Pharmacopsychiatry 19:200–201

Benzodiazepine Receptor Ligands: Tools for Memory Research in Clinical Pharmacology

D. Berenberg[1], R. Dorow[2], T. Duka[2] and N. Sauerbrey[2]

Abstract

In order to study the time course of amnesic effects of the benzodiazepine hypnotic lormetazepam, and their reversal by the benzodiazepine antagonist Ro 15–1788, a combined visual and auditory memory test was developed, which was designed to allow repeated assessments.

Immediate recall as well as delayed free recall and recognition (1 h after drug) were investigated before and after intravenous lormetazepam (0.02 mg/kg) followed 15 min later by intravenous Ro 15–1788 (0.03 mg/kg) or placebo. A third group received placebo followed by Ro 15–1788. Results are based on ten subjects per treatment group and are compared with an age-matched control population ($n=20$) without treatment. Immediate and delayed recall as well as recognition in both visual and auditory tests were impaired abruptly after intravenous lormetazepam. These effects were reversed instantaneously after Ro 15–1788, which had no marked effect on these parameters when given alone. Ratings by visual analog scales (1 h after drug administration) indicated concomitant sedation and impaired concentration after lormetazepam, which was attenuated by Ro 15–1788. By itself, Ro 15–1788 had no effect on these measures.

Interestingly, the performance in delayed free recall of the visual memory test was significantly enhanced in the lormetazepam group prior to administration. Our results suggest that impaired acquisition of new information after lormetazepam is benzodiazepine receptor mediated and may be associated with a drug-induced enhancement of retrieval of information acquired before lormetazepam administration.

1 Introduction

Soon after the discovery of the main therapeutic activities of benzodiazepines in humans, i.e., of their anxiolytic, sedative, muscle-relaxant, and anticonvulsant properties, it was found that these drugs have a major impact on cognitive functions in that they induce memory loss (Clarke et al. 1970; Dundee and George 1976). Although this effect may be of some advantage in the preoperative situation when benzodiazepines are given as premedication, memory loss is a clear disadvantage to patients who take benzodiazepines on a daily basis for the treatment of insomnia or anxiety. A number of studies and extensive reviews of recent literature (Curran 1986; Lister 1985) have reported that dose, type of benzodiazepine (i.e., its physicochemical properties determining pharmacokinetic parameters and hence penetration into brain, distribution in and elimination from body

[1] Department of Internal Medicine, St. Vinzenz Krankenhaus, Altena, FRG.

[2] Research Laboratories, Department of Clinical Neuropsychopharmacology, Schering AG, 1000 Berlin 65, FRG.

tissue, as well as receptor affinity), route of administration, and method of measuring memory performance play a major role in the degree and duration of amnesia found after benzodiazepine intake. The amnesic properties of different benzodiazepines have recently been reviewed in a paper by CURRAN (1986). Diazepam, lorazepam, triazolam, and flunitrazepam are the compounds that have been most extensively investigated in this context. The patterns of amnesia appear to be similar in quality: there is consensus among most experimenters that benzodiazepines have little effect on retention or retrieval of material and that their major impact is on acquisition and consolidation of information presented after drug intake (CLARKE et al. 1970; GHONEIM and MEWALDT 1975). Most investigators have found benzodiazepines to induce anterograde amnesia, i.e., amnesia for events occurring subsequent to application of the drug.

The majority of studies have focused on modal memory models such as that proposed by ATKINSON and SHIFFRIN (1971). This model is based upon a concept of short- and long-term memory and hence implies a time sequence of information processing. By assuming that information first enters short-term memory before passing on to long-term storage, this theoretical concept offers a suitable model for studying the time course and components of this storage and retrieval process and hence lends itself to experimental studies with drugs and other procedures that are likely to interfere herewith. Such theories are without doubt oversimple, yet they have formed a useful basis for developing psychological tests that discriminate between drug effects on short-term and long-term memory.

It has been suggested that benzodiazepine-produced amnesias may be a model for organic amnesias. In comparison with other forms of memory loss, they produce impairments similar to those seen in patients with Korsakoff's syndrome, but not in patients with progressive degenerative dementia like of the Alzheimer type (WEINGARTNER 1985; WOLKOWITZ et al. 1987). In the former group episodic memory is impaired, while the latter group shows severe impairments of semantic memory. The anticholinergic drug scopolamine, another compound with amnesic properties, also impairs the acquisition of new information but additionally affects retrieval of information from semantic memory. It has been suggested as a model for Alzheimer's disease (DRACHMANN and LEAVITT 1974; SUNDERLAND et al. 1986).

2 Benzodiazepine Receptor Ligands

Benzodiazepines are known to act via specific receptors coupled to the GABA receptor/chloride channel complex. They are believed to exert their CNS depressant effects by enhancing chloride conductance in the presence of GABA and thus facilitating GABAergic transmission (STUDY and BARKER 1981; TALLMAN et al. 1980). Recently, other compounds that bind to benzodiazepine receptors and which have minimal effects when given alone (benzodiazepine receptor antagonists) and compounds that have mirrorlike effects to those of the benzodiazepines (hence, inverse agonists) have been characterized (for review see HAEFELY 1985; STEPHENS and KEHR 1985; DOROW et al. 1987a).

Lormetazepam, a benzodiazepine that is marketed in an oral form for use as a hypnotic and in an intravenous form as premedication before anaesthesia, has been thoroughly investigated for its pharmacological and clinical properties (for review see DOENICKE and OTT 1980). Lormetazepam was shown to possess only minor amnesic effects after oral intake of the recommended therapeutic dose when compared with other benzodiazepine hypnotics (ROEHRS et al. 1984; SUBHAN 1984; SUBHAN and HINDMARCH 1984). However, it had potent effects on long-term memory when administered intravenously (OTT et al. 1980). Because of its short half-life of 10–13 h and its safe use in its intravenous form, it was chosen as a probe to investigate amnesia.

The first benzodiazepine receptor antagonist to be tested in humans was the imidazodiazepine Ro 15–1788, which was shown to reverse the typical effects of benzodiazepines while having only sparse effects of its own (HUNKELER et al. 1981; DARRAGH et al. 1981, 1982; DOROW and DUKA 1986). Some authors have suggested that this compound had residual minor inverse agonistic effects as evidenced by an increase of wakefulness (SCHÖPF et al. 1984; FILE and PELLOW 1986; HIGGITT et al. 1986). Furthermore, when a high intravenous dose of Ro 15–1788 was administered 24 h after a single high intravenous dose of a benzodiazepine (lormetazepam or flunitrazepam), symptoms were reported which were interpreted as signs of benzodiazepine withdrawal (DUKA et al. 1986 a). Two studies in animals have demonstrated that Ro 15–1788 blocks or partially reverses acquisition impairments induced by triazolam or diazepam in rodents (BONETTI et al. 1982; THIEBOT et al. 1983).

Ro 15–1788 was recently also applied as a tool to reverse benzodiazepine-induced sedation and amnesia in humans (O'BOYLE et al. 1983; DOROW et al. 1987 b). The authors were able to demonstrate that impairments in a modified word list task were partly reversed after administration of the antagonist. In another study HOMMER et al. (1986) showed that a low dose of Ro 15–1788 could completely block sedative, anxiolytic, and attentional effects but not amnesia. This study suggests that these two qualities may be independent of each other. Thus, benzodiazepine receptor agonists constitute a class of useful probes for investigation into benzodiazepine-induced memory impairments.

3 Interaction Studies with Lormetazepam and Ro 15–1788

An interaction study was performed to estimate the time course (a) of effects on memory produced by a single intravenous dose of benzodiazepines and (b) of the reversal of these effects by Ro 15–1788. A special continuous memory test was developed which allowed a minute-by-minute assessment of the time course of memory performance (DOROW et al. 1987 b). This test investigated memory of both visual material (21 pairs of pictures presented in 2-min intervals) and of auditory material (lists of five semantically related words presented at six times in intervals between the picture tests). Subjects were asked to name the items upon visual presentation to ensure that they had been correctly perceived. Moreover, the test was designed such that medication could be applied intravenously in the intervals, in

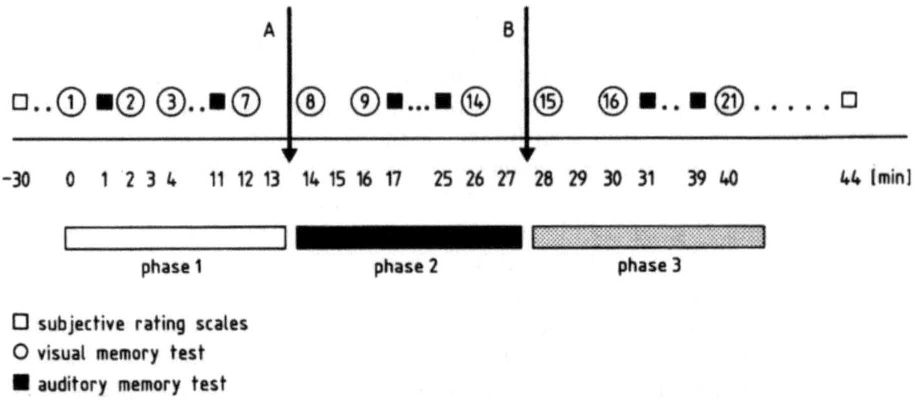

Fig. 1. Schematic representation of the memory test session. *Figures in circles* denote times of visual presentation (1–21); *open squares* denote times of subjective rating scales; *filled squares* indicate times of auditory memory tests; *arrows* indicate times of intravenous application of drugs or placebo (*A*, lormetazepam 0.02 mg/kg or placebo; *B*, Ro 15–1788 0.03 mg/kg or placebo)

this case between the 7th and 8th and between the 14th and 15th pairs of slides, without interfering with the above-described rhythm of material presentation (for details see DOROW et al. 1987b). With a total of 21 pairs of slides, the entire test session thus lasted 42 min. Figure 1 provides a time scale showing when auditory and visual material were presented and when drugs were given. The test was also designed to investigate both immediate recall (20 s after presentation of slides) and delayed recall (1 h after presentation of all pictures). After this, the subjects were randomly presented all 42 slides together with an equal number of previously unseen pictures.

The results were evaluated in two ways: first, the number of slides recalled by the entire group for each of the 21 presentations was recorded; the maximum attainable score for ten volunteers in each group (see below) was thus 20 pictures per presentation. In the second analysis, the entire test session was divided into three phases: phase 1, 14 pictures (7 × 2) shown before first drug treatment; phase 2, 14 pictures shown between the two treatments; and phase 3, 14 pictures shown after the second treatment. The values were then computed as mean scores of volunteers per group and phase. In addition, psychometric scales, i.e., visual analog scales (VAS), the B-L complaints list, and the Bf-S mood scale were applied before and immediately after the memory test session. Blood was sampled at various times before, during, and after the memory tests in order to determine serum cortisol and prolactin levels.

The comparison of the three treatment groups was based on an analysis of variance or on the basis of Kruskal-Wallis tests in each case with multiple comparisons of groups A/B, A/C, B/C. The test procedure was evaluated in a controlled experiment in which 20 subjects (ten males and ten females aged 20–50 years) were challenged without drug treatment.

Results are shown in Fig. 2 as a time course of effects arranged continuously for each pair of presented pictures. It is apparent that almost all slides could be

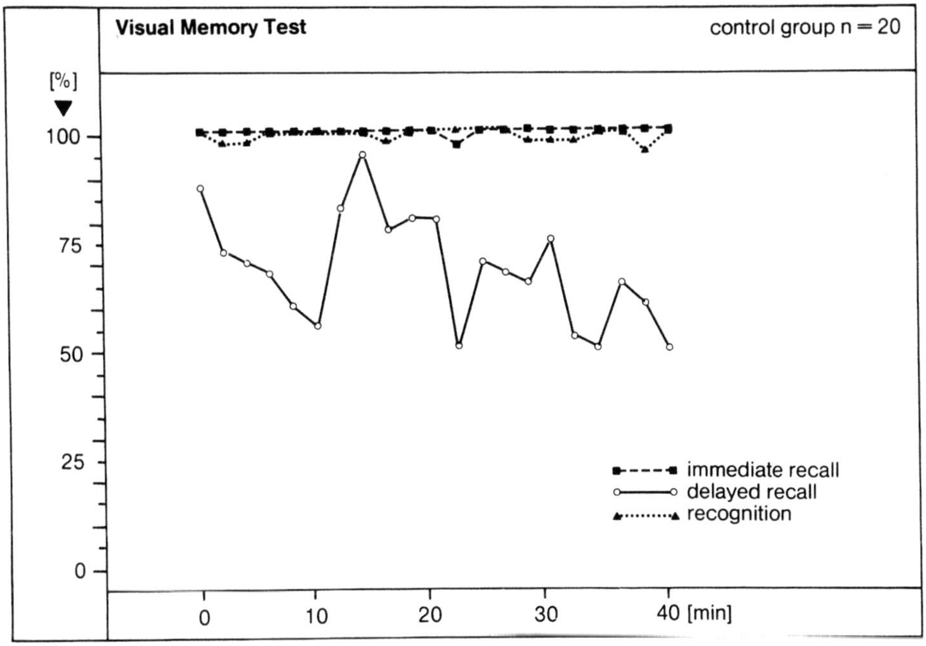

Fig. 2. Time course of performance in memory tests of immediate recall, delayed recall, and recognition of visual material of the control group

Table 1. Performance in the auditory tests (word lists). Figures denote mean percentage (±SD) of words memorized by 10 subjects per group in each phase (maximum 10 items = two word lists)

		Immediate recall	Delayed recall	Delayed recognition
Group A (placebo/ Ro 15–1788)	Phase 1	97.0± 6.8	91.0± 7.4	93.0± 8.2
	Phase 2	100.0± 0	83.0±26.3	90.0±12.5
	Phase 3	96.0± 5.2	81.0±16.0	91.0± 8.8
Group B (lormetazepam/ placebo)	Phase 1	98.0± 4.2	93.0± 9.5	97.0± 4.8
	Phase 2	59.0±33.5[A]	12.0±27.0[A]	20.0±31.6[A]
	Phase 3	77.0±17.0[A]	21.0±28.9[AC]	42.0±30.8[AC]
Group C (lormetazepam/ Ro 15–1788)	Phase 1	98.0± 4.2	92.0±12.3	97.0± 4.8
	Phase 2	54.0±24.1[A]	0 ± 0[A]	5.0±10.8[A]
	Phase 3	89.0± 7.4	75.0±31.0	91.0±12.9

[A, C] Significantly different ($p < 0.05$) from group A or C

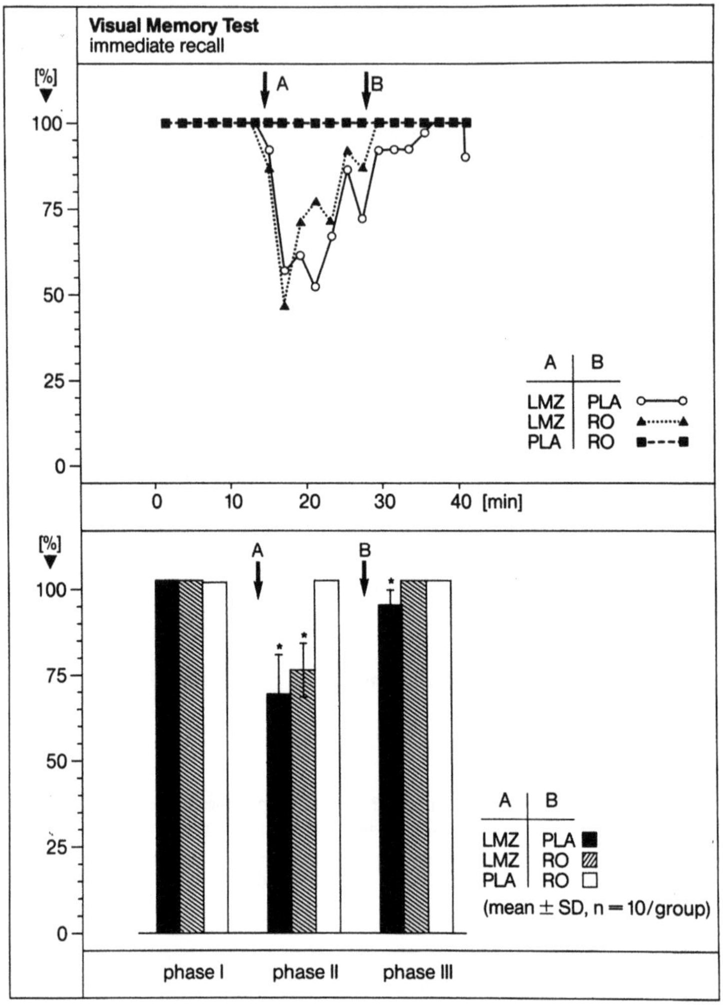

Fig. 3. Time course of performance in tests of immediate recall of visual material. Curves in *top section* show total percentage scores in each group for each presentation ($n = 20$). Columns in *lower section* show the groups' mean percentage scores per test phase (maximum 14 points for each subject = 100%). *Arrows* indicate times of injection. *PLA*, Placebo; *LMZ*, lormetazepam; *RO*, Ro 15–1788; $*p < 0.05$ vs PLA/Ro 15–1788

remembered in the tests of immediate recall and recognition, whereas in the tests of delayed recall only about 75% could be remembered.

In the main study, three groups of ten volunteers each performed the same test procedure, but after seven presentations either lormetazepam 0.02 mg/kg (groups A and B) or placebo (group C) was applied intravenously followed 14 min (7 pairs) later by Ro 15–1788 0.03 mg/kg (i.v.) (group B and C) or placebo (group A).

Lormetazepam injection resulted in immediate amnesic effects as evidenced by an impairment of performance in tests of immediate and delayed recall and rec-

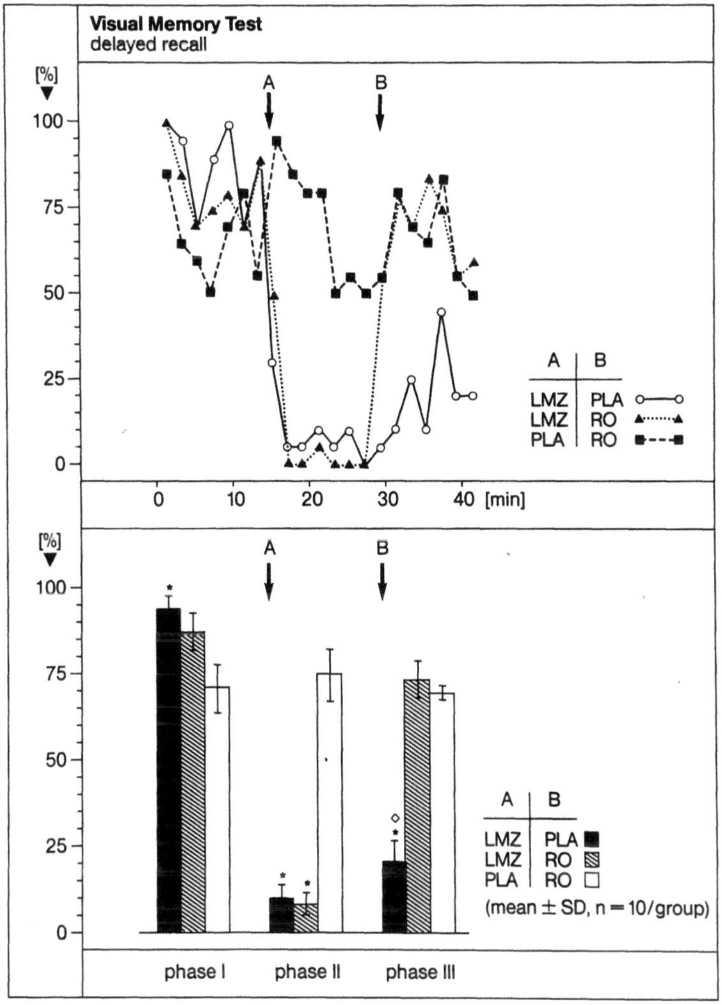

Fig. 4. Time course of performance in tests of delayed recall of visual material. Curves in *top section* show total percentage scores in each group for each presentation ($n = 20$). Columns in *lower section* show the groups' mean percentage scores per test phase (maximum 14 points for each subject = 100%). *Arrows* indicate time of injection. For abbreviations see Fig. 3. $^{*}p < 0.05$ vs PLA/Ro 15–1788; $\diamond p < 0.05$ vs LMZ/Ro 15–1788

ognition. This applies to both memory performance in the picture tests and wordlists (see Table 1, Figs. 3–5). While these effects were still evident in group A up to the end of the test session, they were completely antagonized in group B, which received Ro 15–1788 after the lormetazepam injection. Ro 15–1788 had no effect when given by itself in group C. The amnesic effects were most marked in the tests of delayed recall. Subjects appeared to have less difficulty remembering the word lists than the pictures, which may be due to the lower number of items presented in each phase (ten items in the word lists and 14 items in the picture tests), or to the semantic relationship between the words in each list.

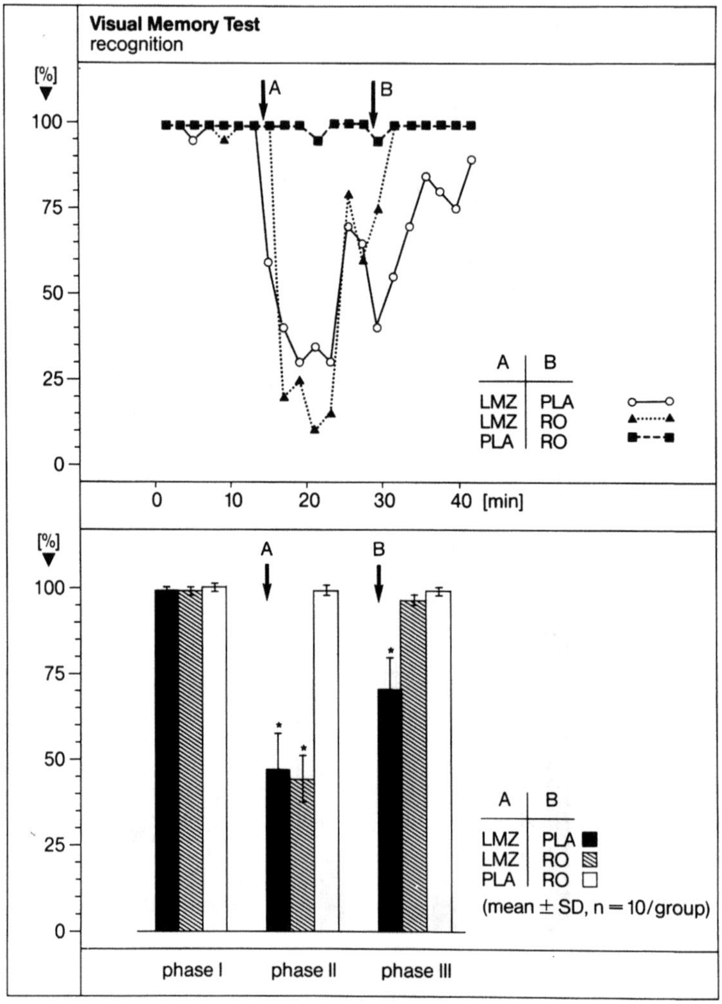

Fig. 5. Time course of performance in tests of recognition of visual material. Curves in *top section* show total percentage scores in each group for each presentation ($n = 20$). Columns in *lower section* show the groups' mean percentage scores per test phase (maximum 14 points for each subject = 100%). *Arrows* indicate times of injection. For abbreviations see Fig. 3. $^{*}p < 0.05$ vs PLA/Ro 15–1788

Seven VASs were completed with the poles high drive–low drive, tired–energetic, sluggish–animated, concentrated–distracted, cheerful–downcast, anxious–relaxed, and restless–composed. Significant differences were only evident in group A: subjects were more relaxed, distracted, tired, sluggish, and had less drive. In Fig. 6 two VAS with the poles tired-energetic and anxious-relaxed are presented. Weakness, dizziness, need of sleep, increased irritability, fatigue, and restless feelings in the legs were symptoms in the 24-item complaints list which were cited more frequently in group A. Sum scores from this list differed between

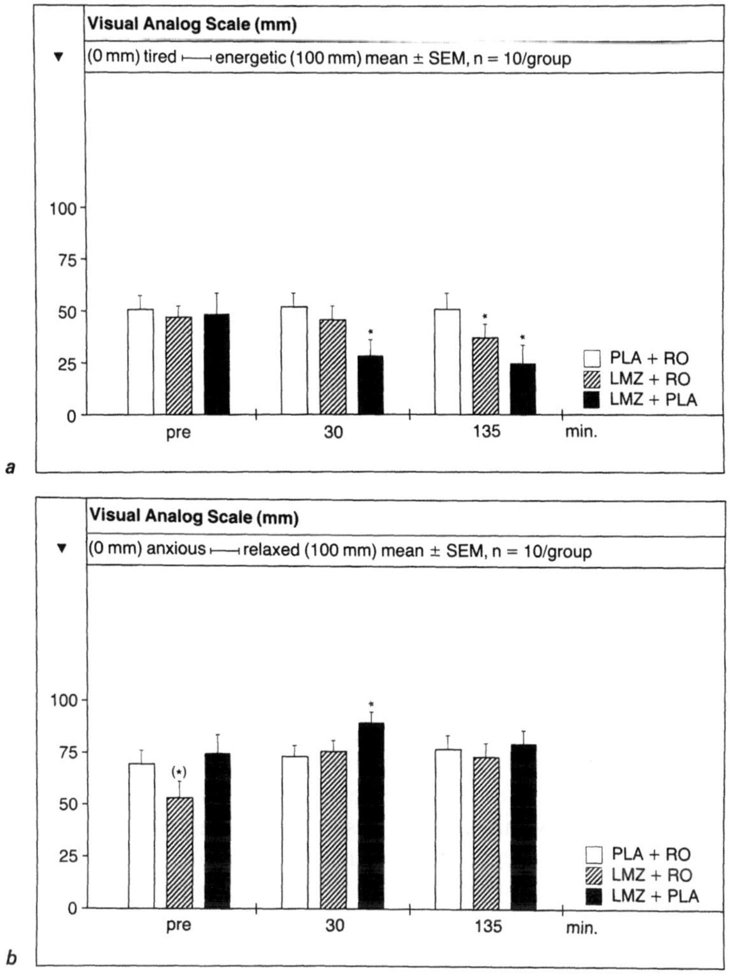

Fig. 6 a, b. Results of visual analog scales tired–energetic (**a**) and anxious–relaxed (**b**) are presented as mean ± SEM of $n = 10$ subjects per group. For abbreviations see Fig. 3. *$p < 0.05$ vs PLA/Ro 15–1788; (*)$p < 0.1$ vs PLA/Ro 15–1788

groups A and C at the $p < 0.1$ level. In the mood scale, the group treated only with lormetazepam indicated significantly better mood than the group treated with placebo and Ro 15–1788 ($p < 0.05$, data not shown; see DOROW et al. 1987 b).

Prolactin and cortisol levels (Table 2) measured 10 min after injection of lormetazepam (group A or B) or placebo (groups C) as well as 15 min after Ro 15–1788 (groups B and C) or placebo (group A) showed no significant differences.

Table 2. Mean (±SD) serum prolactin and cortisol levels before and after intravenous lormetazepam 0.02 mg/kg (groups A, B) or placebo (group C) and after Ro 15-1788 0.03 mg/kg (group B, C) or placebo (group A) administration

Treatment groups	Before injection	After first injection		
		10 min	20 min[a]	30 min[b]
Prolactin (ng/ml)				
A	5.4±4.9	5.6±3.2	5.1±2.0	5.2±1.7
B	4.5±2.2	5.4±2.8	5.6±2.6	5.4±3.1
C	4.9±2.9	4.6±2.3	4.5±1.8	4.3±1.7
Cortisol (nmol/I)				
A	598±323	565±377	534±389	469±309
B	339±133	258± 84	244± 67	214± 77
C	392±305	329±291	293±267	285±286

[a] 5 min, and [b] 15 min after second injection (Ro 15-1788 or placebo).

4 Discussion and Conclusion

In agreement with previous reports, the intravenous administration of benzodiazepines, in this case of lormetazepam, rapidly (<2 min) led to anterograde amnesia as measured in both the word lists and picture presentations. The mode of information presentation did not play a major role with regard to impairments of performance. The effects on immediate recall were less marked and of much shorter duration than those on delayed recall. If it can be assumed that the recall of information within 20 s of presentation primarily reflects the acquisition process and only the beginning of consolidation, then it follows from the above that the drug effects on the acquisition process can be seen to wane as drug plasma – and presumably brain – levels decrease. The main benzodiazepine effect, however, was manifest in the tests of delayed recall, indicating an effect on both acquisition and consolidation of new information.

In the present study no specific test of the effects of sedation on memory was performed and so they cannot be ruled out. Indeed, at high drug levels, the acquisition process could have been subject to further impairment through benzodiazepine-induced sedation. In this context it is interesting to note that only two subjects experienced sedation so strongly that they could not respond to a pair of pictures presented in the visual test. This suggests that the applied dose of lormetazepam did not in general impair the ability to perceive objects and name them. Subjective ratings like the VASs and the B-L complaints lists completed at the end of the memory test session (30 min after lormetazepam) showed a clear impairment of drive, concentration, and wakefulness.

Although this study did not include a group of subjects who were treated solely with placebo, it can nevertheless be stated that Ro 15–1788 had no effect on the present memory tests. This is evident from both a comparison with preadministration evaluations (phases 1 and 2) and from a historical comparison with the control group in which the same test design was applied (Dorow et al. 1987b).

This is in agreement with reports that the compound has little effects of its own (DARRAGH et al. 1981; O'BOYLE et al. 1983) but is in contrast to findings that it induces central activation as measured by EEG parameters and rating scales (SCHÖPF et al. 1984; DUKA et al. 1986a; HIGGITT et al. 1986).

Ro 15–1788 was able immediately and completely to reverse the effects of lormetazepam on memory and psychometric ratings. In all tests of recall and recognition, results in group B were similar to those of group C, which received placebo followed by Ro 15–1788. Signs of withdrawal, as described earlier in both animal and human studies (LUKAS and GRIFFITHS 1982; DUKA et al. 1986a), were not observed. This could be due to the dose and the duration of benzodiazepine action (15 min) before application of the antagonist. In addition, cortisol and prolactin levels remained constant and thus furnished no evidence of withdrawal reactions: cortisol levels would be expected to rise in stressful situations (DUKA et al. 1986a; DOROW and DUKA 1986).

In contrast to the findings of O'BOYLE et al. (1983) and HOMMER et al. (1986) our results provide strong evidence that the amnesic effects of a benzodiazepine can also be antagonized by Ro 15–1788. This discrepancy may be due to the different dose relationships between agonist and antagonist or to the sequence in which the drugs were applied: HOMMER et al. (1986) for instance, first gave Ro 15–1788. It is generally accepted that, depending of course on receptor affinity, antagonists are less likely to be displaced by agonists than vice versa.

Another interesting finding came to light as a result of the design of this memory study. Lormetazepam retrogradely enhanced performance in visual tests of delayed recall. This effect could not be detected in the word lists as groups attained maximum performance (ceiling) in phase 1. Retrograde facilitation of memory performance was observed in earlier studies for material presented before diazepam and alcohol intake (PARKER et al. 1980, 1981; MANN et al. 1984; HINRICHS et al. 1984). Recently, KUBICKI et al. (1986) and OTT et al. (this volume) reported similar findings after oral doses of lormetazepam and flunitrazepam. In our investigation both groups that received lormetazepam remembered more items in phase 1 than the group that first received placebo, although the effect was only significant in group A (lormetazepam plus placebo). That the effect was less marked in group B (lormetazepam plus Ro 15–1788) suggests that the duration of lormetazepam's action on information processing plays an important role, i.e., the prevention of acquisition of new information for 15 min is not enough to fully facilitate consolidation of previous information.

A number of interpretations have been proposed for findings of the retrograde facilitation of information processing by benzodiazepines and related CNS depression. First, benzodiazepines shield ongoing memory processes from new information inputs, thus allowing the interference-free consolidation in long-term storage of most recent inputs (PARKER et al. 1980; HINRICHS et al. 1984). Another explanation offered by OTT et al. (this volume) is that subjects experience a reactive increase of tension with the onset of the drug effects as sedation. This allows them to counterregulate and concentrate more on the presented information. This seems unlikely in our test session as the drug was given intravenously and effects were immediate. Another possible explanation for these findings is that benzodiazepines may have a beneficial effect on recall from memory.

Recently, VENAULT et al. (1986) reported an improvement of memory function in animals after administration of an inverse benzodiazepine receptor agonist, β-carboline carboxylic acid methyl ester (β-CCM). This compound is known to induce convulsions and is thus not suitable for testing in man. Another betacarboline, N'-methyl-β-carboline carboxamide (FG 7142), which was characterized as a partial inverse agonist, could not be studied for its memory effects in humans, since it was shown to produce severe states of anxiety (DOROW et al. 1983). However, another β-carboline, Ethyl-5-isopropoxy-4-methyl-β-carboline-3-carboxylate (ZK 93426), which has been characterized as a benzodiazepine receptor antagonist with some inverse properties (DUKA et al. 1986b; DOROW et al. 1987a), was shown to impair processing of information acquired before administration and to improve delayed recall of picture items presented after administration (DUKA et al. this volume). Interestingly, Ro 15–1788, which is also said to have some residual inverse agonist properties (FILE and PELLOW 1986), did not seem to have any effect in our tests, yet the battery of tests were designed to assess impairments of memory performance, and not improvements. A scrutiny of the time course of delayed recall in the visual memory task shows a 6-min performance dip in phase 2 just before injection of Ro 15–1788 and hence a retrograde impairment.

In summary, these findings suggest that benzodiazepine receptor ligands might prove to be potent tools both in studies of benzodiazepine-induced amnesia and its reversal, and in studies of effects that facilitate acquisition and consolidation of information into memory.

References

Atkinson RC, Shiffrin RM (1971) The control of short-term memory. Sci Am 224:82–90

Bonetti EP, Pieri L, Cumin R, Schaffner R, Pieri E, Gamzu R, Muller RKM, Haefely WW (1982) Benzodiazepine antagonist Ro 15–1788: neurological and behavioral effects. Psychopharmacology 78:8–18

Clarke PRF, Eccersley PS, Frisby JP, Thornton JA (1970) The amnesic effect of diazepam (Valium). Br J Anaesth 42:690–697

Curran HV (1986) Tranquillising memories: A review of the effects of benzodiazepines on human memory. Biol Psychol 23:179–213

Darragh A, Scully M, O'Boyle C, Brick I, Downie WW (1981) Investigation in man of the efficacy of a benzodiazepine antagonist, Ro 15–1788. Lancet:8–10

Darragh A, Lambe R, Kenny M, Brick I, O'Boyle C (1982) Ro 15–1788 antagonises the central effects of diazepam in man without altering diazepam bioavailability. Br J Clin Pharmacol 14:677–682

Doenicke A, Ott H (1980) Lormetazepam. Anaesthesiology und Intensivmedizin:133

Dorow R, Duka T (1986) Anxiety: its generation by drugs and by their withdrawal. In: Biggio G, Costa E (eds) GABAergic transmission and anxiety. Raven, New York, pp 211–225

Dorow R, Horowski R, Paschelke G, Amin M, Braestrup C (1983) Severe anxiety induced by FG 7142, a β-carboline ligand for benzodiazepine receptors. Lancet ii:98–99

Dorow R, Duka T, Sauerbrey N, Höller L (1987a) β-Carbolines: new insights into the clinical pharmacology of benzodiazepine receptor ligands. In: Dahl S, Gram LF, Paul SM, Potter WZ (eds) Clinical pharmacology in psychiatry: selectivity in psychotropic drug action – promises or problems. Springer, Berlin Heidelberg New York (Psychopharmacology suppl 3)

Dorow R, Berenberg D, Duka T, Sauerbrey N (1987b) Amnestic effects of lormetazepam and their reversal by the benzodiazepine antagonist Ro 15–1788. Psychopharmacology 93:507–514

Drachman DA, Leavitt J (1974) Human memory and the cholinergic system. Arch Neurol 30:113–121

Duka T, Ackenheil M, Noderer J, Doenicke A, Dorow R (1986a) Changes in noradrenaline plasma levels and behavioural responses induced by benzodiazepine agonists with the benzodiazepine antagonist Ro 15–1788. Psychopharmacology 90:351–357

Duka T, Höller L, Obeng-Gyan R, Dorow R (1986b) Initial human pharmacology of the neutral benzodiazepine receptor antagonist β-carboline ZK 93426. Br J Clin Pharmacol 22:228P

Dundee JW, George DA (1976) The amnesic action of diazepam, flunitrazepam and lorazepam in man. Acta Anaesth Belg 27:3–11

File SE, Pellow SS (1986) Intrinsic actions of the benzodiazepine receptor antagonist Ro 15–1788. Psychopharmacology 88:1–11

Ghoneim MM, Mewaldt SP (1975) Effects of diazepam and scopolamine on storage, retrieval and organizational process in memory. Psychopharmacology 44:257–262

Haefely W (1985) Pharmacology of benzodiazepine antagonists. Pharmacopsychiatry 18:163–166

Higgitt A, Lacer M, Fonagy P (1986) The effects of the benzodiazepine antagonist Ro 15–1788 on psychophysiological performance and subjective measures in normal subjects. Psychopharmacology 89:395–403

Hinrichs JV, Ghoneim MM, Mewaldt SP (1984) Diazepam and memory: retrograde facilitation produced by interference reduction. Psychopharmacology 84:158–162

Hommer DW, Breier A, Paul SM, Davis M, Weingartner H (1986) Ro 15–1788, a specific benzodiazepine antagonist, blocks the sedative, anxiolytic and attentional but not amnestic effects of diazepam in humans (abstr). 25th ACNP Annual Meeting, p 204

Hunkeler W, Möhler H, Pieri L, Polc P, Bonetti EP, Cumin R, Schaffner R, Haefely W (1981) Selective antagonists of benzodiazepines. Nature 290:514–516

Kubicki St, Rohloff A, Ott H, Fichte K (1986) Vigilanz und Amnesie nach Benzodiazepingabe: Neurophysiologische und pharmakopsychologische Aspekte. In: Schulte am Esch J (ed) Benzodiazepine in Anästhesie und Intensivmedizin. Roche, Basel

Lister RG (1985) The amnesic action of benzodiazepines in man. Neurosci Biobehav Rev 9:87–94

Lukas SE, Griffiths RR (1982) Precipitated withdrawal by a benzodiazepine receptor antagonist (Ro 15–1788) after 7 days of diazepam. Science 217:1161–1164

Mann RE, Cho-Young J, Vogel-Sprott M (1984) Retrograde enhancement by alcohol of delayed free recall performance. Pharmacol Biochem Behav 20:639–642

O'Boyle C, Lambe R, Darragh A, Taffe W, Brick I, Kenny M (1983) Ro 15–1788 antagonizes the effects of diazepam in man without affecting its bioavailability. Br J Anaesth 55:349–356

Ott H, Hemmerling KG, Kugler J, Suttmann H, Doenicke A, Tesch C, Sträßner G (1980) Amnestische Begleitwirkungen nach i.v.-Gabe von Lormetazepam and Flunitrazepam. In: Doenicke A, Ott H (eds) Lormetazepam Noctamid. Springer, Berlin Heidelberg New York, pp 13.1–13.10

Parker ES, Birnbaum IM, Weingartner H, Hartley JT, Stillman RC, Wyatt RJ (1980) Retrograde enhancement of human memory with alcohol. Psychopharmacology 69:219–222

Parker ES, Morihisa JM, Wyatt RJ, Schwartz BL, Weingartner H, Stillman RC (1981) The alcohol facilitation effect on memory: a dose-response study. Psychopharmacology 74:88–92

Roehrs T, McLenaghan A, Koshorek G, Zorick F, Roth T (1984) Amnesic effects of lormetazepam. In: Hindmarch I, Ott H, Roth T (eds) Sleep, benzodiazepines and performance. Springer, Berlin Heidelberg New York, pp 165–172

Schöpf J, Laurian S, Le PK, Gaillard J-M (1984) Intrinsic activity of the benzodiazepine antagonist Ro 15–1788 in man: an electrophysiological investigation. Pharmacopsychiatry 17:79–83

Stephens DN, Kehr W (1985) β-Carbolines can enhance or antagonize the effects of punishment in mice. Psychopharmacology 85:143–147

Study RE, Barker JL (1981) Diazepam and (−)pentobarbital. Fluctuation analysis reveals different mechanisms for potentiation of GABA responses in cultured central neurons. Proc Natl Acad Sci USA 78:7180–7184

Subhan Z (1984) The effects of benzodiazepines on short-term memory and information processing. In: Hindmarch I, Ott H, Roth T (eds) Sleep, benzodiazepines and performance. Springer, Berlin Heidelberg New York Tokyo, pp 173–181
Subhan Z, Hindmarch I (1984) Assessing residual effects of benzodiazepines on short term memory. Pharmaceut Med 1:27–32
Sunderland T, Tariot PN, Weingartner H, Murphy DL, Newhouse PA, Mueller EA, Cohen RM (1986) Pharmacological modelling of Alzheimer's disease. Prog Neuropsychopharmacol Biol Psychiatry 10:599–610
Tallman JF, Paul SM, Skolnick P, Gallager DW (1980) Receptors for the age of anxiety: pharmacology of the benzodiazepines. Science 207:274–281
Thiebot M-H, Childs M, Soubrie P, Simon P (1983) Diazepam induced release of behavior in an extinction procedure: Its reversal by Ro 15–1788. Eur J Pharmacol 88:111–116
Venault P, Chapoutier G, de Carvalho LP, Simiand J, Morre M, Dodd RH, Rossier J (1986) Benzodiazepine impairs and β-carboline enhances performance in learning and memory. Nature 321:864–866
Weingartner H (1985) Models of memory dysfunction. Ann NY Acad Sci 444:359–369
Wolkowitz OM, Weingartner H, Thompson K, Pickar D, Paul SM, Hommer DW (1987) Diazepam-induced amnesia: a neuropharmacological model of an "organic amnestic syndrome". Am J Psychiatry 144:25–29

Résumé

Benzodiazepine Receptor Ligands, Memory and Information Processing: Issues, Comments and Prospects

I. HINDMARCH[1] and H. OTT[2]

1 Introduction

An understanding of memory, or indeed forgetting, is important for explaining both normal and pathological cognitive functions, especially since the capacity for new learning and speed of retrieval tend to be extremely sensitive to many factors that impair the overall functioning of the CNS. Disease, trauma, ageing, psychiatric disorders and the administration of psychoactive drugs can impair the integrity and capacity of information processing abilities with consequent detraction of memory.

Psychopharmacological interest is centred on drugs which have effects in protecting the integrity and stability of cognitive functions and memory processes. While nootropic agents exist, their effects in man are often debatable; however, the recently developed drugs which act as ligands to the benzodiazepine receptor, the β-carbolines, hold promise as promnesic agents in the treatment of memory disorders over a wide range of pathological conditions. They may also have utility as tools to unravel some of the basic concepts of memory and information processing in man.

It is natural to begin with some important riders concerning clinical and experimental work relating to psychoactive drug effects on memory. The actions of benzodiazepine receptor ligands will be directly related to the dose administered and dose-effect relationships might not always have been established, so causing for difficulties in comparing studies using either different dose regimens and the same drug or the same dose regimen and different drugs. The pharmacological profiles of drugs change with respect to time of day, sex and age of patient or volunteer, route and mode of drug administration, interaction with other psychotropics, especially caffeine, alcohol and nicotine, clinical characteristics, pathologies and dose/treatment regimens.

The results obtained from psychopharmacological experiments where memory tests have been applied will, therefore, clearly depend on pharmacokinetic and pharmacodynamic variables and the age and sex of the subject, as well as personality, IQ and mood. They will also depend on extrinsic factors such as time of day of testing, presence of other drugs, nature, duration and task requirements of the test and the naivety of the subject both to the drug and to the test battery. Dif-

[1] Human Psychopharmacology Research Unit, University of Leeds, Leeds LS2 9JT, UK.
[2] Research Laboratories, Department of Pharmacopsychology, Schering AG, 1000 Berlin 65, FRG.

ferences in statistical and experimental designs and in the conceptual bases for protocols can also hinder interstudy comparisons, but some generalisations can be made.

The interactive nature of most of the information presented in this volume necessitates careful reading to obtain a full conceptualisation of the numerous issues. Perhaps also the diversity of methods and measures might encourage experimenters to seek to use common protocols, identical exclusion criteria, similar test batteries and experimental techniques in future studies. The issues, opinions, comments and criticisms raised in the discussions which followed the oral presentation of the papers have been edited into a, hopefully, parsimonious precis of what was said.

2 Concepts, Models and Theories

It is evident that there is no single theory that can fully account for the whole gamut of mental processes and systems subsumed under what cognitive psychologists call memory – be that "short", "long", "working" or "semantic". In addition, there does not seem to be a unique, systematic way to organise the various concepts and models into one theory of memory. While the lack of a unified set of theoretical tenets might seem to be an obstacle to advancing theory and the basic conceptualisation of memory, the range of alternative techniques and models provides for a large number of pragmatic approaches to the study of memory and to the measurement of the effects of psychoactive drugs on decision making, memory and related information processing abilities.

Any experimental study in either patients or volunteers must have clearly defined objectives, sensitive tests, intelligent research strategies and a sound experimental protocol. It must also be decided whether the study is to test retrograde or anterograde amnesia or indeed both. The sensitive tasks performed in a psychological laboratory might have a reduced relevance to reality but the experimental strategies and controls inherent in the laboratory situation enable environmental predictions and generalisations through simulations of information processing systems basic to cognitive behaviour and learning in the real world. In particular, the tasks involved in most standard laboratory learning and recall tests probably reflect a very important component of human memory, namely the capacity to learn and remember new material. Learning is inferred from behaviour change. A proper study of learning must consider both the conditions under which the behaviour change is believed to have occurred and the condition under which the behaviour change is revealed. Conceivably, psychotropic drugs could have their effects as one of the conditions contributing to the behaviour change or as one of the conditions under which the change is revealed. Although the psychological laboratory might be only a simplistic representation of real life, it is possible to control those variables which might have some effect on experimental outcomes. These experimental variables are legion: rehearsal or training time, semantic variables, stimulus complexity, latency and type of recall or retrieval, contextual variables, information load, effects of recency and primacy, learning and acquisition of information. The measures used for assessment are also many: digit span, memory for names and pictures, information processing (critical flicker fusion, digit symbol substitution, car handling, tracking) and temporal integration.

The time on task is an important variable within the general framework of arousal and vigilance. Investigations of learning and memory differ greatly with respect to the amount and complexity of the behaviour to be learned, the familiarity or meaningfulness of the material, the sensory modality through which the material is presented and the behavioural modality by which the learned responses are expressed. Task performance assessments therefore must always take into account the well-known trade off as regards speed and accuracy as well as general influences implicit in the inverted-U model of performance and arousal. Personality, motivation and mood can affect the strategy a subject or patient uses in the test situation and care must be taken in isolating and controlling such sources of variation.

Most tests have a range of values within which they can be claimed to be sensitive and reliable; however, it is often neglected that age alone could place a subject outside the measurement range of a task. Most recent findings suggest, as a preliminary result, that the tests measuring prospective memory, the capacity to remember to do things, may be especially vulnerable to the effects of ageing. Furthermore, many tests are used too infrequently to support generalisations concerning their sensitivity to psychotropic substances. The reliability and validity of a test accrue from the context in which it is used. Validity must be stated with respect to some particular use. If the test does not produce the expected distinction, it may be declared invalid for that use. The manner in which the test is administered, the characteristics of the sample of subjects and the conditions under which the test performance is assessed can greatly influence validity.

Another problem is that memory is usually considered a unitary phenomenon and a single measure is employed to quantify it. Recent findings in cognitive psychology, however, suggest that such an assumption would be a gross oversimplification. Effects on putatively different learning and memory processes must be compared for a proper understanding of psychotropics and their clinical potential.

3 Animal and Human Models

From the cognitive-psychological perspective one of the major issues in animal and human models is whether benzodiazepines actually produce "amnesia" or whether the failure of recall is secondary to sleepiness and sedation. Another point at issue is whether the phenomenon can be subsumed under the category of state dependency.

The number of tests of memory and information processing available to the animal psychopharmacologist is limited, compared with the number and range used in man. One of the great problems in drawing analogies between animals and humans is the non-equivalence of the cognitive systems. For example, a 50% reduction in spatial memory may render an animal completely unable to cope with the immediate environment, whereas a similar reduction could very well force human subjects to off-load information to other cognitive systems or to adopt different strategies allowing them to cope with their environment without apparent loss of performance or cognitive function.

An open mind must be maintained as regards the utility of using animal models since experimental regimens have proven their value in leading to effective pharmacological treatment for use in patients.

4 Methodological and Clinical Aspects

It is evident that a pathological condition will affect psychological behaviour in general and memory functions and information processing in particular. Furthermore, the therapeutic effect of a drug used to treat a pathological state can affect the patients' cognitive system, for better or worse, either by directly altering the efficiency of information processing or by changing the severity of the pathological state.

Traditional neuropsychological theories regarding a specifically localised neural entanglement of mental functions alone are not sufficient to account for the complexity and diversity of memory disturbances associated with neurological and psychiatric conditions, e.g. senile dementia of the Alzheimer type, Korsakoff's syndrome, Parkinson's disease, head injury, stroke, anxiety, depressive disorders and schizophrenia. Although neurotransmitter-based theories of memory have generated some test models (scopolamine challenge etc.) there is not sufficient evidence to identify any neurochemical as a specific "mnesic" agent.

Memory functions and ability presumably depend on an interaction of various neurotransmitter systems, although the possibility should not be discarded that some of them may exert a greater than average influence. Human working memory represents the operation of at least three separate subsystems (a central executive, an articulatory loop, and a visuo-spatial scratch pad), which temselves can further be fractionated. Neuropsychological evidence indicates that any of these three may be impaired independently of the other two, suggesting that any adequate assessment of the influence of a given drug on memory is likely to require several tests.

In any case, memory and information processing abilities are better reflected by considering "whole brain" functions, and many researchers are now using batteries of laboratory tests (behavioural memory, logical reasoning, recall and recognition of visual, verbal and spatial stimuli, memory scanning techniques for numerical and verbal material, digit symbol substitution tests, critical flicker fusion thresholds and divided attention tasks) to measure memory disturbances over a range of clinical pathologies. Rating scales and questionnaires regarding memory lapses and daytime functions, scored by the clinician, the patient, or any of his or her relatives, have a poor intercorrelation and there is little evidence to suggest that a reliable and/or valid relationship exists between subjective questionnaires concerning everyday memory and laboratory task performance. However, a therapist has the advantage of seeing many different patients rather than just one, and hence is likely to give a more balanced and reliable assessment than a relative.

Under these circumstances it is important to remember that test measures which might be sensitive and reliable in young volunteers are not necessarily suitable for elderly or patient populations. Moreover, drugs administered to healthy volunteers may produce paradoxical effects not noticed in patients given the same drug or dose regimen. The ecological validity of laboratory studies must always be considered and correlations determined between experimental results and the effects on patients in their habitual environment. At the same time, clinicians must not disregard the information generated from double-blind placebo-controlled crossover studies in normal volunteers. Indeed, if the effect of a drug on memory per se is to be investigated then this must be done in psychologically normal, physically healthy persons without significant pathology. It is equally important to distinguish between "memory" and general intellectual abilities, particularly in the elderly, as impaired memory can be found in patients with intact intellect; to some extent, the reverse can also be true.

5 Benzodiazepine Receptor Ligands

The benzodiazepines appear to achieve their clinically useful effects through an action at specific receptor sites in the brain. These benzodiazepine receptors are associated with the binding sites for the major inhibitory transmitter γ-amino-butyric acid (GABA). The two receptor sites are subunits of a single membrane protein. A difference from the classical receptor model arises out of the benzodiazepine receptor's function as a modulator.

β-Carbolines exhibit an affinity for the benzodiazepine receptor as high as the benzodiazepines. β-Carboline and some of its derivatives depress but can also enhance GABA-mediated responses in electrophysiological studies. Thus the inhibitory effects of GABA on neurotransmission are either intensified or minimised. However, such ligands are not simply dividable into "agonists" and "inverse agonists" accordingly, but a continuum exists, with the "agonists" and "inverse agonists" representing two extremes. Hence, within the β-carboline series there exist substances with activities ranging from benzodiazepine-like to exactly the opposite.

The basis of this supplement is the aspect of benzodiazepine receptor pharmacology that enables it to disrupt cognitive function, ranging from sedation, i.e. the inability to deal with stimulus input, to frank amnesia. Many psychotropic drugs, including benzodiazepines, have sedative effects which have been associated with a disturbance of memory function as measured on a variety of tests under different experimental protocols, e.g. before and/or after sleep. Generally, sleep itself is considered a time of amnesia. The most likely explanation for this phenomenon seems to be the inability of organisms to transfer information from short-term to long-term memory during sleep. The probability that the process of memory consolidation is inhibited during sleep must, therefore, be considered in all studies evaluating the direct effect of benzodiazepines on memory processes.

Sedation, as measured on subjective rating scales, may be evident following administration of benzodiazepines. Their effect on anterograde amnesia is dose dependent and related to their hypnotic activity. Therefore, investigations of the effect of any variable on memory have also to delineate the influence of that variable on sleep and sleepiness. Since sedatives generally impair learning, the sedative properties of the benzodiazepines may contribute to their amnesic effects even if they are not the underlying cause. But it is not necessarily the sedation or reduction in arousal that is causing amnesia. Some studies have suggested a link between sedation and amnesia, others a dissociation. The reverse is also important in the development of benzodiazepine nootropic or promnesic drugs, viz a better memory performance associated with increased arousal does not mean that the drug has "protective" of "facilitatory" effects on memory unless the receptor characteristics and specificity of the drug's neuropharmacological profile are clearly established.

Benzodiazepine receptor ligands have a broad spectrum of action on memory and information processing, ranging from powerful amnesic effects to cognitive enhancement. It is important, therefore, that the objectives of any study involving these agents should be clearly defined. Amnesia in anaesthetic practice, for surgical, dental or endoscopic procedures following the administration of a particular psychotropic drug, would be desirable but equivalent amnesia without such purpose is regarded an unwanted side effect.

Both promnesic and amnesic properties might be due to state dependency, with the consequence that a change of "state" will reverse or eradicate the observed effects of the drug. It has been found that a benzodiazepine can facilitate learning in anxious subjects and impair it in nonanxious. Thus, before considering whether benzodiazepines, when used as anxiolytics, are likely to alter a patient's ability to learn and remember, it is important to determine whether anxiety alone alters these cognitive functions.

There have been a large number of studies investigating the effects of anxiety on learning and memory. It has been argued that worry and other task-irrelevant cognitive activities associated with anxiety always impair the quality of performance by competing with task-relevant information for space in the processing system. Also, it has been suggested that highly anxious subjects attempt to compensate for these impairments by increasing effort expenditure. Consistent with these suggestions is the observation that the effect of anxiety depends on the difficulty of the task. Important in terms of clinical use of benzodiazepines is the fact that patients receiving anxiolytic therapy or taking benzodiazepines as a hypnotic often have problems when they need to operate machinery or drive a car.

Despite the fact that benzodiazepines are usually taken chronically in the clinical management of anxiety, the vast majority of studies have employed only a single dose of the compound under investigation. Although some tolerance may develop to the impaired acquisition of information, deficits are observed even after patients have taken their medication chronically. With repeated dosing such deficits might ameliorate, but behavioural effects such as poor cognitive and memory functions remain. The development of behavioural tolerance to the rating measures and the existence of pharmacological tolerance to the clinical effects of a drug must be borne in mind when evaluating the long-term therapeutic efficacy of psychotropic agents. Alcohol is a well-known amnesic agent in that it may disrupt memory for a short period of time subsequent to its administration; at the same time it protects the prealcoholic memory state, thereby facilitating recall of prealcoholic information input. This retrograde facilitating effect has also been observed with benzodiazepines commonly found to have a detracting effect on memory. An explanation offered proposes that the benzodiazepine given immediately after input material prevents the acquisition of new material which would interfere with the learning acquired immediately before. From a series of experiments it is concluded that the major reason for benzodiazepine-induced amnesia might be an impaired ability to filter interfering stimuli. Consistent with the overall experience generated from clinical data as well as from studies in normal volunteers is the observation that benzodiazepines primarily seem to impair acquisition processes. Their greatest effect is observed in tests of long-term episodic memory in which impairments result from deficits in the acquisition of new information.

The amnesic effects of the benzodiazepines find a mirror image in the ability of benzodiazepine receptor inverse agonists to improve performance on tasks requiring learning and memory. In fact, the possibility exists that a particular benzodiazepine ligand of the antagonist type may completely reverse the established amnesic activity of a benzodiazepine and act as a memory protector or enhancer. On a neuropharmacological basis it is hypothesised that the nootropic effects of antagonist β-carbolines on cognition are mediated via a direct GABAergic control of cholinergic transmission in certain brain areas. Further possibilities to explore psychopharmacological aspects are offered by β-carbolines which exert bidirectional effects on vigilance. Particularly interesting are compounds displaying a profile of characteristics which lies between either those of the agonists and the antagonists or those of the inverse agonists and the antagonists.

6 Prospects

It is well recognised that memory disturbances are a characteristic feature of many physical and psychiatric illnesses. Clinicians and psychologists have long sought parsimonious measures to assist in the classification, assessment and treatment of these patients. Traditional pencil and paper tests, developed according to principles of objectivity, validity and reliability, have been used with limited success as they were difficult to apply in either a consistent or controlled way. No doubt, these traditional tests will soon be replaced by computer-assisted devices which can control, record and interact with the subject with levels of accuracy, control and replication never before attained and thus will contribute to the reliability of test scores. The validity of these test batteries is still to be established, but they can be adapted to work on both patient and volunteer populations. Individual laboratories and researchers unfortunately use their own, often idiosyncratic, test systems. It would seem that a major advance in measuring memory and effects of psychotropics must await a consensus as to the specifications of a general test battery.

Drugs such as the β-carbolines, with documented benzodiazepine receptor characteristics, could well advance knowledge of the mechanisms underlying aspects of memory and information processing. They can also provide the clinician treating disorders of cognitive functions with a more broadly based armamentarium of therapeutic compounds, and explicit criteria for selecting treatment on the basis of both the behavioural requirements of the patient and the symptom remission given by a particular therapy.

Author Index

Subject Index

MIX
Papier aus verantwortungsvollen Quellen
Paper from responsible sources
FSC® C105338

If you have any concerns about our products,
you can contact us on
ProductSafety@springernature.com

In case Publisher is established outside the EU,
the EU authorized representative is:
Springer Nature Customer Service Center GmbH
Europaplatz 3, 69115 Heidelberg, Germany

Printed by Libri Plureos GmbH
in Hamburg, Germany